D1189576

History of Chemical Engineering

History of Chemical Engineering

William F. Furter, EDITOR

Royal Military College of Canada

Based on a symposium

cosponsored by the ACS

Divisions of History of

Chemistry and Industrial

and Engineering Chemistry

at the ACS/CSJ Chemical

Congress, Honolulu, Hawaii,

April 2–6, 1979.

ADVANCES IN CHEMISTRY SERIES **190**

AMERICAN CHEMICAL SOCIETY

WASHINGTON, D. C.   1980

Library of Congress CIP Data

History of chemical engineering.
    (Advances in chemistry series; 190 ISSN 0065–
2393)

    Includes bibliographies and index.

    1. Chemical engineering—History—Congresses.
    I. Furter, William F. II. American Chemical Society.
Division of the History of Chemistry. III. American
Chemical Society. Division of Industrial and Engi-
neering Chemistry. IV. AC/CSJ Chemical Congress,
Honolulu, 1979. V. Series.

QD1.A355 no. 190 [TP15] 540s [660.2'09] 80–17432
ISBN 0–8412–0512–4 ADCSAJ 190 1–435 1980

TP
15
.H57
1980

Copyright © 1980

American Chemical Society

PRINTED IN THE UNITED STATES OF AMERICA

660.209
F984h
76608

Advances in Chemistry Series

M. Joan Comstock, *Series Editor*

FOREWORD

ADVANCES IN CHEMISTRY SERIES was founded in 1949 by the American Chemical Society as an outlet for symposia and collections of data in special areas of topical interest that could not be accommodated in the Society's journals. It provides a medium for symposia that would otherwise be fragmented, their papers distributed among several journals or not published at all. Papers are reviewed critically according to ACS editorial standards and receive the careful attention and processing characteristic of ACS publications. Volumes in the ADVANCES IN CHEMISTRY SERIES maintain the integrity of the symposia on which they are based; however, verbatim reproductions of previously published papers are not accepted. Papers may include reports of research as well as reviews since symposia may embrace both types of presentation.

CONTENTS

PREFACE

With the publication of this book, the American Chemical Society honors chemical engineering on the occasion of what can be considered, for lack of a precise date, the approximate Centennial of its origin as a distinct profession.

The central theme of the book is the historical identification and development of chemical engineering as a profession in its own right, distinct not only from all other forms of engineering, but particularly from all forms of chemistry including applied chemistry and industrial chemistry.

The volume's twenty-two chapters widely represent both industry and education, and originate in the United States, Canada, Britain, Germany, Japan, Italy, and India. The first chapter, written by F. J. Van Antwerpen, Secretary and Executive Director Emeritus of the American Institute of Chemical Engineers, was the keynote paper of the symposium on which the book is based. Several chapters concentrate on the early emergence of chemical engineering as distinct from chemistry, while others are concerned with more recent developments where chemical engineering has occupied a particularly high profile, as in the case of nuclear energy. Other chapters offer anecdotal and biographical insights into key figures in the history of the profession, and one deals with problems of professional identity. The final chapter, by D. F. Othmer, examines important aspects of the future of chemical engineering.

Earlier reference was made to the lack of any generally agreed date for marking the emergence of chemical engineering as a distinct profession. The reason is that chemical engineering has not one, but two main roots, that developed along differing time schedules. One root was primarily chemical and of mainly European origin, while the other was primarily physical and of mainly North American origin. The former root is characterized by the industrial chemistry approach, which achieved its principal development during the mid-nineteenth century and in which chemists and mechanical engineers teamed to manufacture chemicals on a large scale. The latter is characterized by the unit operations approach, which underwent its major development in the United States during the early part of the twentieth century and in which M.I.T. played a preeminent role. A very important point is that the two concepts fused from the outset in the United States, but did not begin to converge until relatively recently in Europe. The major significance of this factor on the overall history and development of the profession is demonstrated in several chapters.

In the introduction to his famous handbook (1), George E. Davis records that "The first public recognition of the Chemical Engineer seems to have been made in 1880, in which year an attempt was made to found a 'Society of Chemical Engineers' in London." The initial attempt was unsuccessful, but provides a date which is as good as any other for establishing the point of origin of chemical engineering as a unique profession. However even the British claims of origin are disputable. Ivan Levinstein, writing in the Journal of the Society of Chemical Industry in 1886, said "We do not possess what may be termed chemical engineers, or at any rate they are very few indeed in number" (2). E. K. Muspratt, writing in the same issue, stated "It is very difficult to find a manager who has a knowledge of engineering combined with a knowledge of chemistry. Such men must be educated, and it is only now, after the Germans and French have shown the way for forty years, that we are beginning to follow in the path" (3).

Chemical engineering is the fourth most universally popular form of engineering, after civil, electrical, and mechanical. These are known as the "big four" of engineering, and there is neither a consistent nor close fifth. Chemical engineering, at the forefront of scientific advancement not only in the applications of the chemical sciences but also in a much broader range of related fields, must ensure that it does not lose the momentum of the first century of its development to what could be a developing loss of its professional identity.

Among the many previous works on aspects of the history of chemical engineering, the serious historian should be aware of the books (4–7), papers (8–10), and manuscripts (11–13) cited below.

Finally I would like to thank my wife Pamela, and my daughters Lesley, Jane, and Pamela for their many kindnesses during the preparation of this work. I also would like to thank the Petroleum Research Fund of the American Chemical Society for honoring me with the award of a Special Educational Opportunities Grant to organize and conduct the ACS international symposium on which this book is based.

Literature Cited

1. Davis, G. E. "A Handbook of Chemical Engineering," 2nd ed.; Davis Bros. Ltd.: Manchester, England, 1904; Vols. 1 and 2.
2. Levinstein, I. *J. Soc. Chem. Ind., London* 1886, 5, 353.
3. Muspratt, E. K. *J. Soc. Chem. Ind., London* 1886, 5, 414.
4. Van Antwerpen, F. J.; Foudrinier, S. "Highlights, the First Fifty Years of the American Institute of Chemical Engineers;" A.I.Ch.E.: New York, 1958.

5. Dixon, W. T.; Fisher, A. W., Jr. "Chemical Engineering in Industry;" A.I.Ch.E.: New York, 1958.
6. Piret, E. L. "Chemical Engineering Around the World;" A.I.Ch.E.: New York, 1958.
7. Miall, S. "A History of the British Chemical Industry;" Ernest Benn Ltd.: London, England, 1931.
8. Hougen, O. A. "Seven Decades of Chemical Engineering," *Chem. Eng. Prog.* **1977,** *73*(1), 89.
9. Pigford, R. L. "Chemical Technology: The Past 100 Years," *Chem. Eng. News* **1976,** *54*(15), 190.
10. Dodge, B. F.; Guthrie, H. D.; Van Antwerpen, F. J. "The Origin and Development of Chemical Engineering in the United States," presented at the *4th Interamerican Congress of Chemical Engineering, Buenos Aires, Argentina, Apr. 1969.*
11. Trescott, M. M. "Unit Operations in the Chemical Industry: An American Innovation in Modern Chemical Engineering," Univ. of Illinois: Urbana, IL, 1979.
12. Guédon, J.–C. "Chemical Engineering by Design: The Emergence of Unit Operations in the United States," Institut d'histoire et de sociopolitique des sciences, Univ. de Montréal, Quebec, Canada, 1979.
13. Stephenson, R. M. "A Selective Historical Bibliography of Chemical Engineering," Univ. of Connecticut: Storrs, CN, 1979.

Royal Military College of Canada   WILLIAM F. FURTER
Kingston, Ontario K7L 2W3

August 21, 1979

The Origins of Chemical Engineering

F. J. VAN ANTWERPEN

16 Sun Road, West Millington, NJ 07946

Chemical Engineering, born and nurtured in the United States and recognized as a distinct branch of engineering, has had a long, branched road of development. This history on its origins traces the birth of the discipline, being especially concerned with why and how it evolved in the United States and not in Europe where all of the technical building blocks also existed. The article deals mostly with the emergence of the chemical engineering concept and the author only broadly describes later periods of growth and change; no attempt is made to trace the conceptual origins of the later periods.

Any talk on the history of a dynamic subject such as chemical engineering must trace concepts. And how difficult that is to do. The chronicling of chemical processes and their variations and changes, or of chemical equipment, with its myriad new designs, is easier by far, except when we try to isolate and inspect the why and who.

In tracing the origins of ideas the illuminative insights are usually not simple to reconstruct. One cannot, like anthropologists, hope to find the one origin of man, or decide as the ornithologists have that the reptile is the forerunner of the bird, or, for that matter, have a choice of even two beginnings akin to the physicists who are now uncertain whether everything was started by God or a big bang—and aren't really comfortable with the thought that they might be one. Chemical engineering then (wherever you choose to put the then) and chemical engineering now had many, many origins and chemical engineering of the future will trace itself back to concepts we do not foresee now. So a first caveat—this talk on origins of chemical engineering is not an attempt to do more than make a few historical comments on a few things which led to present-day chemical engineering. The second caveat—I will not attempt to describe our history in terms of processes or equipment. We should make

0-8412-0512-4/80/33-190-001$05.00/1

a few distinctions; while chemical engineering is a profession, it is at any one time a collection of facts, assumptions, and design art. The collection is continuously changing—just as our minds change—and just as our impressions of fact and truth change, so chemical engineering changes. Granted chemical processes and the chemical process industries are areas in which chemical engineering is used, but they are not per se chemical engineering. Process equipment, while subject to the careful attention of chemical engineers, is not chemical engineering. This may seem obvious and redundant; but while it is easier to point to a process or an evaporator which has benefited from chemical engineering, indeed may owe its existence as an economic entity to chemical engineering, one is not justified to look on them as anything more than the artifacts of our artifice. Consider the relative ease of tracing the evolution of mechanical things. I have used this illustration before in a talk (1) I gave to celebrate Olaf Hougen's 85th birthday.

> "If one asks generally how Bell invented the telephone most individuals would assume its complete origin was in his mind, and that all that was needed was a fortuitous accident with acid and a call for help to Mr. Watson. Bell had good examples to inspire him; the telegraph was well established; the multiple telegraph, a scheme to allow several telegraph messages to be sent simultaneously over a single wire, interrupting tones of different frequencies, had been thought of; a device, known as a manometric capsule, in which the voice actuated a membrane to produce flame distortations existed; and another device, via a voice-actuated membrane traced (through a stylus) voice patterns on a pane of glass treated with lamp black" (1).

So we know prior mechanical elements of the telephone. The point I am trying to make is that it is far easier to trace the history of a finite object than it is to track concepts. We have countless clocks and precursors; but who first thought of measuring time? Show me a monument to his honor. Only through autobiography will we ever know who was inspired to do what with what idea.

One other example. During the war a major achievement using chemical engineering was the chemical plant which was used to separate plutonium made in the Hanford pile.

Du Pont agreed to design and build the separation plant even before it was certain a pile could be built which would go critical and whether it would make plutonium. Design was well underway before plutonium was produced in any quantity. The problem in the separation plant, which was never met before, was remote handling of plutonium, remote processing, remote control, and, greatly important, remote repair. One illustrative and intriguing concept from this chemical experience follows.

Remote, overhead, shielded cranes could use delicately balanced impact wrenches to disconnect radioactive equipment and could pull it up for burying. But how to guarantee that the replacement would fit exactly in place? The design engineers borrowed a concept from radio and designed equipment with bottom pins, just as on radio tubes, which fitted precisely into designated slots. Perhaps this strikes one as minor (it wasn't though), but it serves to illustrate my point about concept; only through autobiography is it possible to trace concepts and how they are used and reused. If Kekule had not told of his dream about snakes, we would not be only poorer for lack of a charming insight, but we still would be wondering how he ever figured out the structure of benzene.

Who first had the idea of a chemical engineer? We don't know. I mentioned in my history of the first 50 years of the American Institute of Chemical Engineers (2), that the word chemical engineer appeared in 1839 in a *Dictionary of Arts, Manufacturers, and Mines,* and that in 1879 the words were used also on a published drawing. So the idea of an engineer associated with chemical processes existed quite early, in fact only twenty-one years after the formation in England of the first engineering society, the Institute of Civil Engineers, founded in 1818 with culminating organization efforts going back to 1771. A few other dates to put everything in perspective:

- 1608: First chemicals exported from the New World
- 1747: The French founded the first Civil Engineering School
- 1818: First Engineering Society, Civil Engineering, in Britain
- 1836: Civil Engineers try to organize in the United States
- 1843: National Engineering Society formed in Holland
- 1847: National Engineering Society formed in Belgium and Germany
- 1848: National Engineering Society formed in France
- 1848: Boston Civil Engineering Society (BSCE) formed (lasted until merged with ASCE in 1974)
- 1852: First National Engineering Society—the American Society of Civil Engineers (ASCE)—formed in the United States.

The term engineer was not new: our Revolutionary Army had engineering officers and a corps of engineers; in England John Smeaton in 1782 signed himself "Civil Engineer."

In his excellent paper (3) on the evolution of unit operations, W. K. Lewis points out that "Modern chemical industries started with the Le Blanc process in France during the (French) Revolution" and that "the expansion of the chemical industry during the nineteenth century was extraordinary. Moreover, many of its developments were outstanding engineering achievements. The men who achieved these results thought of themselves as chemists . . . rather than engineers" (3). And as pointed out in the paper by Lewis and by others as well, in 1881 a serious attempt was made in England by George E. Davis to begin a Society of Chemical Engineers.

Furthermore, Lewis, who along with Walker and McAdams can be thought of as the earliest and best known proponents of the unit operation concept states that "Hausbrand (a German who published a book on rectification and distillation in 1893) is beyond question the father of the modern chemical engineering treatment of unit operations, in which problems are solved by applying the fundamental principles of physical science to the specific case. He was the world's first process design engineer. Note that Lewis did not say that Hausbrand was the father of unit operations but of "the modern chemical engineering treatment of unit operations" (3). The term "unit operations" was coined by Arthur D. Little in 1915.

But to return to the development of Chemical engineering . . . it is a mystery as to why it was not embraced and fostered in Europe. Certainly the concept was enunciated; Davis in England tried not only to found a society, but he perhaps made the earliest analysis of chemical engineering on record. He did this first in a series of lectures on chemical engineering in 1887 at the Manchester Technical School and then in his *Handbook of Chemical Engineering* published in 1901–1902, and it must have sold well because a second edition was published in 1904.

The conceptual element, the intellectual insight, was all there in England, France, and Germany. Why the failure to capitalize? In answer to a question by the author, E. W. Thiele, at the time he presented two volumes on distillation and rectification for the Library of Historical Chemical Engineering volumes being collected by the American Institute of Chemical Engineers, said that the volumes were the source of data for him and W. L. McCabe at the time they developed the well-known step-by-step construction to determine the number of ideal plates needed to establish a concentration difference in a column. The two volumes which appeared in French were both by Ernest Sorel; one was on distillation and the other on rectification of alcohol. Sorel, by the way, identifies himself as a one-time manufacturing engineer of the State (Ancien de mfg de Etat). Why then the United States?

Alexis de Tocqueville, in his two volumes on democracy in America, a record of observation and opinions made during a nine-month stay in the United States in 1831, had something to say about the character of Americans and their technical strength (3):

> *"The spirit of the Americans is adverse to general ideas; it does not seek theoretical discoveries . . . the observation applies to the mechanical arts.   In America the inventions of Europe are adapted with sagacity; they are perfected and adapted with admirable skills to the wants of the country. Manufacturers exist, but the science of manufacture is not cultivated; and they have good workmen, but very few inventions"* (4).

He warms up to this theme in Volume II in the section entitled "The Example of the American Does Not Prove That a Democratic People Can Have No Aptitude and No Taste for Science, Literature, or Art" (4) and begins:

> *"It must be acknowledged that in few of the civilized nations of our time have the high sciences made less progress than in the United States."*

That sounds worse than it is, for de Tocqueville credits it to an austere religion, a new and abundant country, the spirit of gain, etc., etc.

However, he does credit us for something.

> *"In America the purely practical part of science is admirably understood, and careful attention is paid to the theoretical portion which is immediately requisite to application. On this head the American always displays a clear, free, original, and inventive power of mind.   But hardly anyone in the United States devotes himself to the essentially theoretical and abstract portions of human knowledge.*
>
> *"Nothing is more necessary to the culture of the higher science . . . than meditation . . . and nothing is less suited to meditation than the structure of democratic society . . . everyone is in motion . . . some in quest of power and others in quest of gain"* (5).

Perhaps I have dwelt too long on de Tocqueville—I must admit fascination—but the analysis in 1831 still fits conditions during the development of chemical engineering 50 to 60 years later.

The first course in chemical engineering was offered at M.I.T. when a Professor of Industrial Chemistry, Lewis Mill Norton, founded the now famous Course X—Chemical Engineering.   This was in 1888, one year after Davis' Manchester lecture.   Although it has preeminence, M.I.T. did not claim invention.   The President of the Institution in his December, 1888 report revealed that already 11 members "of the second-year class have already entered upon the course" (6) (M.I.T. had a common first year for engineering students) and then he undertook to explain what it all was about.

"The chemical engineer has been but little known in this country or England, and perhaps not at all, under that name; although his profession is recognized in France and Germany. The chemical engineer is not primarily a chemist, but a mechanical engineer. He is, however, a mechanical engineer who has given special attention to the problems of the chemical manufacture. There are a great number of industries which require constructions, for specific chemical operations, which can best be built, or can only be built, by engineers having a knowledge of the chemical processes involved. This class of industries is constantly increasing, both in number and in importance. Heretofore, the required constructions have, generally speaking, been designed, and work upon them has been supervised and conducted, either by chemists, having an inadequate knowledge of engineering principles and unfamiliar with engineering, or even building practice; or else by engineers whose designs were certain to be either more laborious and expensive than was necessary or less efficient than was desirable, because they did not thoroughly understand the objects in view, having no familiarity, or little familiarity, with the chemical conditions under which the processes of manufacture concerned must be carried on. It was to meet this demand for engineers having a good knowledge of general and applied chemistry, that the course in chemical engineering was established.

"The instruction to be given, while following mainly in the line of mechanical engineering, includes an extended study of industrial chemistry, with laboratory practice. Special investigations into fuels and draught, with reference to combustion, will be a feature of the course. The plan of study has not yet been fully marked out; but a standing committee of the Faculty, consisting of the professors chiefly concerned, will give their attention, throughout the year, to the further development of this department, which, it is believed, will add much to the strength and usefulness of the school" (6).

One wonders if the profession really was recognized in France and Germany, or, are these words of justification for yet another course in the curriculum?

The twig was definitely bent in the direction of mechanical engineering, for in the M.I.T. catalog for 1888–1889 the description for Course X was:

"This course is arranged to meet the needs of students who desire a general training in mechanical engineering and to devote a portion of their time to the study of the application of chemistry to the arts, especially to those engineering problems which relate to the use and manufacture of chemical products" (7).

And later . .

> "The general engineering studies in the course in chemical
> engineering coincide for the most part with the work of the
> students in mechanical engineering" (7).

A look at the curriculum bears this out. Courses included Construction of Gear Teeth, Mechanism of Mill Machinery, and Slide Valve
Link Motion, etc.

Chemistry was analytical: Elements of Organic Chemistry (Perkin
was only 32 years earlier), Industrial Chemistry—Lecture and Laboratory, Applied Chemistry, Thermochemistry, and Applied Chemistry
which included a thesis.

Called chemical engineering then, it is recognizable today as industrial chemistry and perhaps (note the perhaps) was a combination of the
European techniques of a chemist working with a mechanical engineer in
the design of a plant.   The perhaps is in deference to W. K. Lewis' (8)
several conclusions in 1958 that:

> "Davis, [in the] Manchester Lectures thirty years before
> the coining of the term, presented the essential concept of unit
> operations, and particularly an understanding of its value for
> educators; that Davis must be given full credit for the initi-
> ation of the modern chemical engineering profession; that
> Norton's curriculum was differentiating between the chemistry
> of an industry which is always specific to that industry on the
> one hand, and the mechanical and physical operations com-
> mon to many industries on the other" (8).

Further, Lewis concludes that based on the M.I.T. catalogue
description of a fourth year course:

> "'Applied chemistry . . . (which dealt) with materials,
> methods of transportation, evaporation and distillation, etc.,
> etc., devoted to a discussion of the appliances used in manu-
> facturing and applied chemistry considered from an engineer-
> ing point of view . . . this was the first course in unit
> operations ever incorporated in an organized curriculum in
> chemical engineering" (8).

One must respect the opinion and conjecture of W. K. Lewis
because he, Walker, and McAdams were the first to expound the unit
operations—that concept which gave the first distinctive acceleration of
chemical engineering away from industrial chemistry.

My reasons for skepticism about Lewis' sweeping conclusions are
based on the opinion of others from the same era.

A. H. White: "In the spring of 1919 when Colonel William H. Walker and I were still in Army uniform, he told me that he was going to his Maine cottage in June with W. K. Lewis and W. H. McAdams and write a text book on chemical engineering" (9).

C. M. A. Stine in 1928: "The chemical engineer is a comparatively recent product of our industrial development; a couple of decades ago we find but little mention of him. When the American Institute of Chemical Engineers was organized the conception of chemical engineering was rather hazy. What was realized actually was the fact that those engaged in industrial operations needed to supplement the results of the purely chemical research worker in order to adapt these results to use by the manufacturer . . ."

"Perhaps the characteristics which most differentiate the chemical engineer of today (1928) from the earlier activities of those interested in the field is the quantitative treatment of these various unit operations and it is this exact and quantitative treatment of these operations which constitutes the province of modern chemical engineering" (10).

A. D. Little in 1928 after describing Course X and its beginning association with a "general training in mechanical engineering" (11): "Even at that time and for many years afterward there was little distinction between industrial chemistry and chemical engineering. The chemical engineer was still a mechanical engineer who had acquired some knowledge of chemistry" (11).

So I must conclude that an association of Davis and Norton et al. with operations, later included in the concept called unit operation, did not mean that they had discovered them. These operations were not new even at the time of Hausbrand, Sorel, Norton, and Davis and it took the verbal brilliance of A. D. Little to bound the country where our early pioneers would find pay dirt.

For the record, the first—the very first—person entitled to call himself a chemical engineer by virtue of a degree was William Page Bryant, who, after graduating from M.I.T. in 1891, promptly entered the insurance business and spent most of his life as a rating auditor for the Boston Board of Fire Underwriters. Apparently he was an outstanding individual with such a prodigious memory that his fellow students in the year book lauded him as: "W. P. Bryant, the intellectual giant, can repeat word for word almost everything he ever heard" (12).

Other universities in the United States soon entered the field after M.I.T. The second chemical engineering program was offered at the University of Michigan in 1898. The first departments of Chemical Engineering were started at the University of Pennsylvania in 1892, Tulane University in 1894, the University of Wisconsin in 1898, and the Armor Institute of Technology in 1900, where according to White "the first laboratory work in chemical engineering was offered here in 1908 and directed study to such unit operations as evaporation, crystallization, filtering, and drying" *(13)*.

Speculating still further on why chemical engineering developed as it did in the United States, do the remarks of C. M. A. Stine—eventually Vice President of DuPont and President of AIChE—give a better clue than de Tocqueville? Probably.

> *"The formation of AIChE certainly helped. It was the only such society in the world and it gave focus by chemical engineers and for their publication"* *(14)*.

It enlisted the support of the great pioneers such as Little, White, Walker, John Olsen, and McCormack Meade (editor of the publication *The Chemical Engineer*) who apparently deserved special encomia such as that give by John C. Olsen, first secretary of AIChE, when he wrote:

> *"the origin of any human institution or society invariably leads back to some outstanding personality whose initiative and industry were responsible for its early growth and development. In the case of the AIChE that was Richard K. Meade"* *(11)*.

Then too one suspects that the arguments against forming a society of chemical engineers, expressed by the then President of the ACS, Marston T. Bogert, and his insistence that chemical engineers and industrial chemists were the same, gave the soon-to-be-hatched society a goal—to prove we are different. Bogert, it should be pointed out (I knew him: he was a fine courtly gentleman of sincere concern for chemistry in all of its manifestations), tried to assure the founders of AIChE that he was not in any way opposing the formation of a society of chemical engineers.

The formation of AIChE, with its infant pledge to the future, organizationally congealed the dedicated protagonists of the profession in a search first for identity, then for systems and applications which bore out that identity and their claim to uniqueness.

And these new men insisting on the specificity of their calling—that they were not industrial chemists, but rather were engineers—found their way, slowly true, but they found it. And the industry profited.

A primary premise relating to the development of an engineering discipline is that it is required by an established industry. The chemical-process industries in the United States popularly are assumed to have developed after the First World War. Up to that time Germany is credited with being the preeminent chemical power. This is not so. Many developments that we assume are modern were firmly established by 1908, the year AIChE was founded. In that year the United States began the first large-scale chlorination of water; William H. Walker received the Nichols Medal for his work in chemical engineering. The first ten years of the twentieth century saw other notable developments in the chemical industry. J. B. F. Herreshoff developed the first American sulfuric acid contact process in 1900; the Semet–Solvay Company made pure benzene, toluene, and solvent naphtha from coke-oven gas; David Wesson, one of the founders of AIChE, vacuum-deodorized cottonseed oil; A. J. Rossi began the electrolytic manufacture of ferrotitanium at Niagara Falls. The next year the first oil gusher was discovered; Monsanto Company was formed to manufacture saccharin; Diamond Alkali was organized; and the beginnings of the artificial-silk, or rayon, industry were underway. A year later the Hooker Electrochemical Company got its start and J. V. N. Dorr invented the mechanical classifier in 1904.

These were the basic developments that later were to become huge industries. Rubber accelerators were discovered by George Oenslager of Goodrich in 1906 and the cyanamide process for nitrogen fixation was developed in 1905. That same year phenolformaldeyde plastic was developed by L. H. Baekeland, who later became President of AIChE. In the year before the founding of the AIChE, the calcium cyanamide manufacturing process was begun at Niagara Falls; E. L. Oliver produced the first continuous-vacuum filter; and the first kraft paper mill in North America was operated at Quebec. All of this activity testified to the establishment of a huge chemical industry; as a matter of fact, the chemical production in dollars and tons in America in 1910 was greater than the English and the German outputs combined, and it was against this background that chemical engineers came on stage.

But while there was a flourishing inorganic chemical industry, the United States of America had little in the organic field. World War I revealed in dramatic fashion our dependence on Germany for dyes, dye intermediates, pharmaceuticals, and many other organic chemicals, which were largely cut off by a naval blockade. We were also dependent on foreign sources for supplies of nitrogen and potash for fertilizers. There was no synthetic ammonia. Our fixed nitrogen came from Chilean nitrate; a small amount of atmospheric nitrogen was combined with oxygen by the now-obsolete arc process, and the fixation of nitrogen by the cyanamid process was practiced on a small scale. Even though the United States has a large chemical industry by 1917, there were still no high-pressure syntheses for making methanol and ammonia, no synthetic rubber, and no high-octane gasoline. (In fact, octane number hadn't been conceived yet.) The thermal cracking of hydrocarbons had just begun; there were no synthetic fibers, no synthetic detergents, and few organic plastics.

The first synthetic indigo was produced in 1917. Until the development of the Burton process for cracking hydrocarbons in 1913, the petroleum industry had been confined to separating from the crude the compounds that nature had put there. There was no petrochemical industry. Ethylene and acetylene, now produced in enormous quantities as important building blocks for many compounds, were then of minor importance as raw materials for the chemical industry.

Perhaps it is not too immodest to claim that the explosive development of efficient large-scale chemical plants had to await the development of chemical engineering as a distinct engineering discipline with its own methods, literature, research, and practitioners.

Development of Chemical Engineering Curricula

In the meantime, the multiplying process industries needed trained men. What was happening in education since the origination of the first chemical engineering course in 1888? In 1905 the M.I.T. curriculum was reorganized by Professor W. H. Walker to introduce more chemistry. But chemical engineering has gone through several stages of development. These periods detailed here are purely arbitrary divisions based largely on the comprehensive study made by Hougen (*12*). In each of these periods there is a major emphasis on a particular area, but each emphasis gradually shifted to another area of greater insight. The first period was:

- 1898–1915: Industrial chemistry and descriptions, largely nonquantitative, of processes used in industry.
- 1915–1925: The unit-operation concept, chiefly the application of physics, took hold and was the central educational theme. The concept was expanded in a report by the AIChE Committee on Education—at that time under Little's chairmanship—in 1922, seven years after Little's pioneering description of unit operations to the President of M.I.T. The report stated:

"Chemical engineering as a science, as distinguished from the aggregate number of subjects comprised in courses of that name, is not a composite of chemistry and mechanical and civil engineering, but a science of itself, the basis of which is those unit operations which in their proper sequence and coordination constitute a chemical process as conducted on the industrial scale. These operations, as grinding, extracting, roasting, crystallizing, distilling, air-drying, separating, and so on, are not the subject matter of chemistry as such nor of mechanical engineering. Their treatment is in the quantitative way with proper exposition of the laws controlling them and of the materials and equipment concerned in them is the province of chemical engineering. It is this selective emphasis on the unit

operations themselves in their quantitative aspects that differentiates the chemical engineer from industrial chemistry, which is concerned primarily with general processes and products" (16).

In 1922 the chemical engineers still were interring the ghost of a haunting predecessor. Another significant extract from W. K. Lewis' paper, quoted earlier, bears on the historical development of chemical engineering from:

". . . 1910 when there were only 869 chemical engineering students out of a total of 23,241 of all kinds. World War I precipitated a tremendous demand for graduates, reflected in a listing of 5,743 chemical engineering students in 1920, a figure which rose to a sharp peak of 7,054 in 1921–22. This expansion in student numbers resulted largely from establishment of curricula in schools all over the country. The policies of these schools were molded by the educational ideals of Walker and Little" (17).

The development of chemical engineering was the product of several different forces: the need of an industry for specialized engineering talent; the growth of curriculum through a definition of what to teach; and a professional organization formed to promulgate, publicize, and maintain standards—plus of course de Tocqueville's insight on the development predilections of Americans as they strove to found an industrial complex.

The next periods of development in chemical engineering education were as follows:

- 1925–1935: Unit operations were still the dominant theme, but more emphasis was being put on material and energy balances.
- 1935–1945: Applied thermodynamics and process control assumed imortance, but the development does not imply necessarily less emphasis on unit operations.
- 1945–1955: Applied chemical kinetics and process design came to the fore. Unit operations losing its uniqueness as it was consolidated into other concepts.
- 1955–: More and more emphasis placed on engineering science. Rather than emphasizing unit operations, the present trend is to concentrate on the basic engineering sciences; for example, in place of the unit operations of fluid flow, heat transfer, distillation, absorption, drying,

and the like, one uses momentum and mass and energy transfer.

Looking back again over the years during which all of these changes were taking place one realizes with a pang that these developments were the results of insights by chemical engineers building on the achievements of other chemical engineers, and that by and large, their contributions have been chronicled mostly in technical imagery and the human qualities of these teachers, engineers and researchers are preserved as impressions, lovingly retained, only in the minds of students and associates. We mostly need to organize more programs dedicated to those who will be looked on in the year 2000 as ancients worthy of praise.

I have done a partial history on Olaf Hougen, who, with Watson, caused the slow creeping away from unit operations as the dominant theme in chemical engineering to the broad sophisticated exploratory engineering it is today. What better way to end this inadequate, incomplete history of origins than to repeat the words of Professor Hougen at the end of his magnificent Bicentennial Lecture on Chemical Engineering History entitled, "Seven Decades of Chemical Engineering" and published in January, 1977 by CEP "to urge each department of chemical engineering to write its own historical record—for preservation by the American Institute of Chemical Engineers."

We already have about 20 such histories on record; that leaves about 110 to go.

But let us go further and invite each company to do the same, for much of the history and progress of our profession was made when theory met hard practicality. We should be allowed to know what, who, and mostly why.

Literature Cited

1. Van Antwerpen, F. J. "Hougen, Olaf Andreas, His Impact on Chemical Engineering: A Retrospective," to be published.
2. Van Antwerpen, F. J.; Fourdrinier, Sylvia "Highlights of the First Fifty Years of the American Institute of Chemical Engineers;" AIChE: New York, 1958.
3. Lewis, W. K. *AIChE Symp. Ser.* **1959**, 55 (26), 1,3.
4. de Tocqueville, Alexis "Democracy in America;" (Reeve, Bowen, Bradley translation,) Vintage Books, Random House: New York, Vol. 1, p. 326.
5. Ibid., Vol. 2, pp. 36, 43.
6. "Report of the President (MIT) for the Academic Year 1887–1888."
7. "MIT Catalogue, 1888–1889."
8. Lewis, W. K. "MIT Catalogue, 1888–1889," pp. 3, 5.
9. Van Antwerpen, F. J.; Fourdrinier, Sylvia "MIT Catalogue, 1888–1889," p.56.
10. Stine, C. M. A. "Chemical Engineering in Modern Industry," *AIChE Trans.* **1928**, 5.
11. Little, A. D. "Twenty-five Years of Chemical Engineering Progress;" AIChE: New York, 1933.
12. Ferguson, J. Scott, personal communication.

13. White, A. H. "Twenty-five Years of Chemical Engineering Progress;" AIChE:
    New York, 1933.
14. Olsen, J. C. *AIChE Trans.* **1932,** *28,* 299.
15. Hougen, O. *The Chemical Engineer* **1965,** *191.*
16. "Report of the Committee of Chemical Engineering Education of the
    American Institute of Chemical Engineers" 1922.
17. Lewis, W. K. "Report of the Committee of Chemical Engineering Education
    of the American Institute of Chemical Engineers," 1972, p. 5.

RECEIVED May 7, 1979.

Chemical Engineering:
How Did It Begin and Develop?

JOHN T. DAVIES

Department of Chemical Engineering,
University of Birmingham, Birmingham B15 2TT, England

Chemical engineering evolved from a mixture of craft, mysticism, wrong theories, and empirical guesses. The crafts of soap-making and distillation entered Northern Europe from the Mediterranean in the 12th–14th centuries. But improvements were very slow until the Scientific Revolution of the 17th and 18th centuries. Only then were mystical interpretations replaced by scientific theories: though the early theories were often wrong, they nevertheless played a leading role in stimulating thought. Included here are details of the chemical process engineering developed between 1740 and 1913, in particular alkali production, coal carbonization, sulfuric acid manufacture, agricultural fertilizers, and distillation. The origin of the unit operations approach also is discussed.

Applied chemistry has, through the ages, been interesting and useful to man. Dyeing, distillation, metal refining, and the manufacture of wine, glass, soap, and cement have long been practiced in small-scale units. Chemical engineering is the technique of scaling-up such operations and processes, with some of the large-scale plants being operated continuously and with automatic control. So chemical engineering is concerned with keeping costs low by mass production methods, by optimization, and by reducing labor costs. It also is concerned with quality control through better instrumentation related to automatic control systems.

How did all of this begin? How did chance discoveries, magic formulae, superstition, and religion give rise to (*i*) the scientific revolution, and (*ii*) the possibility of scaling-up chemical processes and controlling them closely?

Ancient Greece to the 17th Century Scientific Revolution

The ancient Greeks were occupied with forms and shapes. They also loved enquiry, reason, and knowledge for their own sakes. In their

0-8412-0512-4/80/33-190-015$07.25/1

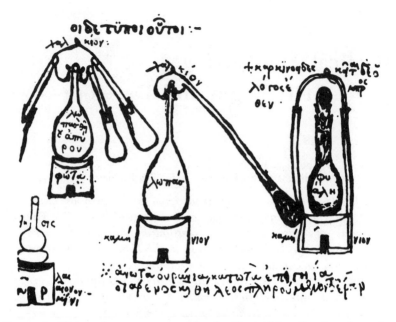

*Figure 1. Chemical apparatus used by the Alexandrian chemists about the
1st century AD (1)*

philosophy, the intellectual interest of a theory counted much more than profit or utility; a vision of the whole was more important than an analysis into separate components.

For example, Epicurus (who flourished ca. 300 BC) valued theories solely for providing naturalistic explanations of phenomena which superstition was attributing to the agency of the gods. If there were several possible naturalistic explanations, Epicurus held that there was no point in trying to decide between them.

Thus theories, though numerous in the ancient world, were tested little: the lack of instruments admittedly provided little opportunity but there was even less interest. Nor were theories generally related to the known crafts: Aristotle (384–322 BC) mentioned that pure water can be made by evaporating sea water, but provided no theory of this. Pliny (in the 1st century AD) described a primitive method of condensation in which the oil obtained by heating rosin is collected on wool placed in the upper part of the apparatus. Typical stills of the 1st and 4th centuries are shown in Figures 1 and 2; simple stills are described also in Arab texts of the 7th and 8th centuries AD.

From the 7th century AD, "Greek fire" was used in warfare, particularly to set enemy ships on fire. It made a major contribution to the Byzantine naval victories of those times. There has been much speculation as to the nature of "Greek fire". A recent study (3) concludes that the flame-thrower used hot crude oil, or perhaps a distillate from it.   The liquid was forced (by air pumped into the top of the heated feed tank) through a nozzle and emerged as a turbulent jet which was ignited.   The whole process could be described as early chemical engineering.

In the 1st century AD, Pliny recorded the empirical use of oil for calming a rough sea, and Plutarch also mentioned this practice.   A little of this empirical knowledge may have passed directly into the so-called Dark Ages of Europe, though the interpretation of the effects then became rather mystical.   For example, Bede, in his famous history of 731 AD, records that in the seventh century holy oil was used to calm the seas in stormy weather.

The calming of the sea then was attributed to the holiness of the oil rather than to its physical properties.   However, holy oil consisted mainly of olive oil, which now is known to spread well and to be very effective in calming a rough sea.   It thus appears that the knowledge of the wave-calming properties of oils may not have been lost completely during Europe's Dark Ages.

But in Europe religious and personal authority reigned supreme over the crafts and skills practiced in the Middle Ages (see Figure 3). Indeed about 2,000 years were to elapse from the discussions of the ancient Greeks before the unprecedented confluence of cultural, religious, and technological changes produced an environment in which

Institute of Petroleum

Figure 2. A still of the 4th century AD (2). Note the sand bath immediately below the flask.

Figure 3. Alchemists distilling the "quintessence"; a rather mystical search for a fifth essence (after earth, air, fire, and water). The quintessence was supposed to be the purest nature of created things, making up the heavenly bodies, and latent in all things.

extensive testing of theories by systematic experiment and observation became accepted practice, so that theories which were too vague (or too mystical) or too complicated to be tested were no longer given serious consideration.

The first stirrings of this modern outlook can be traced back dimly into the 12th and 13th centuries. It was then that Aristotle's ideas of the dignity and self-confidence of man began to be revived: the philosophies of law, government, and the physical universe began to be examined. Peter Abélard, the French Catholic philosopher of the early 12th century, was optimistic concerning the power of human reason to achieve knowledge of the natural and the supernatural. He maintained that by doubting we come to questioning, and by questioning we perceive the truth. He believed in the power of reasoning—that healthy skepticism was a stepping stone to understanding. His questioning attitude extended even to the Apostles and the Holy Fathers, whom he thought could err sometimes: only the Scriptures, he said, were infallible.

Also about the middle of the 12th century the production of alcohol, via the distillation of fermented substances, was discovered in Salerno. The influence of the Arab alchemists was strong in southern Italy (*see* Figure 4), and many of their writings were being translated into Latin.

By the 13th century the craft of soapmaking (described in the Bible and by Pliny in the 1st century AD) had reached England.

In the 14th century, the distillation of wine to produce alcohol became a minor industry, and strong alcoholic beverages were being prescribed for various ailments. At the same time the use of sweetened strong drinks based on alcohol (liqueurs) was introduced into northern Europe from Italy (*4*).

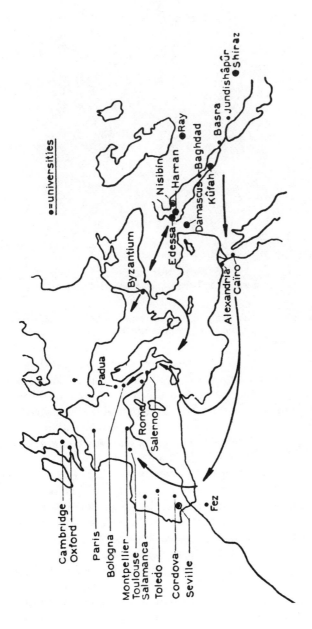

Figure 4. The routes by which classical science came to Western Europe

Figure 5. Leonardo da Vinci's drawing of eddies and bubbles in water which has flowed from a square hole into a pool (5)

By the 15th century, the Renaissance was in full flood, sweeping Italy and then making its impact felt north of the Alps. Included in the Renaissance was an excited interest in the rediscovered world map and the geography of Ptolemy, and in the mathematical traditions of Plato and Pythagoras whose famous theorem involved a reasoned deduction from postulated axioms. The intellectual power of man was being rediscovered but in a new context—that of Christianity. This religion involved a belief in a governing Lord, leading directly to a belief that there were governing laws. Also during the 15th century, technology and craft skills were being improved; for example, the distillation of wine and beer to obtain alcohol became more popular, and a cooling tube on the still in the form of a coil or serpent was introduced.

But even through the 16th century real scientific progress remained rather slight and much of the classical tradition was retained. Leonardo da Vinci in his drawing (1509) of the eddies and bubbles in turbulent water (*see* Figure 5) still is concerned with forms rather than with testing a theory. However, some theories were being discussed. Copernicus was questioning the accepted theory of planetary movement and was proposing that the sun, and not the Earth, was the center of the planetary system.

Technology and crafts (such as distillation, *see* Figures 6 and 7) likewise were continuing to progress through the 16th century. Glass-making was expanding to meet the growing demands for window glass, bottles, and table ware. In alchemy, Paracelsus mentioned the various "operations" which were used, including calcination, sublimation, dissolving, putrefaction, distillation, coagulation, and coloration: "Whoever shall now ascend and pass these seven steps, he shall come to such a wonderful place that he shall see and experience many secret things in the transmutation of all natural things" *(6)*. This idea, relating to laboratory operations, was perhaps the first intimation of the concept of "unit operations." Salt manufacture was concerned with coagulation of the impurities in the brine, evaporation, and crystallization. Sugar processing involved pressing, evaporation, and crystallization (*see* Figure 8).

At the beginning of the 17th century, clocks and optical instruments capable of testing scientific theories were being improved still further. At the same time the ideas and conflicts of the Reformation were spreading and taking root. This repudiation of human authority and the freedom of individuals to indulge in critical discussions of each other's ideas were related perhaps to the rather unstable structure of European society. This instability doubtlessly was accentuated by the frequent waves of epidemics that swept across Europe (the Great Plague of London [1664–1665] was not an isolated phenomenon). In these circumstances of drastic social upheaval, intelligence and increasing literacy had a greater scope than up to this time *(9)*. Whatever the causes of the Reformation, however, the attitude of questioning established authority was essential before a critical experimental science could flourish.

E. J.Brill

Figure 6. Serpent coolers on a 16th century still (7)

British Museum

Figure 7. Chemistry, particularly distillation, in the 16th century (8)

The Scientific Revolution of the 17th century (associated with such names as Kepler, Gilbert, Galileo, Torricelli, Pascal, and Newton) was thus a result of the confluence of three factors (9):

- the excitement of the Renaissance (and the Christian context of this);
- the antiauthoritarian ideas of the Reformation;
- the improved optical, timekeeping, and metallurgical techniques.

In particular, the more accurate instruments such as clocks, telescopes, thermometers, and compound microscopes allowed the testing of scientific theories to be much more exact and extensive than ever before. Such was the origin of the flowering of scientific activity which occurred in the 17th century; the technology and new theories began (and are still) advancing hand in hand, each strengthening the other. Galileo began his experiments on the period of oscillation of pendulums in 1581, which led to the pendulum clock of much improved accuracy.

During the early decades of the 17th century, the English philosopher Francis Bacon wrote of the importance of experiment and observation. Knowledge, he maintained, was useful in giving man sovereignty over nature. He advocated a keen exchange of intellectual views and thought that destructive criticism was particularly important. He

British Museum

Figure 8. Sugar processing in the 16th century (8)

further emphasized the importance of generalizations and of a fundamental approach, in that theories being tested should be applied also in new circumstances.   For such testing, he stated, certain observations are especially valuable in that they allow one to decide between two rival theories.   "We must put nature to the rack to compel it to answer our questions" (10).

Theories now were becoming public rather than private; they were disseminated more widely by the relatively newly invented printing press.   Confidence in the new scientific method was now increasing rapidly: many of the new theories (e.g. Newton's theories of 1666–1687) were confirmed by experiments.   In particular, the new experiments and observations were showing decisively that the current theories were superior to those of the ancients—Aristotle's physics (that bodies move only if they are being pushed) was wrong, and Ptolemy's maps clearly had been in error.

With this new self-confidence in Europe, scientific activity forged ahead; whereas before the 17th century basic experimental science had not been worthwhile either in intellectual or monetary terms, suddenly the new theories, with their elegance, sharp testing with the new instruments, and their utility in navigation and in waging war, made science very important.   Theories which made accurate predictions and stood up to repeated testing were clearly the best theories in this society.   The Scientific Revolution had occurred; the scientific approach was established as a philosophy.

The distinctive constraints of this philosophy are the four criteria (9):

- The scientist uses words and symbols in a relatively explicit, formal manner, in contrast to the vague, emotive words used in poetry and religion (love, grace, redemption, etc.)
- He has a strong aesthetic appreciation of the elegance of basically simple (though perhaps rather abstract and mathematical) general theories.
- He makes sharp predictions, and carries out precise experimental tests to see whether or not theories work accurately over wider and wider ranges (i.e. a scientific theory is potentially falsifiable).
- He is ready, in the face of criticism and refutation, to exchange or modify his theory in favor of a better one.

China and Arabia

One well may ask why there was no comparable spectacular development of science in ancient China or Arabia between the 9th and 12th centuries.   Why was there only one scientific revolution?   The threefold coincidence which initiated the revolution has been discussed above. There was never a comparable situation in China, where the Confucian religion (with its great respect for the records of antiquity, a dominant interest in social and moral problems, that the maxim the cautious never err), the inflexible written language (a new symbol is necessary for each new concept), and the lack of the competitive spirit (in contrast to that between the different European states), all conspired against any scientific activity sufficient to make a breakthrough.

Moreover, the ancient Chinese recruited their best scholars for State service, removing them from the practical affairs of life. Such a system effectively preserved the status quo. The result was that the technological feats (such as the extensive walls and canals and the high crop yields through composting) were achieved without abstract theories: the technique was that of a million men each with a teaspoon. There was never great interest in inanimate forces, nor was local initiative encouraged (the irrigation systems were controlled by the central authorities). Thus, in spite of skill, tenacity, and frequent outbreaks of plague in China, the intellectual climate remained quite unlike that of post-Renaissance Europe.

In Arabia, science was strongest between the 9th and 12th centuries. The tradition was continuous from late antiquity, based on Greek science and crafts. There was no sudden exciting rediscovery as in Europe at the Renaissance. Although there were considerable achievements in mathematics and astronomy, the crafts and intellectual attitudes changed very slowly; indeed alchemists still practiced their art in the ancient city of F×es (Morocco) until about 1956. Even today some of the stills used there for distilling perfumes are of a form dating back to the early centuries AD, when the Arabs improved the distillation apparatus by cooling (with water) the tube leading from the head of the still. They discovered a number of essential oils by distilling plants and juices, and also distilled crude oil to obtain white spirits (4). But the alchemists, seeking the transmutation of baser metals into gold, were trying to take a single great leap forward. They did not know (as we do now) how to shuffle forward by testing, modifying, or replacing their theories with better ones. Consequently, the alchemists continually fell back defeated in their main purpose. The Arab religion was not sympathetic to worldly knowledge for its own sake, nor to improvement through change. Such critical aspects as Arab science had in its great days were replaced gradually by more mystical attitudes, with the emphasis in alchemy on spiritual exercises rather than on scientific experiments. However, it was from the Arab world, via Sicily and Spain, that the crafts of the Alexandrian Greeks were transmitted to Europe in the 12th–14th centuries (*see* Figure 4).

Chemistry: The Modern Atomic Theory (1661–1811)

In the late seventeenth century chemistry, in spite of the empirical knowledge of the alchemists, and also of the tremendously improved understanding of physics, was developing only slowly and with great difficulty.   The birth of chemistry as a science perhaps can be dated from Robert Boyle's publication in 1661 of *The Skeptical Chymist*.   In this book, Boyle severely criticized the approach of the alchemists, whose researches were directed to the practical objectives of making gold and medicines, and whose rather mystical theories of matter were so vague and ambiguous as to cover all sorts of phenomena discovered subsequently.   Boyle maintained that the pursuit of chemistry should be related less closely to spectacular immediate objectives; but that systematic experimental tests were required.   This spirit of inquiry led him to analyze various substances, and he rejected the then-current theory of the four elements being earth, air, fire, and water.   Instead he advanced the opinion that only substances from which nothing different could be obtained by decomposition should be regarded as the elements of matter.

During the first half of the 18th century there was considerable interest in alkaline bases and earths, and in the isolation of new metals. Among the famous chemists of that time was the Frenchman H. L. Duhamel.   In 1736 he published his work comparing sea salt (i.e. NaCl) and digestive salt (i.e. KCl), showing that the acid was the same in both, but that the bases were different.   A year later he established the identity of the elemental bases of common salt and alkali (sodium carbonate) by heating common salt with sulfuric acid to obtain sodium sulfate, then reducing this with charcoal, treating the product with acetic acid to give sodium acetate, and finally igniting this to produce sodium carbonate *(11)*.

Lavoisier, beginning in 1772, experimented with various combustible substances, and in 1789 interpreted his results in terms of Boyle's concept of elements, among which he included many metals and solid nonmetals, and oxygen, hydrogen, and nitrogen gases.   The concept of chemical affinity was developed about this time as an empirical classification of substances according to their relative chemical reactivities.

Even at the beginning of the 19th century, chemistry still was not quantitative; it had to wait for the leading role of a theory.   John Dalton's Atomic Theory (1803–1808) served this purpose admirably.   Dalton visualized the atoms of any particular element as all being exactly alike, while those of different elements had different, but characteristic, weights. Compounds, according to this theory, are formed by the union of atoms of different elements in simple proportions.   From this theory Dalton predicted that chemical changes between substances would occur only in certain simple weight ratios, and confirmed this prediction experimentally.

With the theory of Avogadro (1811) which stated that the particles of gaseous elements normally consist of several atoms combined into groups (which he called "molecules") and his further hypothesis that equal volumes of all gases contain the same number of molecules, the atomic theory was virtually complete.

Chemical Processes and Distillation (1740–1913)

Higher agricultural productivity at home, and later the availability of cheap food imported into Britain from the New World, made possible both a great increase in population and a movement of people from the countryside into the towns. This concentration of labor and wealth in turn made possible the industrial revolution, i.e. craft processes could be practiced on a much larger scale, albeit rather empirically. For example, in the United Kingdom in 1785 there were 971 soapmakers, producing an average output of 16 tons each, but by 1830 there were only 309 soapmakers, with an average output of 170 tons per year each (12). In North America, soapmaking remained a household craft until 1800, with few industrial developments.

Alkali. The soapmakers, as well as the textile and glass manufacturers, required increasing quantities of alkali for their processes, and in Britain it was obtained (until about 1806) entirely from the ashes of certain plants and seaweeds. But towards the end of the 18th century, supplies of alkali were becoming inadequate and expensive—the scarcity is reflected in the fact that Lord Macdonald of the Isles was making £10,000 a year (an enormous sum in those days) from his share in the profits from burning Scottish seaweed to yield sodium carbonate. Alkali ashes also were being imported into Britain from America and Spain.

In France the supply position was worse. By 1776 the political and financial situation there was making the continuity of the imports of ash to that country (particularly from Spain) doubtful, and a prize was offered by the French Academy of Sciences for a new, commercial process in which soda alkali could be produced from common salt. Duhamel's reactions (mentioned earlier) were, of course, completely uneconomical, but it had been established clearly from such studies in pure chemistry that common salt, sodium sulfate, and sodium carbonate were related through the element sodium, and that a commercial process might, therefore, be achieved. It did not prove easy, however, and it was 1789 before Nicolas Le Blanc devised his process (described later) for making alkali from common salt. He did not base his process on the then-current theory of chemical affinity, which suggested that iron should be used to produce alkali from sodium sulfate because of the great affinity of iron for sulfate (13, 14). Indeed, the theory of the precise chemistry of the Le Blanc process remained obscure until about 100 years later, and Le Blanc well

Imperial Chemical Industries, Mond Division

Figure 9. Late Victorian "black-ash revolver," a revolving furnace used in the second stage of the Le Blanc process (18)

may have devised his process by means of a simple but fallacious analogy with the smelting of iron ore *(13, 14)*.

He patented *(15)* the process in 1791, and in 1794–1795 a small factory was operated at Franciade (near Saint-Denis, on the Seine in France) by Le Blanc, Dizé, and Shée *(16)*. The process proved uneconomical, however, and not until 1808 did it flourish, when there was a special remission of the salt tax and when the supplies of alkali ashes were proving to be seriously inadequate because imports of foreign alkali were discouraged actively *(13)*. In 1810 there were several alkali factories operating in France; that year the one at Marseilles produced 1000 tons. In 1814 its production was 3500 tons, and in 1820 it was 9000 tons *(17)*.

In Britain the Le Blanc process was introduced between 1802 and 1806, but Britain was trading very widely and still was importing much alkali ash. Consequently, the Le Blanc process did not develop rapidly there. It wasn't established firmly until 1823; but by 1840 so much alkali was being manufactured in Britain by the Le Blanc process that there was enough being made to supply the local market as well as creating a surplus for export to America.

The Le Blanc method (used on a commercial scale in Britain until 1885) made alkali by first heating salt and sulfuric acid in batches to give Na_2SO_4 and HCl gas. The solid sodium sulfate was heated then in revolving furnaces (*see* Figure 9) with a mixture of limestone and coal to produce a solid product (black ash) from which sodium carbonate then was extracted. A very dirty calcium sulfide remained. Disposal of this was a problem, but the HCl gas was even worse. At the original Le Blanc factory at Franciade the small amount of HCl gas produced was partly passed into the atmosphere, but some was converted to ammonium chloride by collecting the HCl in a lead chamber and then introducing ammonia vapor (15, 16). But in Britain in the years following 1823, the quantities of hydrogen chloride from the Le Blanc process were so vast that the treatment with ammonia was not feasible: the HCl was passed directly into the atmosphere (*see* Figure 10) causing great damage to crops and trees in the vicinity of the works. This necessitated considerable financial compensation (19).

In 1836 William Gossage had the happy inspiration of using a derelict windmill, which happened to stand near his works, as an absorption tower to remove the HCl gas. He filled the old windmill with gorse and brushwood from the surrounding countryside, and irrigated this packed tower with a downward-flowing stream of water, introducing the HCl gas at the top (11). This simple absorption tower worked very well, and so was born empirically what is still a standard chemical engineering operation. Gossage in his patent of 1836 stated clearly the importance of using extensive surfaces over which water is caused to pass in the same direction as the smoke and gas. Towers of stone or brickwork, packed with twigs, broken bricks, or coke, soon were used widely to absorb the HCl gas from the Le Blanc process, though the HCl solution flowing from the bottom still had to be disposed of, usually by dumping it into a convenient river.

A more direct route to alkali from common salt is the ammonia–soda process (1863 +) involving a chemical reaction between CO_2 and a concentrated aqueous solution of NaCl saturated with ammonia. This reaction was discovered in 1811 by A. J. Fresnel, who had shown in the laboratory that $NaHCO_3$ could be precipitated from saturated NaCl solution in the presence of ammonium bicarbonate (21). But repeated efforts to scale-up the reaction to commercial production were, for many decades, all frustrated because of the difficulties in recovering and conserving the ammonia and in obtaining sufficiently pure CO_2 and using it under pressure. Not until 1861 did the Belgian, Ernest Solvay, approaching from an engineering standpoint the problems of efficient CO_2 absorption and of distilling off and recycling the ammonia with minimal losses, achieve a commercial process. So important for success was the engineering (particularly the high-efficiency carbonating tower (21) 80 ft

Figure 10. Muspratt's factory for making soda by the Le Blanc process, near Liverpool, England, about 1830. Note the tall chimney stack for dispersing the vast amounts of HCl gas, and also the windmill nearby (20).

high with the ammoniated brine entering at the top and the CO_2 gas at the bottom, and containing plates and bubble caps) that the ammonia–soda process was given Solvay's name. In 1873 the Solvay process was introduced into England by Ludwig Mond and John Brunner. By 1885 it rapidly was replacing the Le Blanc process, over which the Solvay process had three advantages: (1) an easier separation step (filtration to remove the precipitated $NaHCO_3$); (2) the absence of a dirty and difficult-to-dispose-of by-product; and (3) continuous operation. Thanks to the Solvay process, the price of sodium carbonate fell from about $80 a ton in 1870 to about $24 a ton in 1900. A typical Solvay plant is shown in Figure 11.

Figure 11. A typical Solvay plant at Winnington, England (18)

Imperial Chemical Industries, Mond Division

Coal Carbonization. Coal was first carbonized on a practical scale at the beginning of the 18th century to obtain coke for smelting iron ores, but not until the end of that century were any by-products recovered. Coal gas was produced for lighting by William Murdoch in 1795, and a few years later he was manufacturing coal gas on a sufficient scale to light a factory in Birmingham (in the Soho area of that city) by gas flames. By 1823, in London alone, 250×10^6 cu ft of coal gas per year were being produced, mainly for illumination, from coal carbonization plants. Associated with the gas making were the by-products ammonia, coal tar, and coke. The ammonia was needed for nitrogenous fertilizers such as ammonium sulfate (and later for the Solvay process), and the tar (after distillation) gave a variety of useful products. The coke was used as a fuel and in metallurgy (and also for packing absorption towers).

Sulfuric Acid. This was required in increasing quantities for many developing industries (e.g. textile treatment, fertilizers, and alkali manufacture) and as Liebig pronounced in 1843, that it was no exaggeration to say that we may judge fairly the commercial prosperity of a country from the amount of sulfuric acid it consumes.

As long ago as the 1730's, Joshua Ward had begun manufacturing it in small batches by burning sulfur-containing substances and saltpeter (KNO_3) above a shallow layer of water under a glass bell. In 1746, Roebuck and Garbett in Birmingham (England) scaled-up the reaction, making a reaction chamber (about 6 sq ft) from lead sheets supported on a wooden framework. Other small plants soon followed, each making a few tons of acid per year. The effect of the larger scale of the lead chamber process on the price of sulfuric acid was striking—from £280 a ton in 1746 to £50 a ton a few decades later.

Freed now from the limitations of glassware, the size of the acid chambers soon increased from a few hundred cubic feet to chambers each with the capacity of a large concert hall, several being used in sequence.

Quite early in the development of the chamber process (about 1800) it was made continuously by blowing the gases and steam through the chambers, and it also was shown (in 1806) that the saltpeter served as an important intermediary. Before this it had been thought that sulfuric acid resulted from the simple combustion of sulfur in air, the saltpeter being supposed merely to accelerate the burning of the sulfur, and that a cheaper method of promoting rapid combustion would be to burn the sulfur in a strong current of air. But these attempts had ended in failure: the theory was wrong. In 1806 Clément and Desormes in France showed that the action of the saltpeter was really to decompose into nitrogen oxides, which then catalyzed the formation of sulfuric acid—the first clearly characterized example of a catalytic reaction *(16)*. The catalyst had been found entirely by chance.

Towards 1810 it was becoming clear from chemical theory that one part of sulfur should furnish about three parts of concentrated acid.

Figure 12. Batchwise concentration of sulfuric acid at Clermont–Ferrand, France, in the middle of the 19th century (24)

Chemical science thus could assist the progress of chemical technology. With better understanding and control of the process, yields rose dramatically from about 30% to 80% or even 90% (16). The next step was to recover the rather expensive nitrogen oxides, and about 1830 Gay–Lussac (in France) devised his absorption tower to recover these oxides from the gases leaving the chambers. The tower was packed with coke (or later, stoneware in molded forms), over which sulfuric acid flowed. But the tower proved rather difficult to operate in practice (22) and was not adopted generally until 1869.

Concentrated sulfuric acid was required for many processes, and this concentration was carried out in small batches in the mid-19th century (*see* Figure 12). In 1859, however, Glover (following Gossage's earlier design of a tower for concentrating sulfuric acid (23)) had developed his packed tower to concentrate the acid using the incoming hot gases from the sulfur burners, and to recover the nitrogen oxides dissolved in the acid coming from the Gay–Lussac tower.

One problem Glover had to face was the high operating temperature, but he finally solved this by making his tower of firebricks set in molten sulfur, with fireclay tiles for packing (25). Another was that a tower designed to be a good denitrator is not necessarily a good evaporator (23). After about 1869 when both the Gay–Lussac and Glover towers were in common use, higher concentrations of nitrogen oxides could be used economically, and the size of the lead chambers (for a given output of sulfuric acid) was halved (26).

In the year 1820 the United Kingdom produced 3,000 tons of sulfuric acid, but by 1860 (after the Le Blanc soda process had become a major user of sulfuric acid), production had risen to 260,000 tons. In 1900 it had reached nearly 1,000,000 tons (a quarter of the world's output) (27).

Agricultural Fertilizers. Manure and fertilizers are essential for improving crop yields. Using farmyard manure is as old as agriculture itself, and the use of wood ash was known to the Celtic people in ancient times, and also to the Arabs. The properties of manures and composts were described by the Arabs in the 10th century. Waste wool and bones were used also in olden times to improve crop yields. In England in the 16th century, marl (calcium carbonate-rich clay) was used to improve crops, and by 1718 ground mineral phosphates were being applied also to the soil. Ground bones were being applied more widely also in the 18th century (28) and yields of grain crops per acre had, by then, increased to twice those of medieval times. By 1827 some 40,000 tons per year of bones were being imported for use as agricultural phosphate fertilizer. But the efficacy of ground bones was limited by their very low solubility, and following preliminary suggestions by Kohler in Austria and by Liebig, J. B. Lawes in 1841–1842 established a factory for making the bone phosphate more soluble by treatment with sulfuric acid to give superphosphate. Acid from the chamber plants (about 70% H_2SO_4) had to be concentrated to 80% before it could be used for solubilizing the bones, this concentration being affected in lead pans supported by cast-iron plates protected from the fire by refractory bricks (see Figure 12).

Lawes soon found that the supply of bones was becoming limiting, and in 1842 he extended his operations to include mineral phosphates, later imported from Norway, Belgium, and the United States. The superphosphate industry became the largest user of sulfuric acid, and by 1861, 40,000 tons a year of phosphatic materials were being solubilized.

A synthetic nitrogenous fertilizer, ammonium sulfate, was manufactured first in the United Kingdom in 1815, using ammonia from the coal gas plants and sulfuric acid; by 1879 the national output of ammonium sulfate had reached 40,000 tons a year. With superphosphate and the nitrogenous fertilizers being produced in quantity, grain yields in England now rose to 3.3 times those of the Middle Ages. But ammonia production remained limited to the coal carbonization plants until 1913 when the Haber process was developed. Although Döbereiner (1823) had found traces of NH_3 when hydrogen was burned in air, the industrial production of synthetic ammonia was impossible until suitable catalysts had been discovered and the engineering techniques were available for working at the necessary pressures of 200–300 atm. Partly due to chemical fertilizers, the crop yields per acre are today about 5 times those of the Middle Ages.

The composting of organic waste materials, long practiced by the Chinese, changed very little over the centuries, being essentially a small-scale batch operation.   With the adoption of the process by the Western world in the present century, progress has been made in understanding the fundamental chemical reactions involved and in applying composting to large-scale waste treatment *(29)*.

Distillation. The practice of distillation was introduced into Northern Europe from the Arab world, via Spain and Italy, during the 12th–14th centuries.   The distillation of alcoholic drinks from the 14th century onwards has been described above.   By the 17th century a little natural crude oil also was being distilled commercially, and at Broseley (in the Midlands of England) oil extracted from the local bituminous rock was being distilled to give a turpentine-like distillate and a pitch residue. In 1746 a patent was granted in the United Kingdom for the distillation of oils from coal tar *(30)*.   Later in the 18th century, wood tar was distilled in England to produce pine oil, and by 1822 coal tar was being distilled (also in Britain) to yield a light oil (naphtha), which was used as a lamp oil and a solvent.   When Macintosh in 1823 required a light oil in which to dissolve the rubber used for his waterproof garments, he obtained it from a small tar-distilling firm in Leith, which obtained its tar from Birmingham.

In 1838, in Birmingham, heavy tar oils (creosotes) were being used for preserving railroad cross ties and other timbers *(30)* and by 1850 many more tar distilleries were in operation, and tar oils were being exported from Britain.   In 1860 the first petroleum refinery was built in Pennsylvania.   However, these early tar and petroleum refineries were extremely simple, since no close fractionation was required.   The products simply were collected (condensing the vapors in a water-submerged coil) on the basis of specific gravity, which varied with the time and temperature of the distillation of the batch of oil.   Though some of these oil stills later were connected in series and operated continuously, they were, from a thermal point of view, quite inefficient.   They also produced only wide-boiling fractions.

Alcohol distillation demanded better equipment, and in the early years of the 19th century several stills were designed in which the vapors passed through cylinders which were divided into compartments by perforated plates.   These horizontal stills operated by partial condensation *(4)*.   In France in 1818 J. B. Cellier devised a still for making brandy from large volumes of dilute aqueous solution, using a vertical column with bubble plates.

In 1830 Aeneas Coffey of Dublin designed a still which operated continuously and gave good alcohol separation *(31)*.   He fed the preheated mixture of water and alcohol (from fermentation) into a vertical series of shallow chambers placed one on top of the other, separated from each other by perforated plates, heated by live steam, and using reflux *(see*

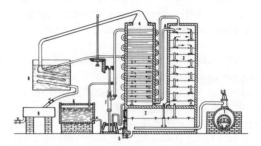

Figure 13. The Coffey still of 1830 for distilling alcohol from fermented mash, using perforated plates. 1, Boiler. 3, Stripping column. 4, Rectifying column. 6, Feed. 8, Condenser.

Figure 13).   In such a still 85% ethanol could be produced from an initial 5% of ethanol.   The Coffey still was the forerunner of the modern column, and its efficiency in separating close-boiling fractions led later to its adoption for hydrocarbon separations, e.g. Coupier's benzole still (1863) for purifying benzene from toluene.   Partial condensation also improved the separation achieved in the Coupier still.

In the tar stills at Oldbury, Birmingham, an improvement was made in 1869 in the overall thermal efficiency.   Vapors from one still were passed through a coil immersed in the tar being prepared for the next distillation, warming it and boiling off any associated water *(30)*.   But not until the end of the 19th century and the early decades of the 20th century were the great advances made which now characterize the highly efficient and selective distillation operations designed by chemical engineers.   For example, it was 1900 when fractionating columns containing perforated trays were introduced into the tar stills at Oldbury. Today it is routine to fractionate close-boiling mixtures such as *o*- and *p*-xylene by distillation.   A modern fractionating tower for crude petroleum is shown in Figure 14.

Chemical Engineering Knowledge Up to 1915

During the 19th century, the emphasis was on process development. The successful development of new processes generally had been achieved with fairly simple empirical machines and structures—furnaces (sometimes revolving, e.g. the black-ash revolvers (*see* Figure 9) used in the Le Blanc process after 1853), and various types of pots, ovens, and mixing vessels in which chemical reactions were brought about.   A few industries, however, required considerable engineering (e.g. the columns and towers of the Solvay process).   The importance of such Chemical Engineering was recognized clearly by 1880, in which year an attempt was made to found a Society of Chemical Engineers in London *(26)*.   At this stage a chemical engineer was regarded as a mechanical engineer with some knowledge of process chemistry. By the end of the century, the Coffey still was becoming used more widely, though the approach to equipment generally remained very empirical. But separations operations, then as now, were by far the most expensive steps in most processes, and with the need for increasingly pure chemicals (e.g. aromatics for dye synthesis), greater interest began to be taken in separation equipment.

The influence of basic science was far from strong in this field. Whereas in chemistry itself, the theory of elements had suggested the

Esso

Figure 14. Modern fractionating tower for separating the components of crude oil, Fawley Refinery, England (32)

possibilities of making Na_2CO_3 from NaCl, and later the atomic theory helped greatly in the synthesis of various new dyes (e.g. Kekulé's theory of the benzene ring in 1865), the separations equipment (e.g. absorption towers) remained empirical. The mechanical and structural engineering (and indeed the plumbing) of the chemical industry scarcely were affected by such pure science as the Second Law of Thermodynamics (1854), theories of heat transfer, Reynold's characterization of fluid flow (1883), or the dimensionless group theory.

Materials of construction of large tanks were posing problems. Glass was unsuitable for the scaled-up processes, and such chemically resistant materials as lead, ceramics, high-silicon iron, enameled pans, and rubber linings (the latter first suggested by Le Blanc (15) for containing sulfuric acid) were being considered and used on a trial-and-error basis (33).

Chemical Engineering Education (1887–1915)

A hundred years ago George E. Davis was the Alkali Inspector for the "Midland" region of England. His function was to monitor the pollution coming not only from the alkali factories but also from the many other types of chemical plants in the region. He thus obtained access to a wide variety of chemical processes, incidentally forming in his mind the rudiments of modern Chemical Engineering education.

After resigning from the Alkali Inspectorate in 1884, Davis became an independent consultant, and in 1887 he gave a course of lectures at the Manchester Technical School (England) in which he analyzed the various contemporary chemical processes technologies into a series of basic operations (now called "unit operations"). Davis (34) pointed out that all of the diverse chemical process plants were largely combinations and sequences of a comparatively small number of operations such as distillation, evaporation, drying, filtration, absorption, and extraction. Davis thus rediscovered the steps (or operations) of Paracelsus, typical of laboratory chemical preparations, but not familiar to the mechanically minded chemical engineers of Victorian times. The dazzling successes of the 19th century chemical processes did not blind Davis to the importance, for plant design, of the operations approach in the many and varied chemical industries of which he had experience. He published these ideas, from his 1887 lectures in the *Chemical Trade Journal* during the next few years. In 1901 he systematized this approach in his *Handbook of Chemical Engineering*. Davis was motivated by a concern for industrial competition in chemicals from the United States and Germany (26), and by a realization that the scaling-up of a chemical plant required a new sort of chemical engineer. The book was such a success that a second, enlarged edition of over a thousand pages appeared in 1904 (23, 26).

In the Preface, Davis wrote:

"The object of this Handbook is not to enable anyone to erect a works of special character . . . but to illustrate the principles by which plant of any kind may be designed and erected when certain conditions and requirements are known. We cannot make the best use of our abilities unless we are taught to investigate the principles underlying the construction of the appliances with which we have to work" (26).

In the 1904 edition there is, for example, a sample calculation of the heat balance on a Glover tower treated as an evaporator, which shows how inefficient it was then ("what a heat waster it is" (23)). There is also a discussion on the efficiency of various packings, explaining in terms of surface areas why coke is 1.5 to 2 times more efficient than bricks (23, 26). But in general, Davis' approach was still empirical; the operations are described as procedures of practical utility, and are not based on fundamental physics. Neither the work of Osborne Reynolds nor dimensionless group theory had been assimilated yet into the profession.

Davis' idea, it is interesting to note, was adopted in the United States much later; George E. Davis "presented the essential concept of unit operations and particularly an understanding of its value for education" (W. K. Lewis (35)). At MIT, Arthur D. Little in 1915 coined the term "unit operations", and W. K. Lewis' textbook was organized on the basis of Davis' system. The oil and petrochemical industries, developing on a large scale in the United States, undoubtedly accelerated the striking and rapid growth of the unit operations approach in that country.

In England also, Davis' concept was being accepted. In 1907 the University of Birmingham started its mining degree course, and in the 1910–1911 session, there were included in this course such basic operations as crushing, conveying, pumping, and hydraulic separations. There were also lectures on fluid flow, including flow through closed channels, orifice flow, and the behavior of free-falling particles. In 1911–1912 a lecture course on the refining of petroleum was added, and in 1912 the latter topic was part of a new degree course entitled "Petroleum Mining." The associated laboratory experiments included the distillation of crude petroleum. Over the years, the refining side of the course steadily assumed more importance, and in 1922 a separate Department of Petroleum Engineering was established in Birmingham. It is interesting to note that some of the research work in this department was concerned with the hydrogenation of coal—60 years ago the era of massive imports of cheap oil was still to come and go! The Birmingham Department of Petroleum Engineering was later (1946) renamed the Department of Chemical Engineering.

Another early course in England was that at the Battersea Poly-technic (London).   In the 1914–1915 session a course entitled "Chemical Engineering" was set up, the basic operations being dealt with more explicitly than in Davis' handbook *(36, 37)*.

Modern Chemical Engineering

Continuous processes became more common in the 1920's and 1930's. To operate these and scale-up a process efficiently, the flows and recovery of heat had to be understood.   Thus thermodynamics, material and ther-mal balances, heat transfer, and turbulent flow (particularly the behavior of eddies at interfaces *(38)*), as well as reaction kinetics and catalysis became (and are still) the foundations on which chemical engineering rests.   Dimensionless group analysis, used much earlier by physicists, became used more widely by chemical engineers.   Of course materials of construction (e.g. stainless steels, titanium, and Teflon coatings) are im-portant too, as are automatic control, computer programs, operations research, and critical path planning.   But the latter techniques may continue to change, and thus thermodynamics, material and thermal balances, turbulent flow, and reaction kinetics of continuous flow systems must constitute a hard core of chemical engineering knowledge.   Allied with this core are the technologies of modern separations processes (including liquid–liquid extraction and distillation).   These are particu-larly important because the separations equipment in a typical chemical plant costs many times more than the chemical reactor itself.

The Philosophy of Chemical Engineering

Besides differences in the scale of their operations, there can be different motives for theorizing and experimenting between chemical engineers and pure chemists.   For the pure chemist the motives are usually curiosity and a desire to see a simplifying pattern relating ap-parently disconnected phenomena.

The motive of the engineer, on the other hand, is to make something which will operate satisfactorily, and the engineer usually has less choice than the pure scientist in the systems to be studied.   Many useful and important chemical engineering processes involve very complex mate-rials, including mixtures containing many components, and liquids with anomalous flow properties.   To deal quantitatively with these systems so that the effects of variations can be predicted precisely, the chemical engineer needs to alter as many of the variables as widely as possible. Preferably this is done by studying them one at a time, though often this procedure is not physically possible.

Often, in dealing with a complicated practical situation, the engineer arbitrarily reduces the number of variables in his theory by combining them into dimensionless groups, of which a well-known example is the Reynolds number characterizing the flow of fluid through a pipe. Such dimensionless groups are evaluated in the laboratory, and are used then for predicting the behavior in a large-scale chemical plant. But this procedure reduces somewhat our confidence in our predictions; though the group as a whole may have varied widely in the laboratory experiments, one or more of the variables within the group may have been virtually unchanged. Because of this reduced confidence in using dimensionless groups in scaling-up predictions, the chemical engineer usually builds a pilot plant, intermediate in size between the laboratory system and the proposed full-scale production plant, so that he can check whether the scaling-up predictions of his simplified theory are working sufficiently accurately.

If, however, the chemical engineer can evaluate his variables separately (i.e. can go basic), thus putting forward a sufficiently complex but nevertheless precise theory to predict the behavior of his complicated practical system, he can eliminate the pilot-plant stage, proceeding to scale-up directly from laboratory studies to the design and construction of the full-scale production plant. This saves considerable expense and time.

In recent years the ready availability of computers to handle the algebraic calculations, and the widespread use of the methods of operations research, have made it much easier for the engineer to use more complicated mathematical correlations on which to base his predictions. This approach is, however, less fundamental than is a theory in chemistry or physics which links previously unrelated concepts.

Just as there are often several possible ways of designing a given laboratory experiment in pure science (e.g. in the detection of fundamental particles, or in the synthesis of some substance), so there are usually many possible ways of designing a chemical plant. The chemical engineer, for example, designing a plant to produce a new polymer, can arrange the required sequence of mixers, reactors, coolers, and other pieces of equipment in various spatial relationships to one another, and the plant still will work. Design is thus partly an art, though theoretical considerations will dictate to the modern chemical engineer whether, for example, he should use several chemical reactors in series or a single larger reactor with recycle.

The Scientific Society

A scientific approach to problems such as food supply, depletion of resources, and pollution is more necessary than ever. Projections of trends into the far future are always unreliable because of man's great ingenuity in changing the trend, through his conscious wish or need to do so, particularly through the application of new discoveries. Who, for example, in 1930 would have predicted transistorized computers, electricity from nuclear power stations, or penicillin? Who today would venture to predict what completely new devices, resources, chemicals, high-yield strains of crops, and power supplies will be available in the year 2030? Will it become feasible, for example, to modify the bacteria in soil to "fix" atmospheric nitrogen (to make it available to growing plants) thus avoiding the need for applying chemical fertilizers?

For 400 years mankind has enjoyed the intellectual endeavor and the material fruits of the scientific approach; indeed his application to ordinary life of new scientific discoveries has been faster and faster. Thus man clearly has wanted the scientific society, and is prepared to pay the prices of rapid change, uncertainty, and a certain amount of industrial pollution. On balance, for most people (particularly in a world of rapidly increasing population) the advantages of the scientific society considerably outweigh the disadvantages. But the disadvantages can be, and should be, reduced to a low level by studying them scientifically (9).

The freedom (through engineering and science) from hunger, cold, and disease in the developed countries has given man a new confidence and a choice—the first real choice since the end of the Middle Ages. Western man now can decide between working hard to achieve a higher standard of living, or enjoying more leisure at the existing standard of living. But even to maintain this existing standard in the face of a rising world population, depletion of resources, and the threat of increasing pollution, still will require a sustained chemical engineering research effort (9, 39).

Acknowledgement

The author is grateful to J. R. Harris for his friendly help on the history of the Le Blanc process, and to the Associated Octel Company for their interest and assistance.

Literature Cited

1. Taylor, Sherwood F. *Chem. Ind. (London)* **1937**, 38–41.
2. Ellis, S. R. M.; Mohtadi, M. F. in "Modern Petroleum Technology," 3rd ed.; Inst. of Petroleum: London 1962; p. 272.
3. Haldon, J.; Byrne, M. *Byzantinische Zeit.* **1977**, 70, 91–99.
4. Forbes, R. J. "Short History of the Art of Distillation"; E. J. Brill: Leiden, 1948.
5. da Vinci, Leonardo Drawing 12660 Verso, 1509, from the Royal Library, Windsor Castle, England.
6. Paracelsus "Von Naturlich Dingen"; 1527.
7. Serpent cooler, according to 16th century sources, copied, for example, by Forbes (4).
8. van der Straet "Nova Reperta"; distillation and sugar processing in the 16th century.

9. Davies, J. T. "The Scientific Approach," 2nd ed.; Academic: London and New York, 1973.
10. Bacon, F. "Distributio Operis," prefixed to the "Instauratio Magna"; see "The Philosophical Works of Francis Bacon"; Robertson, J. M., Ed.; Routledge and Sons: London, 1905.
11. Hardie, D. W. F. "A History of the Chemical Industry in Widnes"; I. C. I. Ltd.: Liverpool, England, 1950.
12. Hardie, D. W. F.; Pratt, J. D. "A History of the Modern British Chemical Industry"; Pergamon: Oxford, 1966.
13. Gillispie, C. C. "The Discovery of the Leblanc Process," *Isis* **1957**, *48*, 152.
14. Gillispie, C. C. "The Natural History of Industry," *Isis* **1957**, *48*, 398.
15. LeBlanc, N. French Patent 1791, (9) 170.
16. Smith, J. G. "The Origins and Early Development of the Heavy Chemical Industry in France"; Clarendon Press: Oxford, 1979; pp. 63–66.
17. Barker, T. C.; Dickinson, R.; Hardie, D. W. F. "The Origins of the Synthetic Alkali Industry in Britain," *Economica* **1956**, 158.
18. Photographs 64/1127P and 70/756/1.
19. Barker, T. C.; Harris, J. R. "A Merseyside Town in the Industrial Revolution. St. Helens, 1750–1900"; Frank Cass & Co.: London, 1959.
20. Muspratt's chemical works, Vauxhall Road, Liverpool, England, about 1830.
21. Reader, W. J. "Imperial Chemical Industries, a History"; Oxford University Press: London, 1970; Vol. I.
22. Firth, G. "The Northbrook Chemical Works, Bradford, 1750–1920," *Ind. Arch. Rev.* **1977**, *2*, 52.
23. Davis, G. E. "A Handbook of Chemical Engineering"; Davis Bros.: Manchester, 1904; Vol. II, pp. 204–207, 271–3.
24. Maw, W. H.; Dredge, J. *Engineering (London)* **1876**, *21*, 145–150.
25. Campbell, W. A. "The Chemical Industry"; Longman: London, 1971.
26. Davis, G. E. "A Handbook of Chemical Engineering"; Davis Bros.: Manchester, 1904; Vol. I. preface, pp. 3, 12–13.
27. Fleck, A. "The British Sulphuric Acid Industry," *Chem. Ind. (London)* **1952**, 9, 1184.
28. Todd, Lord "Chemistry and Agriculture," *Chem. Ind. (London)* **1978**, 357.
29. Gray, K. R.; Biddlestone, A. J.; Clark, R. "Review of Composting, Part 3: Processes and Products," *Process Biochem.* **1973**, *8* (10), 11.
30. *M.T.D. Magazine* (Midland Tar Distillers Ltd., Oldbury, Birmingham) **1965**, *48*, 19–42.
31. Rothery, E. J. "Aeneas Coffey, 1780–1852," *Chem. Ind. (London)* **1969**, 1824.
32. Modern still for refining crude oil, Fawley Refinery, England, an Esso photograph.
33. Swindin, N. "Engineering Without Wheels"; Weidenfeld and Nicholson: London, 1962.
34. Swindin, N. "The George E. Davis Memorial Lecture," *Trans. Inst. Chem. Eng.* **1953**, *31*, 187.
35. Lewis, W. K. *Chem. Eng. Prog.* **1958**, *54*(5), 51.
36. Peck, W. C. "Early Chemical Engineering," *Chem. Ind. (London)* **1973**, 511.
37. Tailby, S. R. "Chemical Engineering Education Today," *Chem. Ind. (London)* **1973**, 77.
38. Davies, J. T. "Turbulence Phenomena"; Academic: New York, 1972.
39. Davies, J. T. "Energy and the Environment: Educational Needs," in "Energy and the Environment"; Walker, J., Ed.; University of Birmingham: England, 1976.

RECEIVED May 7, 1979.

Conceptual and Institutional Obstacles to the Emergence of Unit Operations in Europe

JEAN–CLAUDE GUÉDON

Institut d'histoire et de sociopolitique des sciences, Université de Montréal, C.P. 6128, Succursale "A", Montréal, P.Q., H3C 3J7, Canada

The constitution of a discipline such as chemical engineering depends on a complex process of negotiations between universities, engineering associations, and industries. In the United States, universities thoroughly reorganized their curricula around unit operations, and chemical engineers were fitted easily within industry. In Germany, by contrast, industries reorganized their division of labor and formed teams of chemists with engineers. In Britain and France, chemical industries never succeeded in formulating their needs clearly. Teaching institutions were out to serve themselves and not the industry. Thus, institutional characteristics led each of these countries in directions other than the American one. Furthermore, in Germany, an alternate conceptual scheme came to occupy a place functionally similar to that of unit operations in the United States.

The notion of unit operation is a wholly American achievement. It also represents an entirely original solution to many of the problems confronting chemical industries. In another work *(1)*, we have attempted to show that this notion of unit operation came to be, not as the result of A. D. Little's creative genius, but as the output of a complicated process involving chemical industries, educational institutions, and professional organizations. In effect, it can be argued—and this is what we tried to do—that industries, universities, and engineering associations tacitly negotiated, and that they did so from whatever source of strength they could muster. The notion of unit operations—and we take it that this notion is familiar to readers of this chapter—can be approached then

0-8412-0512-4/80/33-190-045$07.75/1

as an organizing principle incorporating the compromises inherent in the process of negotiations just alluded to. In particular, unit operations, at the time of its inception, had no rival as a successful management strategy. As David Noble has stated, A. D. Little had done for industrial chemistry what Taylor had done for mechanical engineers (2). In fact, the tenets of scientific management, particularly those concerned with a redistribution of the division of labor along modular lines, seem to have influenced deeply A. D. Little's conception when, in 1915, he first defined unit operations.

Meanwhile, in Europe, events proceeded very differently. If we survey the period running roughly from 1800 to 1925, we see that each major European nation—namely Great Britain, France, and Germany— witnessed a process of negotiations affecting the development of their chemical industries, just as the United States did. However, the partners, so to speak, involved in these negotiations, while similar in name, were quite different in nature. British chemical industries differed markedly from its German counterparts and significantly from its French competitors. Likewise, educational systems and professional organizations were quite different from one country to another.

The thesis of this chapter is simple to state, if not to demonstrate: it is that the organizing principles guiding European chemical industries evolved with a great deal of specificity coinciding with national boundaries. As a result, each country came to a kind of modus vivendi which maintained itself as long as international competition remained on a commercial plane. When the First World War broke out, however, only Germany could rest satisfied with the basic principles which formed the underpinnings of its chemical industries, while Britain and France had to come to terms with the fact that their chemical industries hid deep organizational weaknesses.

In the remainder of this chapter, we will attempt to identify the significant partners in this process of negotiation which we see as central in order to understand the evolution of chemical industries in Britain, France, and Germany. For the sake of clarity, we will limit the use of chemical engineering to the unique process of negotiation leading to the notion of unit operations in the United States. In the case of Europe, where different processes of negotiations took place, we will use the phrase "industrial chemistry" until the American model is adopted. We will do this even though titles such as "chemical engineers", "ingénieurs chimistes," or its German equivalent were used in all three countries concerned. By arguing that European chemical industries each evolved along specific lines, we hope to show that the solution provided by the notion of unit operations to American chemical industries could not emerge in Europe.

The Case of Great Britain

As recently as 1962, D. W. F. Hardie wrote lines which reveal a great deal about one of the lasting tendencies of British chemical industries:

> "It has been widely and readily assumed that increase in the scale and complexity of its operations, and the greatly increased and increasing applications to them of scientific data and methods, have transformed chemical technology into something other than what it has hitherto been. . . . The operational and process sides of chemical technology are still carried on very much in the empirical tradition which I have sought to define and exemplify" (3).

Curiously, Hardie's emphasis on the empirical side of chemical technology and his downplaying of the scientific elements which make up modern chemical engineering sound like faint echoes of heated debates taking place in Great Britain in the latter part of the nineteenth century and the first decades of the twentieth.

For many years, Britain had led the field in chemical industries as she had done in practically every other aspect of industrial life. The Tennants and Muspratts, to mention but a few names, had conceived large-scale industrial production of bleaching powder and soda (4, 5, 6, 7). Rather rapidly and, at any rate, by the 1850's, Britain had cornered the market for what was to be called heavy chemicals, i.e. sulfuric acid, soda, bleaching powder, and the like. Through a series of maneuvers which never have been told in detail, Britain managed to displace French chemical industries from their commanding lead and began to expand and conquer new markets such as the rapidly growing market of the United States (8).

By and large, this growth of the heavy chemical industry took place without much reference to either scientific discoveries or even training. The personnel, on the whole, learned what it needed on the spot and the industry's evolution, although rapid, was gradual enough to permit in-house foremen and technicians not only to adapt but even to contribute to its progress. These improvements rarely involved chemical changes; more often than not, progress took the shape of new vessels and machinery, larger batch sizes, and so on. Yet, they were essential as long as competition among companies took the simple form of selling ever larger quantities of products at ever lower prices. For example, the transition from Sicily sulfur to pyrites which occurred ca. 1830–40 was important for manufacturing sulfuric acid. However, it took place in haphazard fashion (9) and the crucial improvement took the shape of a good roasting device needed to extract the sulfur, and not in a new chemical process. In other words, progress had much more to do with the design of new equipment than with the invention of new chemical processes.

On a more fundamental level, it is easy to see that chemical industries were troublesome to the English to the extent that they could not be fitted readily into their recently acquired understanding of "mechanical manufactures." Chemical industries did not conform to the principles governing the organization or the mechanization of mechanical manufactures. In 1835, Andrew Ure, himself a chemist, clearly stated why this was the case:

> "A mechanical manufacture being completely occupied with one substance, which it conducts through metamorphoses in regular succession, may be made nearly automatic; whereas a chemical manufacture depends on the play of delicate affinities between two or more substances, which it has to subject to heat and mixture under circumstances somewhat uncertain, and must therefore remain, to a corresponding extent, a manual operation" (10).

Ure was not alone in making an exception of chemical industries. Babbage also maintained this distinction. However, Babbage was interested in the fact that industries in general could benefit from the input of science and, somewhat curiously, used chemistry to make his point, which he meant to be valid for all manufactures.

> "The method now practised, although not mechanical, is such a remarkable example of the application of science to the practical purposes of manufactures, that in mentioning the advantages derived from shortening natural operations, it would have been scarcely pardonable to have omitted all allusions to the beautiful application of chlorine, in combination with lime, to the art of bleaching" (11).

Both Ure and Babbage can be taken as fair representatives of British theoretical concerns for the organization and management of industries. As such, they display the difficulties inherent in chemical industries. In particular, how could they benefit from the intervention of the "engine man" soon to become the engineers (12)? As for science, its role within chemical industries was to remain miniscule in Britain for the greater part of the nineteenth century, and, as Babbage had pointed out, the pursuit of science was to be the preserve of wealthy men, probably sons of manufacturers, seeking intellectual status through this newly respected facet of high culture (13). This is in fact exactly what happened to James Sheridan and Edmund Knowles Muspratt (14). And science may have been seen as a source of wealth by Perkin, the creator of the synthetic dye industry, but it must not be forgotten that Perkin sold his business as soon as he had made enough money to pursue his own research in "pure" chemistry *(15).*

Given this state of affairs, the British process of negotiations affecting the development of chemical industries involved but one real participant around 1880, or more exactly one buyer—the industrialist. Scientists and engineers did not look all that attractive to this unique buyer and, therefore, they could not have asked for much even if they had been in a position to imagine what they should ask for in the first place. But this was hardly the case, as British professional organizations were just beginning to emerge at that time and were largely under the control of manufacturers, and not of practitioners. (G. E. Davis, one of the promoters of the Society of Chemical Industry, would have liked to have seen the phrase "Chemical Engineer" used in the title, but he finally gave in to the manufacturer's title.) Manufacturers were in control and intended to remain in control.

D. S. L. Cardwell skillfully has sketched this situation in a kind of thought experiment well worth reproducing here:

> *"To summarize the position reached, let us invoke (in imagination) a manufacturer of (say) 1880. Let us suppose that he is a progressively minded man, a supporter of the technical education movement and of the new university college. . . He will favour extended education for all classes and may even have good ideas on secondary education. But if he is asked why he does not use science—research—and scientists in his enterprise he may well reply along the following lines: 'the suggestion that scientists be employed in industry is absurd; as well as Mme. Schumann to teach my daughters to play the pianoforte. A man of science cannot be constrained to follow any prescribed path; he cannot produce discovery to order, neither is it desirable that he should be expected to do so. . . Also we know that although great benefits flow from science, it may take many years before such discoveries are of use, and even so we cannot predict just what use they will be" (17).*

Cardwell's hypothetical industrialist probably stands in need of some nuances, but as a capsule of a general trend, he looks fairly convincing. The emergence, about 1880, of the so-called "Endowment of Research Movement" provides supporting evidence for Cardwell's imaginative effort. Support for research, at that time, did not mean support for applied research; on the contrary, it favored "pure" science with the result that the state fellowships actually contributed to bias the pursuit of science by inserting "a wedge between the high-fliers (who usually chose academic research) and the rest who often settled for posts in industry" *(18, 19).*

That Cardwell's imaginary industrialist needs to see his picture somewhat altered can be drawn from Cardwell himself. "To blame the industrialist," he writes, "for failure to appreciate science and research at a time when the higher educational institutions of the country did little to forward original work was both unjust and foolish" *(20)* In this, he is supported by the historians of English technical education, Michael D. Stephens and Gordon W. Roderick *(21)*. In fact, it is clear from the record that industrialists in general, and chemical manufacturers in particular, sought to make some use of technically trained personnel. Doing so, however, involved several major difficulties. As an example, let us see what Edmund K. Muspratt wrote as early as 1870:

> *"Having so many new processes in hand, I thought it advisable to engage a well educated chemist as head of the laboratory at Widness. It was difficult to find at that time a suitable man in England, and in Germany, owing to the war, which had taken so many young men for service in the army, it was difficult to find a young chemist willing to come to this country. After some correspondence with Professor Knapp of the Brunswick Polytechnic, I engaged a Dr. Jursich, who had been educated at Berlin" (22).*

In the same line of thought, it must be remembered that the organization of the Society of Chemical Industry had been pushed in 1881 by Merseyside industrialists and that the motive behind this move was to examine "what steps could be taken to bring into closer touch professional scientific chemists and those engaged in industry" *(23)*.

A decade later, when German industrial successes no longer could escape anybody's attention and when 45 hard-pressed soda companies reorganized themselves collectively to form the United Alkali Company, one of the first consequences of this huge merger was the creation of a central laboratory *(24)*. In short, British industrialists cannot be grossly condemned for indifference to science. As Cardwell states, "the reason for such condemnation seemed to lie in the belief that a 'demand' for science should somehow arise from the heterogeneous mass of industry" *(25, 26)*. There is profundity in his remark, for, indeed, neither the form nor the content of this demand or of this science can be expected to preexist the specific historical context from which it is about to emerge. In effect, both the demand and the science had to be invented and the inventors had to be the very people who somehow could think of a way to integrate the multifaceted and often conflicting demands of many, many industries into one single voice. Furthermore, it was necessary to reinterpret this synthetic request into curricular terms so as to make them comprehensible to teaching institutions. Finally, to complicate matters further, these teaching institutions were nearly as varied as the industries themselves and were not interested necessarily in looking favorably upon industry's call for help. No wonder, therefore, that so many people and so many Royal Commissions focused on the problems of technical education (e.g., 1868, 1872–1875, 1882–1884, 1889, 1891, 1894, 1895, 1896, 1897, 1898, and 1902) *(27, 28)*. No wonder, also, that industrialists themselves took initiatives in order to try and fill their needs, and the Muspratts, once again, provide a ready example of one of the possible roads manufacturers could follow: they set up their own teaching institution which eventually became the Royal College of Chemistry *(29)*.

Integrating scientifically trained personnel within an industry which had never used them required nothing less than the reworking of the division of labor involved in its productive process. Not only did such a reworking require leaving room for scientific knowledge, but, more importantly, it had to incorporate some notion of the relationship prevailing between theory and practice. In short, much had to be invented.

Now Britain found itself in the peculiar position of possessing a chemical industry which, while being still the largest in the world until late in the nineteenth century, also displayed many signs of age and obsolescence. The growth of British chemical industries was not the result of some technical push, as it were; on the contrary, it was but the consequence of Britain having undergone her industrial revolution earlier and on a greater scale than anyone else. Thanks to the powerful development of mechanical manufactures, chemical works had to respond to a broadly based demand. Textile industries were avid consumers of dyeing and bleaching materials; glass making required large amounts of soda which, in turn, generated a huge demand for sulfuric acid. It can be said, therefore, that the development of the British chemical industry is in large part due to the general movement toward industrialization and it arose as a sort of annex to it. Little wonder, therefore, that men like Ure and Babbage could not incorporate chemical industries in their theoretical discussions of manufacturing.

With the advent of synthetic dyes, the British chemical industry, like any other, had to assimilate the fact that from now on chemistry as knowledge must somehow be incorporated into the chemical industries. Competition had changed. It now involved innovation as a major factor. A new form of productive process had to be invented. In this regard, the traditional advantages of British chemical industries were no longer sufficient. It had cheap raw materials in the form of abundant supplies of coal tar. It had capital and controlled the richest market in the world. But it did not have much of a new resource—namely the chemically trained brains of able men.

Nor was this the only disadvantage of the British chemical industries. Their very size proved inimical to change in at least three ways. First, reference to past successes and prior examples of foreign competition being beaten off could only help the conservative cause advocating more of the same. It turned out to be insufficient to cope with the competition of the new production processes that were emerging at the same time in Germany. The second cause is perhaps a little more elusive but can be stated as follows: international industrial exhibits starting with the London initiative of 1851 quickly led to direct statistical comparisons of all sorts between countries. As a result, the British nation periodically could observe that it was losing ground to German dye industries, but the relative decline of the whole chemical industry of Britain was far less dramatic, so that national pride, even though it was hurt in a particular sector, could find solace in the fact that global figures still testified to England's superiority. Little more is required to restore a feeling of complacency in a people, any people. Finally, and most importantly, the size of British chemical industries and their structure meant that a fair amount of time was needed before investments could be recouped.

Britain, however, never suffered from a lack of Cassandras who, sensibly enough, used the German example to try and shake British apathy. It is in this context that a second culprit often is identified—namely education (30). But accusing British education is no better than accusing British industries. It is not false, but it cannot account by itself for the misfortune of British chemical industries. A good example of this thesis is provided by Edwin Williams' book *Made in Germany*, which first appeared in 1896. The reaction to this book was very heated and it even was discussed in the French press (31). Placing the guilt squarely on British education, Williams went so far as to say that, compared with what existed in Germany, it was like a candle to an electric lightbulb (32).

Further striking statistics came in at the 1902 meeting of the British Association for the Advancement of Science. A survey of industrial chemists had been answered by 502 individuals, among whom only 107 were graduates, and of these only 59 had studied in Britain (33). Even if one multiplied the figures by three, as Dewar did in a flight of unsupported optimism, it yielded only 1,500 chemists in British industries while Germany, at the same time, employed at least 4,000 in its industries. Furthermore, 84% of the German chemists were graduates while the corresponding figure for Britain amounted only to 34% (34, 35).

Yet, all of this is not sufficient to indict education alone. In the years preceding the First World War, higher education underwent very rapid growth and the number of science graduates increased tremendously (36). But the increase in the number of applied scientists did not keep the same pace (37); on the contrary, the vast majority of science graduates perceived themselves as future teachers either at the primary or secondary level *(38)*. In fact, Cardwell estimates that in 1913–1914, "the total number of post-graduate students in the grant aid colleges who were either studying for higher examinations or doing research in science was 172 in England and 17 in Wales" *(39)*.

The gist of all of these figures is clear: by 1910 or so, British education provided many science graduates, although apparently not enough to satisfy the needs of the rapidly expanding secondary schools. Clearly, these science graduates could have envisaged pursuing a career in industry, but apparently very few did so; neither did they try much for higher degrees of research. For reasons which will not be elucidated here, they turned to teaching.

In summary, it is not false to say that industry did not make very large demands for science graduates. Despite some notable exceptions—and E. K. Muspratt has been quoted earlier in this context—industries seem to have remained content with low technical input from trained personnel. Conversely, it is not false either to put the blame on British education or the lack of it. However, when science graduates finally appeared in large numbers, it was in response not to industry's demands, but rather to the needs of a quickly expanding educational system. Consequently, both accusations constitute half-truths at best. Each, by itself, does not resolve the problem at hand.

The solution, in our opinion, hangs with this question of division of labor which already has been raised earlier. Clearly, industrialists are not a priori opposed to integrating trained personnel into their manufacturing, especially if they can be shown that this will increase their profits. As things stood in the nineteenth century, they were not convinced that they should. And if they were not convinced, it is because they did not know exactly what they wanted or what they needed.

Characteristically, it is perhaps in chemical manufacturing that the introduction of scientifically trained personnel went furthest. But what is remarkable in this trend—a trend, incidentally, which is anything but clearly marked—is that it copies a model, that of Germany. Not only do chemical manufacturers of Britain copy the organization of German industries (when they decide to change at all), but, as we have seen, they even buy their personnel from Germany, either in the shape of Germans (or Swiss) who move to Britain, or in the shape of degrees bought by Englishmen going over to study in Germany. In short, British industries tended to model themselves after Germany (which we will examine later) whenever their productive structure was analogous to that of the dominant German industries. But they did so with delays and hesitations and sometimes only halfway. It must be remembered that Britain found it very difficult to move from the Le Blanc to the Solvay process and that the United Alkali Company formed in 1891 turned out to be a real dinosaur when it attempted to save the Le Blanc process at a time

when most of the Continent had switched to the superior Solvay method
to produce soda. The creation of a central laboratory was not sufficient
to stave off disaster (40).

It takes little thinking to understand why British industries should
face such difficulties: the situation is not specific to Britain. Like most
human organizations, chemical industries find it difficult to overcome
their own organizational inertia and when they must do so, they find it
easier to copy someone else than to invent something new. However,
copying always involves risks, if only because the borrower must assess
whether the local situation is sufficiently analogous to that taken as model
to insure a successful transfer.

It is at this junction that the nature of the British chemical industries
becomes crucial. As we have noted before, it was an industry of heavy
chemicals whose growth had been stimulated by the first industrial revolution. By and large, it required very little scientific input and whatever
technical input was needed came mainly from mechanical engineers.
British dye industries, in effect, when they started, attempted to extend
the well-known model of heavy chemicals. Perkin, for example, seems
to have been the sole source of ideas for his own firm, thereby demonstrating how little understanding he then had of the role of scientific
knowledge in what later would be called science-based industries. But
that is not surprising. What is surprising is that a conception of science-
based industries emerged elsewhere—namely Germany—and quickly
came to threaten the new British dye industries. By the time the lesson
had been absorbed, German dye manufacturers had taken such a commanding lead that British industries, by and large, were thrown out of
business or just barely managed to survive. In any case, they were
hardly in a position to experiment with new divisions of labor.

Enough has been said now to provide a suitable backdrop to the
question at hand—namely, what factors shaped British chemical engineering in such a way as to preclude its ever evolving like its American
counterpart? We now can turn to the epilogue of this story, which takes
us to just after the First World War, at a time when British industrial
chemists had taken full measure of their inadequacies.

The Institution of Chemical Engineers emerged in Britain only in
1922, that is fully fourteen years after its American older sister. A sister
association it was indeed, for close ties were established between the two
societies from the very start:

"... a member of one Institution should have the privileged
membership of the other when in the country of the latter, and
also . . . a member of one Institution should receive the
publications of both" (41).

Some of the discussions following the birth of this new professional
society are worth recounting as they reveal the state of uncertainty which

still prevailed in Britain in 1922 regarding the definition and status of a chemical engineer. Just like the AIChE, the Institution of Chemical Engineers immediately focused on educational matters *(42)*. Definitional problems associated with educational matters had to be solved: that was obvious to everyone.

> *"Many times in the short existence of the Institution have*
> *the Council endeavoured to describe a chemical engineer. . .*
> *We have come to the conclusion that a chemical engineer as*
> *such does not in reality exist today" (43)*.

It must be noted that on the inaugural meeting of May 2nd, 1922, a definition for a chemical engineer had been agreed upon which in retrospect looks so broad as to include just about anybody:

> *"A chemical engineer is a professional man experienced in*
> *the design, construction, and operation of plants in which*
> *materials undergo chemical or physical change" (44)*.

Quite clearly, would-be chemical engineers in Britain had but the faintest notion of what they were or wanted to be. It is no wonder, therefore, that the English should react so positively to the appearance of the famed *Principles of Chemical Engineering* by Walker, Lewis, and McAdams:

> *"It would appear that the United States are far ahead of*
> *this country in the study of chemical engineering. The book*
> *under review is in itself evidence of the fact" (45)*.

The war, of course, had brought the Americans into prominence, and, already in 1916, some were saying that the United States ". . . was developing the training and production of chemical engineers on an enormous scale" *(46, 47)*.

That the new Institution was looking very much toward the United States can be demonstrated easily by pointing out that, in 1925, an American delegation of the AIChE was invited to visit its English colleagues. Charles L. Reese, leader of the American delegation, even gave the Presidential Address for that year. In the course of this address, he broached the question of the definition of the chemical engineer and thus underlined the importance of unit operations to his British audience.

> *"Our friends from the Massachusetts Institute of Tech-*
> *nology have, perhaps, done more to establish the idea of what*
> *a chemical engineer is than anyone else, and their idea is that*
> *a chemical engineer is one who understands and knows and*
> *has learned the fundamentals of the elementary processes such*

as distillation, filtration, precipitation, flow of gases and
liquids, heat transmission, and so on" (48).

Reese then followed with a description of several documents well known
to historians of American engineering, such as A. D. Little's report on
education prepared in 1922 and R. T. Haslam's article on *The School of
Chemical Engineering Practice of the Massachusetts Institute of Technology*.   All in all, the London-based Institution could claim no longer
that it ignored American developments and in fact British chemical engineers went a good deal beyond simply acknowledging familiarity with
the American model.   In his reply, Sir Arthur Duckham flatly stated:
"We in our Institution have taken the American Institute as our model . . ."
(49).

In a sense, this statement can end the first part of our analysis, for it
shows that in 1925 British chemical engineers had come to recognize the
superiority of American chemical engineering and, presumably, were
doing all they could to promote unit operations in their own country.   At
the same time it shows that Britain had been unable to construct such a
notion itself, despite the jolt provided by the First World War and the
clear insights it provided on the backward state of British chemical
industries.   As a result, change in Britain took the shape of borrowing:
after being tempted by the German model, Britain finally settled for its
American rival.   Showing how this American model came to dominate in
Britain is another story which goes well beyond the ambitions of this
particular chapter.

The Case of France

The evolution of industrial chemistry in France is no less interesting
than that of Britain.   The French did not know any better than the
British about how to integrate scientific knowledge into chemical industries and the solution that eventually would prevail in this country would
result also from the complex interplay of local factors.   Once again the
nature of French chemical industries and that of the educational facilities
were to play a large role in the shaping of industrial engineers in that
country (50).

Of all the European countries, France was undoubtedly the first to
initiate technical education on a large scale.   Without going back to the
eighteenth century, which nevertheless had witnessed the emergence of
such engineering schools as Ponts-et-Chaussées, the creation of the Ecole
Polytechnique during the revolutionary period had provided a model
which was to be envied and imitated all over Europe for several decades
(51, 52, 53).   In the course of the nineteenth century, France built a
remarkable system of Grandes Ecoles which have remained as the back-

bone of her technical education up to this day.   Besides the two schools already mentioned, one can add the Ecole des Mines, the Ecole Centrale des Arts et Manufactures and, on a somewhat lower level, the Conservatoire des Arts et Métiers.

At first sight, it looks as if France, unlike Britain, had solved the problem of training her technical personnel.   A closer look, however, reveals a somewhat more complex situation.

An extensive pamphlet calling for the creation of a school of practical and industrial chemistry published by the Paris Chemical Society appeared in 1891 *(54)*.   Although clearly aiming at convincing the powers-that-be, and therefore quite biased, this document nevertheless provides an interesting picture of French activities in applied chemistry in the last decade of the nineteenth century.   Its most striking feature is the accusing finger it points in the direction of the Grandes Ecoles and particularly the Ecole Polytechnique:

> *"The Ecole Polytechnique which, from its beginnings, and because of the needs of that time, has been derailed from its prime objective, has always been a school for theoretical sciences"* *(55)*.

The anonymous author of this pamphlet then proceeds to survey existing resources in Paris.   The Conservatoire des Arts et Métiers teaches industrial chemistry, but only in lectures and without any opportunity for practice *(56)*.   Likewise, the Ecole Centrale, although the product of an initiative taken by industrialists, quickly came to perceive itself as the rival of Polytechnique.   As a result, it turned to theoretical sciences *(57)*.   The case of Centrale is quite instructive for another reason.   For many years, it had been graduating engineers with titles such as ingénieur chimiste, ingénieur mécanicien, but the value of these distinctive titles quickly can be gauged from the fact that all of the graduating students took the same final examination, regardless of their alleged specialty.

As for the recently created Ecole Municipale de Physique et Chimie Industrielles, funded by the city council of Paris, our author contends that it trains chemists for the laboratory rather than for industry *(58)*: he may not have been entirely wrong when we note that Pierre Curie taught there and Langevin studied there before embarking on a doctorate *(59)*. Finally, and quite ironically, the only place where one may get serious training in both applied and theoretical chemistry in Paris is at the Pharmacy school.   But, alas, the orientation of this school had very little to do with industry, as could be expected *(60)*.

On the side of the Facultés, which were reorganized into universities only in 1896 *(61)*, little had been taking place.   Between 1835 and 1839, only 49 students graduated on the average with a *Licence-ès-sciences* each year; almost 30 years later, this figure had not doubled yet and it is only ca. 1885–1889 that the number of science graduates increased significantly to reach an average yearly total of 351. Yet, even this figure is small when one considers that it represents the training capacity of a major nation in all of the sciences (62).

Despite such unpromising beginnings, it was in the universities that interesting developments ended up taking place. This unexpected turn of events was the result of a complex interplay of institutional factors, as we shall see.

In examining the French educational system, it never must be forgotten that it obeys strict hierarchical rules. In technical matters, it is clear that Polytechnique and Centrale were the two top schools. Access to these schools is obtained through competitive examinations open to students who, upon having completed their Lycée and passed the Baccalauréat, prepare for these contests in so-called classes préparatoires. The form of these rigorous entrance contests is extremely important in that it emphasizes some sectors of scientific knowledge at the expense of others. In particular, the future engineers of Polytechnique and Centrale were (and still are) submitted to doses of abstract mathematics probably unequalled anywhere else on earth for applied scientists. Physics, taught in a highly formalized fashion, also is granted a good degree of prestige. Chemistry, on the contrary, because of its low degree of mathematical formalization, is viewed as a somewhat inferior science and its weight in entrance contests to Grandes Ecoles is relatively low. With this situation clearly in sight, it is easy to understand why industrial chemistry had difficulties finding room in engineering schools of that kind, however paradoxical that may seem at first sight. Over time, the imbalance between scientific subjects grew to influence more than just the preparatory classes. It biased the whole scientific training of Lycées, as the structure of the Baccalauréat exam tended to approximate the trends ruling over the entrance contests.

Several authors pointed out this defect (63) but the Ecole Polytechnique had a firm grip on power as many of its graduates, although nominally engineers, saw their careers culminate in high civil servants' positions. And the tendency for the other Grandes Ecoles was to vie for prestige with Polytechnique, Centrale, and the other schools by moving in the direction of ever more abstract science at an ever higher level.

Meanwhile, faculties of science had been developing gradually in several French cities, particularly after the Minister of Education, Victor Duruy, had granted them the right to train future teachers (64)—a role hitherto limited to Normal schools. Until then, they had acted mainly as examination offices for secondary schools while their courses and conferences had attracted mainly amateurs and mondains (65).

The new role granted to the faculties of science did not solve their problems. At first, few students were attracted, and further measures had to be taken, if only to provide more opportunities and accelerate career prospects of faculty members *(66, 67)*. In other words, internal rather than external institutional pressures seem to have been the major factor in the evolution of the faculties of science. It is within this somewhat improbable context that these faculties of science resorted to creating courses in applied science. Not only did it help to create opportunities for teachers, but it also gave great political advantages to faculties of science which, from then on, could display proudly their usefulness to government and industries alike.

The data indicate that this trend was general throughout France and it all took place at about the same time. Dijon saw its industrial chemistry program separate itself from chemistry proper in 1900 after having been created in 1898 *(68)*. The same year (1898) saw Besançon start teaching applied chemistry *(69, 70, 71)*. Schools of applied science or chemistry also sprang up in Nancy, Lille, and Bordeaux, while courses or programs in industrial chemistry appeared in Caen, Montpellier, and Rennes. All of these courses or programs appeared either in the last decade of the nineteenth century or the beginning of the twentieth century. *(72, 81)*. And neither Paris nor Lyons should be forgotten *(82, 83)*. Toulouse also stands out, thanks to Paul Sabatier *(84)*.

From town to town, the training of industrial chemists took on a distinct flavor. In some places, like Marseille for example, an engineering school worked next to, but outside of, the faculty of science *(85)*. In other places the school, while incorporated within the faculty of science, worked like an autonomous whole. Nancy, under the able direction of Albin Haller, went probably furthest in that direction and eventually became one of the leading training centers in industrial chemistry in France. Yet another solution was offered by Montpellier where industrial chemists were students already holding a licence and taking extra training in applied chemistry. In effect, Montpellier copied Polytechnique in a rough way by using its licence program to give the general training in science which they felt the future engineer had to know. Then the special program in chemical technology which followed played the role of an Ecole d'Application. In fact, Montpellier offered the most ambitious program to be found in a French university at the time. It had but limited success, however, probably because its teaching strategy required efforts similar to those needed to make it through Grandes Ecoles without granting anything like the prestige and status which these Grandes Ecoles could give to their graduates.

The somewhat chaotic development of programs in applied chemistry in several faculties of science in France is quite suprising in a country well known for its Jacobin-inspired centralism. Yet variations were not limited to organizational matters; with geographical disparities came also the concern to respond to local needs. For example, Bordeaux created a laboratory that specialized in resins because of the large forests in nearby Landes (86), while Grenoble took on paper making and electrotechnology and Nancy soon incorporated a special brewery school into its programs. Wine making also stimulated chemical research in several southern French universities, particularly in Bordeaux and Montpellier.

This enormous diversity led to one major consequence: although many of these schools or programs awarded the title of ingénieur chimiste, the meaning and value to be ascribed to this title was very difficult to ascertain. As late as 1925, a well-informed observer of French industrial chemistry could write:

> "Another and more essential defect of our schools of chemistry is that, as they have been generally created on the initiative of universities, or sometimes of cities, they go too much for a kind of regional competition which is as adverse to the national interest as is Parisian centralism in other circumstances. Their activities are not coordinated." (87).

The author of these lines, Matagrin, is only reflecting a professional concern then being voiced in France. In May 1919, a Syndicat Professionnel des Ingénieurs Chimistes (88) had been formed finally in France and, the following year, the Union Nationale des Associations d'Anciens Élèves des Ecoles de Chimie held a meeting during which a number of resolutions were passed. It is significant that they should have asked for the uniformization of curricula in the different schools of chemistry as well as that of the diploma of ingénieur chimiste which, moreover, should be awarded by the state (89).

The demands made by industrial and applied chemists in 1920 were caused directly by the war experience of course, as well as inspired by analogous German associations which had appeared earlier. But beyond these direct causes, one also must look for factors which had been at work for a much longer period of time. These have to do with the relative status of ingénieurs chimistes and ingénieurs des Grandes Ecoles within industries. Once again, Matagrin can serve as our guide here. Speaking of the large companies where a division of labor is indispensable, Matagrin notes that the general director is rarely a chemist: "rather one chooses a technician endowed with managerial qualities . . .—a graduate from the Ecole Centrale, the Ecole des Mines, and sometimes even a Polytechnicien; . . ." (90). As for the technical director, more often than not, he is not a chemist either, but an engineer versed in questions of tools, uses of energy, and so on *(91)*. And Matagrin concludes that it is regrettable to find chemists placed hierarchically below technicians who have little or no knowledge of their field *(92)*.

A last problem has to do with the competition coming from abroad. Despite remarkable progress, the French educational institutions were unable to satisfy the industrial needs of the country *(93, 94)* with the result that foreign personnel was called into France as it was in Britain *(95)*. Quite clearly, these two countries offered sizeable opportunities for young graduates out of German universities or Technischen Hochschulen. In fact, Graebe, Professor of Chemistry in Geneva, symptomatically complained to a French visitor that if France proceeded to train her own industrial chemists, then many positions would be lost for foreigners *(96)*. Moreover, these foreign doctors (for many had earned this title) were apparently ready to work for quite moderate salaries by French standards and thus contributed to keeping the chemists' wages at relatively low levels within French industries *(97)*. Finally, the introduction of chemists trained in Germany or in Switzerland contributed to introducing into France a model of integration of trained personnel which had evolved within the German context—which we shall examine shortly—and which, therefore, did not necessarily fit the French context. It thus reinforced a wide tendency in France—namely the tendency to look up to Germany for technical lessons, and not to other countries like the United States, for example. In fact there were very few reasons why the French should have looked toward the United States before the First World War: if they admired the rapid economic growth and the potential of the young Republic, they were far less impressed by the general state of its science and technology *(98)*. It must be remembered, for example, that the *Principles of Chemical Engineering* by Walker, Lewis, and McAdam appeared in French only in 1933, under the misleading title of *Principes de Chimie Industrielle*. It seems, moreover, that its reception in France was the result of a misunderstanding. What the French seem to have retained from this volume is not the modular nature of unit operations, but rather their theoretical content *(99)*. The degree of mathematical formalization inherent in unit operations could be used to make industrial chemistry look like physics. As a result, it could fit better into the general scientific style characteristic of Grandes Ecoles education.

In conclusion and summary, it can be said that the French situation could not lead to the emergence of a formula such as unit operations to solve its problems in industrial chemistry. As we have seen, French industrial chemistry and even ingénieurs chimistes were condemned to remain below so-called engineers from the Grandes Ecoles. As a result, it was graduates of Grandes Ecoles who had a voice in the managerial organization of the chemical industries, not chemists or ingénieurs chimistes. Consequently, their role was tailored from above and not negotiated as between partners of roughly equal standing. The division of labor prevailing in most French chemical industries was set up with very little input from industrial chemists.

Traumatized by the 1870 defeat and worried by the rapid industrial growth of Germany, French ingénieurs were far more attentive to German developments than to American progress. The influx of German chemists into French industries, even if, seemingly, it was moderate—and no figures on this point are available to our knowledge—was significant enough to be noted by several observers. The presence of these foreign specialists within several French industries could not but reinforce the prestige of the German model. However, the German model was not to be imported easily into France for reasons similar to the British case. Like the latter, French chemical industries were fairly old and their main strength lay in heavy chemicals. And it was just as unable to formulate a well defined demand to universities or technical schools as its British counterpart.

As for French chemists, they did not constitute a single, well-organized group. Having been trained in ways which varied greatly from place to place, which is something difficult to imagine within the French educational system, they found it impossible to form into an effective lobby as their German or American colleagues were able to do. Consequently, they never were in a position to make themselves heard in a strong, simple, and coherent fashion.

Finally, we saw that universities took initiatives favoring applied chemistry, but these emerged much more to solve internal career problems of faculty members than in response to the nation's fundamental worries. Each university reacted as if it were alone and, consequently, each evolved its own particular solution, often dictated by local needs and opportunities. Since no professional group existed, no one could help to harmonize university programs with industry's needs. In short, the French situation was so disorderly as to preclude the production of a specific and coherent solution by industrial chemists themselves. In practice, decisions remained very much in the hands of the managers and the ingénieurs from the Grandes Ecoles. The model would have to come from the outside and somehow be fitted within rather rigid and partially incompatible institutions.

The Case of Germany

Compared with the previous two cases, Germany presents us with a success story. We have repeatedly seen that both the British and the French looked to Germany for inspiration and models to emulate. While the industries of these two countries displayed only very moderate rates of growth, Germany, on the contrary, posted impressive figures which quickly placed her in a position to threaten British hegemony: only the United States had comparable rates of growth.

Such industrial success drew international attention and observers tirelessly repeated a very similar message: Georges Blondel *(100)*, Raphael–Georges Lévy *(101)*, and Victor Cambon *(102)* all made a point of praising German higher technical education and contrasting its flourishing state with the semipoverty afflicting much of French academic research. In Britain, similar views prevailed. We already have mentioned Edwin Williams' *Made in Germany*, but just as telling is this letter to the Editor found in the professional journal *Engineering*:

> *"So many discourses have been delivered on these two texts [German chemical industries and technical education] by speakers more or less qualified, from Chief Justices down, that the public mind, perhaps, if only by default, admitted the necessity of providing 'technical education' and also that the rise of the chemical industry in Germany is one of the pieces of evidence chiefly relied on by those about to demonstrate such a necessity"* *(103)*.

But what was so peculiar about German higher education that it should so impress foreigners? Two variables may be singled out in answering this question. First, a very extensive technical education system existed besides the well-developed universities; second, the number of students involved in German technical education dwarfed anything other European countries could show.

To obtain full figures about German public education, one should turn to W. Lexis' *Das Unterrichtswesen im Deutschen Reich* or, alternatively, its English summary *(104)*. Just a few examples will suffice to demonstrate how far ahead of her neighbors Germany was at the turn of the century. A French observer of the German scene notes that between 1878 and 1902, the number of students attending the nine (soon to be eleven) Technischen Hochschulen of Germany grew from 5,474 to 16,826 and if this number then decreased a little in the following years, it was because standards of admission had been raised *(105)*. His figures reproduce Lexis', from whom he probably got them *(106)*. All in all, Germany produced about 3,000 engineers per year in the decade preceding the First World War *(107)*.

Meanwhile, the 22 universities of the German Empire harbored 37,677 students in 1903, out of which 15,205 were in the philosophical faculty where science was taught *(108)*. Ca. 1908, this number had grown to 47,000 students *(109)* and in 1911 it had reached 59,000 of which 10,000 hailed from Berlin *(110)*.

Not only was German technical education well developed in terms of students, but it also was supported generously by public funds. As a result, most technical schools benefited from material facilities unequalled anywhere, save perhaps the United States. A quick glance at the annual budgets of the Technischen Hochschulen is enough to gain an idea of the enormous financial efforts made by Germany on behalf of her technical education. In Berlin, for example, the total intake of Charlottenburg was 1,744,366 marks for the year 1902–1903 (111). It was by far the largest technical school, but annual budgets in the vicinity of 500,000 or 600,000 marks were not at all uncommon. And, of course, all of this money translated into very good equipment which never failed to impress foreign visitors. Dantzig, for example, was the tenth technical school to be built, and it opened in 1904. To a French visitor, its equipment and laboratories appeared truly luxurious (112). Germany, in short, suffered from no shortage of educational facilities; on the contrary, it invested heavily in them in order to be able to respond to the ever-growing needs and demands of industry.

On the other side of the coin, German industries had learned to use this personnel on a scale unknown anywhere else. Leverkusen, a model factory created from scratch by Carl Duisberg, then working for the Bayer Company, is perhaps the best example of what German chemical industries were like at the beginning of this century. It was an amalgam of American ideas about plant layouts with some very original ideas of Duisberg himself (113). It is worth quoting in this respect the awed reaction of Victor Cambon when he visited it around 1908:

> "The perfection reached in this respect still surpasses that encountered within universities and polytechnical schools. Reagents, gases, heat, cold, vacuum, pressure, electricity, motive power at any speed are provided to the operator, without his doing anything more than pushing buttons or opening faucets. Therefore, he has no reagent to prepare: a central service takes care of that; there is no solution to shake or precipitates to extract and dry: machines take care of that; there is no glassware to clean: two assistants do this work. The chemist, always a PhD, stands there like a pope set within a sanctuary closed to the masses" (114).

In this particular factory, there were 203 operating chemists. In comparison with such figures, Britain and France could but agree that Germany indeed had found a way to integrate scientists into its industries and, moreover, that it paid handsomely to do so.

That a new division of labor had been worked out was obvious to all, and it constituted the goal to be reached by Germany's envious neighbors. However, it was much easier to describe this division of labor than to implement it. The French repeated the wise but general words of the prominent chemist, Caro, for example *(115)* and the English essentially discovered the same reality from their own observers *(116)*, but words remained words.

What is of particular interest to us is the process leading German manufacturers to invent a mode of production allowing for a significant scientific input. In other words, how did they manage to integrate science successfully into their chemical industries when other European countries could not do so?

As in the case of Britain and France, one of the crucial factors resided in the nature of the chemical industry. In the case of Germany, we confront a situation quite different from the British or French ones. In Germany, traditional heavy chemicals had never developed very much, precisely because it had been unable to beat back the competition of the larger and older British industries. This meant that the spear head of the German chemical industries, unlike Britain or France, gradually was built around synthetic dyes to which were progressively added other branches of industrial chemistry also benefiting from a high level of scientific input, such as pharmaceuticals.

German dye manufacturers did not invent their new process of production and their integration of scientific input into industry overnight. Actually, just like their British and French colleagues, they tended to be rather conservative. However, a very loose patent situation prevailed until German unification came into force, and this allowed them to reap the benefits of imitation at first.

The transition to a large scientific-industrial complex, aptly symbolized by Duisberg's technical utopia, evolved only very gradually and not before several obstacles had been overcome. We so far have emphasized the conservative nature of chemical industries. However, we must remember that no industry is ever so conservative as to stand absolutely still. On the contrary, innovations always take place and foremen play a large part in this respect. Separating out the innovative process from the production process meant that foremen, and others as well, saw their activity circumscribed more narrowly and, therefore, their role was impoverished by this refined division of labor. Not suprisingly, they resisted this trend and continued to do so until about the mid 1870's in the case of Germany *(117)*.

John J. Beer recounts that, in the case of the Bayer Company, the hiring of university-trained personnel took place in the decade between 1874–1884. This was done not because managers wanted to introduce a pattern of systematic innovation—that point had not dawned yet upon them—but rather to try and improve the traditional production process *(118)*. In fact, the chemists hired by Bayer were integrated within the chemical works on the basis of what could be learned from the introduction of the Solvay process of soda production, and not as the result of an imaginative leap into the future.

A crucial push toward the new industrial pattern apparently was provided by the new Patent Act of 1876. Having started their industries by imitating, and later on by improving upon their competitors' products, German chemical manufacturers asked for a patent act allowing them to make the most out of their recently gained ability. This is the reason why the Patnet Act of 1876 did not protect the finished product, but rather the process of production. This meant, in turn, that a company could never be sure that their benefit-earning products could not be produced some day in a different way by a rival firm. As a result, competition between companies quickly brought the point home that having highly trained chemists involved (as well as sharp patent lawyers) improved markedly one's chances.

One of the first attempts to integrate scientific knowledge into the production process consisted of extending the links which had been established rather early between university professors and industries (119). This method was tried because early attempts to use PhDs within industry had not succeeded:

"... the company failed to weld these [doctors in chemistry] into a research team, preferring to attach each man to a specific production division where his work became largely analytical and routine in nature ..." (120).

So, it was thought to create a sort of postdoctoral fellowship allowing a young PhD to work under the direction of a good chemist on problems related to the company's interest. In this fashion, several advantages were obtained: the sponsoring chemist was drawn on indirectly as a consultant and the company did not have to invest in the construction of costly laboratories (121).

However promising, the scheme devised at Bayer did not work out if only because the industry asked questions which could not be answered readily (122). In effect, the Bayer Company was discovering empirically that scientific research is extremely hard to plan, particularly when both time and resources are of the essence. Moreover this particular strategy could have worked only as long as the needed scientific input remained small. It is difficult to imagine how the massive and systematic investigations of organic syntheses which were to characterize the dye industry in later years could have been carried out in several small laboratories with a very limited staff. Clearly, Bayer Industries was devising an approach to the integration of science into manufacturing which could be termed artisan in nature at best.

One of the guinea pigs of the Bayer experiment happened to be Carl Duisberg, and he found a way to organize thoroughly the integration of chemists within the Bayer works.   In 1896, he explained to an American audience what his ideas were in this respect, and they are so central to our thesis that they bear being discussed at some length.   In his New York address of May 18, 1896, Duisberg said:

> "We have found it most satisfactory if we ourselves intro-
> duce the young chemist to our special field.   For that purpose
> every chemist must first pass through our experimental dye
> and print laboratory, so that he learns dyeing and printing and
> becomes aware of the requirements of the dyers' industry as
> regards dyeing properties and the fastness of colours.   When
> the chemist has finished in this laboratory, he is introduced to
> the scientific laboratory, the function of which is to keep us and
> our chemists informed of everything new that appears in the
> field of our manufactures in literature, patents, etc. and it is at
> the same time principally the laboratory of inventions" (123).

Duisberg's strategy consisted of making chemists circulate through the different chemical divisions of the plant until they were familiarized with all aspects of chemical activities in it.   Then he could join his own personal slot, that which was best adapted to his abilities.   It is at this point that Duisberg comes out strongly in favor of a solution which essentially precludes the emergence of an American-style chemical engineer.

> "In opposition to many of my friends I place myself . . .
> on the standpoint . . . that the chemist does not require
> [engineering] as a necessity.   Nothing, in my opinion, is
> worse than to make of a chemist an ingénieur-chimiste as is
> done in France, or chemical engineer as is very often done in
> England.   The field of chemistry which the chemist has to
> master is at the present so enormous that it is practically
> impossible for him to study at the same time mechanics which
> is the special field of the engineer.   Division of labour is here
> absolutely necessary.   I leave to the engineer and to the
> chemist their respective sciences, but I desire that both work
> together" (124).

Duisberg did not use the phrase "division of labor" lightly: for him, it clearly was the central tenet of his basic conception of chemical works.   It was, in fact, on the basis of American familiarity with extensive division of labor that he thought the German approach to be best suited to American chemical industries.

In organizing his industrial research laboratories, Duisberg retained the collective bench and work area devised several decades earlier by Liebig (125).   However, he separated each work space from the others by shoulder-high shelves.   In this manner, neighboring chemists still could communicate easily with one another, but they could not see exactly what the others were doing. In other words, each chemist knew somewhat what the others were doing. He also knew that his own work fitted somewhere into this collective effort, but he never knew enough to reconstruct the complete grid. In this fashion, each chemist had but limited opportunities to pass on laboratory secrets to another firm. Conversely, the incomplete privacy, while fostering a congenial research atmosphere, ". . . made it difficult for a chemist to hide important discoveries as Böttiger had done with Congo red . . . in the hope of leaving the company and selling them to someone else" (126).

Chemical laboratories of German dye industries have been described often as tedious, meticulous, and endless (127). These adjectives are interesting because they bring to mind the drab process of chain work. In fact, the connotative power of these words reveals itself accurately enough if one thinks of this chain not as a place where workers are distributed, but as a chain of successive reactions carried out in order to explore the maze of organic syntheses in a collective and systematic manner. Somewhat elliptically, Victor Cambon wrote that "From the moment they began to be Americanized, the Germans . . . became fertile inventors" (128). Actually, he seems to have hit the nail fairly squarely, because Americans did place great emphasis on division of labor, and Germans found a way to apply division of labor in such a way as to integrate scientific research in an industrial production process.

Remember that all of this gigantic effort was needed. Around 1900, Hoechst, for example, tested about 3,500 colors per year and roughly 18 of these reached the market. A few years later, 8,000 colors were tested within a single year and only 29 could be used commercially (129). In the end, dye manufacturers also became knowledge producers as patents discovered by one company became commodities to be traded, exchanged or leased (130).

Curiously, if the German dye industries, and later on pharmaceutical industries, benefited well from this work arrangement which had succeeded in using efficiently the university-trained chemist, the same is not true to the same extent for the other chemical industries. As a matter of fact, criticisms were expressed even by chemistry teachers who saw dye industries biasing the whole of chemical education owing to their relative weight in the whole field of industrial chemistry: curricula tended to emphasize the teaching of organic chemistry at the expense of other fields (131). It was pointed out in 1900, that if Germany had completely overtaken all of the other nations in the tar-derived dyes, this was not true of many other branches of chemistry. Britain and Belgium, it was argued, still led the way in soda production. Modern explosives had been invented by Nobel and several French chemists. France also led the way in the chemistry of fats and it was Pasteur and no German who had elucidated the fermentation processes.   Finally, Austria had built the largest sugar plants in the world while the United States could supply water gas (a mixture of hydrogen and carbon monoxide) in large quantities *(132).*   The solution, it was argued by Ost in particular, was to lay greater stress on practice and, characteristically, to parallel industrial specialization with similar specialized education *(133).*

Ost's call for a return to practice, for a new stress to be placed on the teaching of inorganic chemistry, and for greater specialization did not, however, threaten Duisberg's concept of a strict division of labor between the chemists and the engineers.   On the contrary, it provided the means to extend the lessons learned within dye plants to all chemical industries, thus allowing a simple transfer of the "Duisberg model" to take place from one plant to the next and from one branch of the chemical industry to another.

The end of our story is now at hand.   Obviously Duisberg did not create the German approach to industrial chemistry all by himself; however, like A. D. Little in the United States, he found effective ways to communicate what he meant.   For this reason, we have called the German team approach Duisberg's concept, if only for the sake of convenience. But no matter what the name, it constituted an entirely original solution to the problem of integrating chemical knowledge within an industrial context.   At the same time, the Duisberg approach, by its careful distinction between chemists and engineers, precluded the possibility of seeing a mixed kind of technician emerge, such as the American chemical engineer. It also included a distinct advantage over the American concept.   German industries could take what the universities and Technischen Hochschulen had given and integrate them within their works through an internal reorganization entirely separate from academic concerns.   The American chemical engineer, on the other hand, not only required some internal reorganizations of chemical works, but he depended for his existence on the creation of new curricula.   Obviously, it was easier to move industries in Germany and universities in the United States.

As time passed, the soundness of Duisberg's approach never was challenged seriously.   The creation of the Interessen–Gemeinschaft der Farbenindustrie in 1925 responded to the higher demands of international competition characterizing the world after World War I.   Carl Duisberg was chosen as its new director *(134).*   His strategy was simple. He extended the methods which had served Bayer so well before the war as far as he could.

Many years later, it was observed that the process of innovation at I. G. Farben always rested on strict obedience to the principle of a division of labor between chemists and engineers which had been elaborated in the prewar years:

"When a result susceptible of being exploited on an industrial scale was confirmed in the laboratory, the inventor himself generally was put in charge of testing his invention on a semi-industrial scale. To do so, competent mechanical or electrical engineers were put at his disposal, along with the necessary trained personnel" (135).

All of this means that the division of labor which had been tested within dye industries had become the rule for the greater part of German chemical industries.   It then acquired a status which some were willing to call a doctrine:

"In terms of other doctrines which could be comparable to I. G. Farben's, we only see the American doctrine of 'chemical engineering;' that of I. G. Farben can then be characterized as being the doctrine of 'pure chemistry'" (136).

Kahan was writing this in 1949, but even more recent examples tend to show that Germany, unlike most of the world, has remained faithful to Duisberg's viewpoint.   As recently as 1955, the British chemist, manager, and historian of chemistry, Harold Hartley still could say that the only country in the world which had not yet accepted the notion of chemical engineering as it had emerged in the United States was Germany (137).   Only in very recent years have American theoretical inputs made any significant headway in Germany, but these reach well beyond the notion of unit operations (138).

Conclusion

We argued at the outset that the notion of unit operation was the result of complex negotiations between industries, universities, and professional organizations, and we used this hypothesis to explore three different European settings.   It must be clear by now that each country surveyed followed a distinctive path in elaborating a form of industrial chemistry.   That in itself should be interesting to anyone interested in exploring the question of national technological styles.   But this was not our central concern.   What interested us most was to demonstrate that each country evolved a form of industrial chemistry on the basis of guiding assumptions involving educational facilities, kinds of industries, and the professional men themselves.   In some cases, as in Germany, these guiding asumptions were stated explicitly once they had been recognized.   In the case of France or of Britain, these guiding assumptions never became explicit since the various institutions involved could never find a way to listen to each other and kept on responding to their own internal logic. However, explicit or not, these guiding assumptions, themselves the results of negotiations, took the place occupied in the United States by unit operations. As a result, European events in the domain of industrial chemistry were not so much obstacles to the notions of unit operation as they were substitutes for it.

A second point to be stressed is that not all of the local solutions survive forever. That is obvious, of course. What is less obvious is that the survival value of a particular solution depends on factors reaching well beyond purely technical questions.

Finally, it is clear that successful models can be exported sometimes but not always. It would be interesting, in this respect, to ask whether American chemical engineering was easier to export than its German rival. Intuitively, we are tempted to answer by saying that it would depend on the relative degree of flexibility exhibited by universities and industries, but justifying such an answer would take us too far afield. Conversely, successful models are integrated too well within their social, economic, industrial, and educational contexts to be displaced easily by another model originating from a different base. The German example is very telling in this respect.

Acknowledgments

We would like to take this opportunity to thank A. King for his kind and perceptive comments.

Literature Cited

1. Guédon, J.-C. "Chemical Engineering by Design: the Emergence of Unit Operations in the United States," submitted for publication in *Technology and Culture.*
2. Noble, D. N. "America by Design"; A. Knopf: New York, 1977; p. 194.
3. Hardie, D. W. F. "The Empirical Tradition in Chemical Technology"; Loughborough College of Technology: 1962; p. 30.
4. Haber, L. F. "The Chemical Industry During the Nineteenth Century"; The Clarendon Press: Oxford; 1958.
5. Haber, L. F. "The Chemical Industry (1900–1930)—International Growth and Technological Change"; The Clarendon Press: Oxford, 1971.
6. Hall, Marie Boas. "La croissance de l'industrie chimique en Grande-Bretagne au XIXe siècle," *Rev. Hist. Sci. Leurs Appl.* **1973,** *26,* 49–68.
7. Stephens, Michael D.; Roderick, G. W. "The Muspratts of Liverpool," *Ann. Sci.* **1972,** *29,* 287–311.
8. Haber, L. F. "The Chemical Industry During the Nineteenth Century"; The Clarendon Press: Oxford, 1958; chapters 2, 4.
9. Ibid., pp. 21–22.
10. Ure, A. "The Philosophy of Manufactures or an Exposition of the Scientific, Moral and Commercial Economy of the Factory System of Great Britain"; Frank Cass and Co: London, 1967; p. 2.
11. Babbage, C. "On the Economy of Machinery and Manufactures"; Carey and Lea: Philadelphia, 1832; p. 39.
12. Ibid., p. 31.

13. Ibid., pp. 275–6.
14. Stephens, M. D.; Roderick, G. W. "The Muspratts of Liverpool," Ann. Sci. 1972, 29, 310.
15. Cardwell, D. S. L. "The Organisation of Science in England"; Heinemann: London, 1972; p. 173.
16. Reference withdrawn.
17. Ibid., pp. 248–9.
18. Russel, C.; Coley, N. G.; Roberts G. K. "Chemists by Profession—The Origins and Rise of the Royal Institute of Chemistry"; The Open University: Walton Hall, Milton Keynes, 1977; p. 190.
19. MacLeod, R. M. "The Support of Victorian Science: the Endowment of Research Movement in Great Britain, 1868–1900," Minerva 1971, 9, 197–230.
20. Cardwell, D. S. L. "The Organisation of Science in England"; Heinemann: London, 1972; p. 157.
21. Stephens, M. D.; Roderick, G. W. "The Muspratts of Liverpool," Ann. Sci. 1972, 29, 310–311.
22. Ibid., 306.
23. Ibid.
24. Ibid., 307.
25. Cardwell, D. S. L. "The Organisation of Science in England"; Heinemann: London, 1972; p. 173.
26. Drew, W. N. Trans. Inst. Chem. Eng. 1924, 2, 18.
27. Cardwell, D. S. L. "The Organisation of Science in England"; Heinemann: London, 1972; Chapters 5, 6.
28. Ibid., p. 111.
29. Stephens, M. D.; Roderick, G. W. "The Muspratts of Liverpool," Ann. Sci. 1972, 29, 294.
30. Ibid., 311.
31. Barine, A. "La fin de Carthage," Revue des deux mondes 1896, 137, 362–377.
32. Ibid., 373.
33. Russel, C.; Coley, N. G.; Roberts, G. K. "Chemists by Profession—the Origins and Rise of the Royal Institute of Chemistry"; The Open University: Walton Hall, Milton Keynes, 1977; p. 191.
34. Ibid.
35. Cardwell, D. S. L. "The Organisation of Science in England"; Heinemann: London, 1972; pp. 204–207.
36. Ibid., pp. 209–217.
37. Russel, C.; Coley, N. G.; Roberts, G. K. "Chemists by Profession—the Origins and Rise of the Royal Institute of Chemistry"; The Open University: Walton Hall, Milton Keynes, 1977; p. 190.
38. Cardwell, D. S. L. "The Organisation of Science in England"; Heinemann: London, 1972; p. 217.
39. Ibid., p. 215.
40. Landes, D. S. "The Unbound Prometheus"; The University Press: Cambridge, 1970; p. 273.
41. Trans. Inst. Chem. Eng. 1923, 1, viii.
42. Trans. Inst. Chem. Eng. 1924, 2, 8.
43. Duckham, Sir Arthur. "Presidential Address," Trans. Inst. Chem. Eng. 1924, 2, 15.
44. Trans. Inst. Chem. Eng. 1923, 1, ix.
45. Davidson, W. B. "Book Review of 'Principles of Chemical Engineering,'" J. Soc. Chem. Ind. 1923, 42, 1117.
46. Donnan. J. Soc. Chem. Ind. 1916, 35, 1190.
47. Elliott, M. A. Trans. Inst. Chem. Eng. 1924, 2, 19.
48. Reese, C. L. "Presidential Address," Trans. Inst. Chem. Eng. 1925, 3, 13.
49. Duckham, Sir Arthur. "Presidential Address," Trans. Inst. Chem. Eng. 1924, 2, 16.

3. GUÉDON   *Unit Operations in Europe*   73

50. Thépot, A., submitted for publication in *Actes de Colloque Gay–Lussac*.
51. Léon, A. "Histoire de l'éducation technique"; Presses universitaires de France: Paris, 1961.
52. Léon, A. "La Révolution française et l'éducation technique" Société des études robespierristes: Paris, 1968.
53. Monteil, F. "Histoire de l'enseignement en France"; Sirey: Paris, 1966.
54. Anon. "De la nécessité de la création d'une grande école de chimie pratique et industrielle sous le patronage de la société chimique de Paris"; Dupont: Paris, 1891.
55. Ibid., p. 9.
56. Ibid.
57. Ibid.
58. Ibid., p. 12.
59. Lauth, C. "Rapport général sur l'historique et le fonctionnement de l'Ecole municipale de physique et de chimie industrielles"; Imprimerie générale La Hure: Paris, 1900.
60. Anon. "De la nécessité de la création d'une grande école de chimie pratique et industrielle sous le patronage de la société chimique de Paris"; Dupont: Paris, 1891; p. 10.
61. Ponteil, F. "Histoire de l'enseignement en France"; Sirey: Paris, 1966; p. 314.
62. Mouton, M. R. "L'enseignement supérieur en France de 1890 à nos jours (étude statistique)" in "La scolarisation en France depuis un siècle"; Mouton: Paris, 1974; p. 176.
63. Leclerc, M. "La formation des ingénieurs à l'Etranger et en France"; A. Colin: Paris, 1917; p. 102.
64. Mouton, M. R. "L'enseignement supérieur en France de 1890 à nos jours (étude statistique)" in "La scolarisation en France depuis un siècle"; Mouton, Paris, 1974; p. 193.
65. Ibid., p. 177.
66. Le Châtelier, H. *Rev. Int. l'Enseignement* 1909, 58, 45.
67. Gosselet, J. "L'enseignement des sciences appliquées dans les universités"; *Rev. Int. l'Enseignement* 1899, 37, 97–107.
68. "Enquête sur l'enseignement technique dans les universités françaises," *Rev. Int. l'Enseignement* 1909, 57, 252.
69. Ibid., 257.
70. Genvresse, P. "La chimie appliquée à l'Université de Besançon," *Rev. Int. l'Enseignement* 1898, 35, 32–34.
71. Genvresse, P. "La chimie industrielle à l'Université de Besançon," *Rev. Int. l'Enseignement* 1902, 44, 311–312.
72. "Enquête sur l'enseignement technique dans les universités françaises," *Rev. Int. l'Enseignement* 1909, 57, 333–334.
73. Bichat, E. "L'enseignement des science appliquées à la Faculté des sciences de Nancy," *Rev. Int. l'Enseignement* 1898, 35, 299–307.
74. "Enquête sur l'enseignement technique dans les universités françaises," *Rev. Int. l'Enseignement* 1909, 57, 343.
75. Buisine. "Les laboratoires de chimie appliquée à la Faculté des sciences de Lille," *Rev. Int. l'Enseignement* 1897, 34, 11–15.
76. "Enquête sur l'enseignement technique dans les universités françaises," *Rev. Int. l'Enseignement* 1909, 57, 423.
77. Vezès, M. "Une foundation régionale—le laboratoire des résines à l'Université de Bordeaux," *Rev. Int. l'Enseignement* 1901, 41, 118–26.
78. Gayon, U. "L'enseignement de la chimie appliquée à la Faculté des sciences de Bordeaux," *Rev. Int. l'Enseignement* 1898, 36, 397–400.
79. "Enquête sur l'Enseignement technique dans les universités françaises," *Rev. Int. l'Enseignement* 1909, 57, 343.
80. Ibid., 424–431.

81. Cavelier, J. "L'enseignement de la chimie appliquée à l'Université de Rennes," *Rev. Int. l'Enseignement* **1899**, *38*, 490–493.
82. Gernez, D. "L'Ecole de chimie industrielle annexée à la Faculté des sciences de Lyon," *Rev. Int. l'Enseignement* **1895**, *30*, 39–42.
83. Friedel, Ch. "La chimie appliquée à la Faculté des sciences de Paris," *Rev. Int. l'Enseignement* **1898**, *36*, 481–487.
84. Partington, J. R. "A History of Chemistry"; MacMillan: London, 1964; Vol. IV, p. 858.
85. Barthelet, E. "L'Ecole d'ingénieurs de Marseille," *Rev. Int. l'Enseignement* **1901**, *42*, 218–223.
86. Vezès, M. "Une fondation régionale—le laboratoire des résines à l'Université de Bordeaux," *Rev. Int. l'Enseignement*, **1901**, *42*, 119.
87. Matagrin, A. "L'Industrie des produits chimiques et ses travailleurs"; Doin: Paris, 1925; p. 328.
88. Ibid., p. 349.
89. Ibid., p. 329.
90. Ibid., p. 308.
91. Ibid., p. 318.
92. Ibid., p. 319.
93. Friedel, Ch. "La chimie appliquée à la Faculté des sciences de Paris," *Rev. Int. l'Enseignement* **1898**, *36*, 482.
94. Lévy-Leboyer M. "Le patronat français a-t-il été malthusien?," *Le Mouvement social* **1974**, *(88)*, 3–50.
95. Friedel, Ch. "La chimie appliquée à la Faculté des sciences de Paris,' *Rev. Int. l'Enseignement* **1898**, *36*, 482.
96. Genvresse, P. "Relation résumée d'un voyage dans des universités françaises et étrangères," *Rev. Int. l'Enseignement* **1898**, *36*, 298.
97. Matagrin, A. "L'Industrie des produits chimiques et ses travailleurs"; Doin: Paris, 1925; p. 56.
98. Haller, A. "L'industrie chimique"; J. B. Baillière et fils: Paris, 1895; p. 44.
99. *Chim. Ind.* (Paris) **1934**, *31*, 812.
100. Blondel, G. "L'essor industriel et commercial du peuple allemand"; Librairie de la Société du recueil général des lois et des arrêts: Paris, 1900.
101. Lévy, R.–G. "L'industrie allemande," *Rev. Deux Mondes* **1898**, *145*, 806–838.
102. Cambon, V. *L'Allemagne au travail* **1909**, 13–29.
103. E. H. "To the Editor," *Engineering* **1901**, *72*, 263.
104. Lexis, W. "A General View of the History and Organisation of Public Education in the German Empire"; A. Asher: Berlin, 1904.
105. Schoen, H. "L'enseignement supérieur technique en Allemagne," *Rev. Int. l'Enseignement* **1909**, *57*, 217.
106. Lexis, W. "A General View of the History and Organisation of Public Education in the German Empire"; A. Asher: Berlin, 1904; pp. 135–136.
107. Cambon, V. *L'Allemagne au travail* **1901**, *72*, 263.
108. Lexis, W. "A General View of the History and Organisation of Public Education in the German Empire"; A. Asher: Berlin, 1904; 217.
109. Schoen, H. "L'enseignement supérieur technique en Allemagne," *Rev. Int. l'Enseignement* **1909**, *57*, 217.
110. Pyenson, L.; Skopp, D. "Educating Physicists in Germany circa 1900," *Social Studies of Science* **1977**, *7*, 332.
111. Lexis, W. "A General View of the History and Organisation of Public Education in the German Empire"; A. Asher: Berlin, 1904; 137.
112. Cambon, V. *L'Allemagne au travail* **1901**, *72*, 16–29.
113. Beer, J. J. "The Emergence of the German Dye Industry"; University of Illinois Press: Urbana, Illinois, 1959; p. 146n.
114. Cambon, V. *L'Allemagne au travail* **1901**, *72*, 184–185.
115. Lévy, R.–G. "L'industrie allemande," *Rev. Deux Mondes* **1898**, *145*, 811.
116. Beilby, G. T. "Chemical Engineering," *J. Soc. Chem. Ind.* **1915**, *34*, 771.

117. Beer, J. J. "The Emergence of the German Dye Industry"; University of Illinois Press: Urbana, Illinois, 1959; p. 76.
118. Ibid., p. 77.
119. Ibid., p. 65.
120. Ibid., p. 76.
121. Ibid., p. 79.
122. Ibid.
123. Duisberg, C. "The Education of Chemists," *J. Soc. Chem. Ind.* **July 1931,** 173–174.
124. Ibid., 174.
125. Beer, J. J. "The Emergence of the German Dye Industry"; University of Illinois Press: Urbana, Illinois, 1959; p. 146 n.
126. Ibid.
127. Ibid.
128. Cambon, V. *L'Allemagne au travail* **1901,** *72,* 50.
129. Beer, J. J. "The Emergence of the German Dye Industry"; University of Illinois Press: Urbana, Illinois, 1959; p. 89.
130. Ibid., p. 108.
131. Ost, H. "Die chemische Technologie an den Technischen Hochschulen," *Zeitschrift für angewandte Chemie* **1900,** *27,* 660.
132. Ibid.
133. Ibid.
134. Flechtner, H.-J. "Carl Duisberg—vom Chemiker zum Wirtschafts-führer"; Econ–Verlag GMBH: Düsseldorf, 1960; *passim.*
135. Mahan, J. "L'essor de l'I. G. Farben et le recrutement de ses cadres supérieurs," *Chim. Ind. (Paris)* **1949,** *61,* 503.
136. Ibid., 504.
137. Fletchtner, H.-J. "Carl Duisberg—vom Chemiker zum Wirtschaftsführer"; Econ–Verlag GMBH: Düsseldorf, 1960; *passim.*
138. Bucholz, K. "Verfahrenstechnik (Chemical Engineering)—Its Development, Present State and Structures," *Social Studies of Science* **1979,** *9,* 33–62.

RECEIVED May 7, 1979.

4

The Improbable Achievement: Chemical Engineering at M.I.T.

H.C.WEBER

Department of Chemical Engineering, Massachusetts Institute of Technology, Cambridge, MA 02139

Occasionally in almost any field there develops, from a small beginning, an achievement which grows rapidly and to a state of excellence hardly expected. Chemical engineering education, which started at M.I.T. almost 90 years ago, is an outstanding example of such an achievement. There must have been certain unusually favorable conditions, and a fortunate choice of personnel, not only to make such a result possible, but to enable this Chemical Engineering Department to hold its position of world-wide prestige throughout the years. The possible reasons for the department's success, together with the personalities and accomplishments of many of its members, will be traced in the pages which follow.

The Beginning

In 1888, Professor Lewis Mills Norton founded the course in Chemical Engineering in M.I.T.'s Chemistry Department. This course today probably would be called Industrial Chemistry. It was mainly descriptive, and consisted of a series of lectures describing the commercial manufacture of chemicals used in industry. Material was taken in considerable degree from German practice. The early development of student interest in chemical engineering is attested to by the granting in 1891 of seven Bachelor's degrees in that new area, 11 in the more traditional Chemistry. Norton died in 1893, but chemical engineering continued to develop under the leadership of Professor Frank Hall Thorpe, '89, whose *Outlines of Industrial Chemistry* was first published in 1898.

---

[1] Current address: La Hacienda, 10333 W. Olive #F179, Peoria, AZ 85345.

0-8412-0512-4/80/33-190-077$05.00/1

Arthur A. Noyes indirectly but very significantly affected the further development of chemical engineering at M.I.T. Noyes graduated from the Institute with a Masters degree in Chemistry in 1887. After graduate study in Germany, he returned to M.I.T. and began a brilliant career in physical chemistry. Later he was to serve as acting president of the Institute until a permanent head was found. Doctor Noyes had firmly in mind the establishment of a Research Laboratory of Physical Chemistry, even going so far as to offer to support a laboratory of this type for three years from his own funds. On September 20, 1903, such a laboratory was opened and under Noyes' direction attained world-wide recognition. It must be remembered that, at this time, little in the way of chemical research was being done in the United States.

William H. Walker was an associate of Noyes in the Chemistry Department. Walker, perhaps from close contact with the Research Laboratory of Physical Chemistry, saw the desirability for a Research Laboratory of Applied Chemistry and founded a laboratory of this type in 1908. Walker assumed charge of Chemical Engineering in 1912 and held that position until 1920.

Another name closely associated with chemical engineering during the early years of its development was that of Arthur D. Little, M.I.T. '85. Little was not a member of the faculty, but he was interested deeply in chemical engineering education. For several years he and Walker carried on the firm of Little and Walker in Boston, one of the early industrial consulting firms. Later the firm of Little and Walker became Arthur D. Little, Inc. Both Walker and Little had extraordinary vision and a capability for clear and forceful expression. They enunciated four fundamental concepts which formed the foundation upon which chemical engineering at M.I.T. has been built.

These are:

1. The grouping together of certain steps common to most industrial processes, i.e., heat transfer, distillation, fluid flow, filtration, crushing and grinding, and crystallization, to mention a few. These they called unit operations, to be studied as such irrespective of the particular industry of which they were a part. This idea unified and greatly simplified the study of industrial chemical processes.

2. The formation of a Research Laboratory of Applied Chemistry where industry could have specific problems worked on by a team of experts, faculty members, and research associates or research assistants, for a fee. This idea was broadened later to include all departments at M.I.T. by formation of the Division of Industrial Cooperation. The name of the latter was changed later to the Division of Sponsored

Research. One of the later outcomes of this early organization to serve industry was the Industrial Liaison Office, which in time would bring in $2,000,000 per year in unrestricted funds.

3. A School of Chemical Engineering Practice where students in an industrial plant, but under the guidance of a professor, carried out chemical engineering tests and studies significant to the operation of the plant.

4. An important concept, developed by Noyes and strongly advocated by Walker and Little, was the use of the case-and-problem method in class instruction instead of the then more usual lecture system.

In 1901, Warren K. Lewis entered M.I.T. as a student in mechanical engineering, but a year later transferred to the Chemical Engineering Course being offered by the Chemistry Department. After graduation Lewis studied at the University of Breslau in Germany, returning to the United States with a PhD degree in 1908. In 1910, after a brief period with industry, he was attracted back to M.I.T. as an Assistant Professor in Chemical Engineering, largely because of Walker's prodding. (By 1909 the number of degrees granted in the Chemical Engineering option had passed those in Chemistry). In 1920, Chemical Engineering became a separate department. Warren K. Lewis became its first Head, a post he held until 1929 when he resigned to devote his full time to teaching.

Even with such an outstanding nucleus, it is doubtful if chemical engineering at M.I.T. could have attained its position of prestige without the help of an administration, under President MacLaurin, which was most sympathetic to the aims of education and research. President MacLaurin wholeheartedly backed Walker and Lewis and with the aid of Arthur D. Little, was successful in having George Eastman donate $300,000 to establish the first School of Chemical Engineering Practice.

The Pioneers: Walker and Noyes, The Fundamentalists; Lewis, the Builder

Walker and Noyes were entirely different personalities. Walker was a qualitative, imaginative thinker. He was concise and to the point, almost abrupt in his manner. He was loyal to a fault with his associates as long as they were truthful and sincere. On one occasion, Walker had some question on Raoult's Law. He knew Noyes would know the answer, but he was reluctant to ask Noyes. Instead he sent Lewis, then a young professor, to Noyes' office. When Lewis returned Walker asked him if Noyes had taken some time to answer. When Lewis said Noyes did not answer immediately, Walker said, "That is good, I am not so stupid after all" *(1)*. Walker disliked wasting time. He often said, "An engineer who plays a moderately good game of pool has used judgement; if he plays an excellent game he has wasted his time" (1). Walker repeatedly accused wordy talkers of having a diarrhea of words and a constipation of ideas. He could express himself clearly and with a beauty in the choice of words that was remarkable. Because of his ability he was in great demand as an expert witness in patent suits. Naturally, Walker was a superb lecturer. After his departure from M.I.T., each revision of the Walker, Lewis, and McAdams text lost something in the parts which were rewritten without Walker's help.

Noyes, on the other hand, was an outstanding and methodical experimenter of great technical ability and imagination. He was kind, retiring, and very much a gentleman. He was most patient with students. His text *Principles of Physical Chemistry*, written with Miles Sherrill, was outstanding in its brevity and coverage. This book, together with the notes on chemical engineering being developed by Lewis and Walker, made life anything but easy for Course X students in that period from 1914 to 1923. Students of this period vividly remember these notes, hard to read, duplicated on an early mimeograph machine, with stencils typed on a typewriter which made all small letter e's and o's the same—just a solid round black dot. The notes contained numerous errors, and often considerable class time was spent by Lewis interpreting what the notes were supposed to say. These notes were revised and used in class over a period of years. About 1920, Walker, Lewis, and McAdams together with some graduate students spent the summer at Walker's summer home in Maine. They returned in the fall with essentially a complete manuscript. This resulted in the so-called "Bible," *Principles of Chemical Engineering*, which without a doubt laid the world-wide foundation for quantitative chemical engineering.

Through the years it became increasingly evident that Walker and Noyes, two strong men with such contrasting personalities, could not continue to work smoothly in the same department. Doctor MacLaurin met Walker in an Institute corridor one day and remarked, "Doctor Walker, I understand there is some friction in your department". Walker answered, "Doctor MacLaurin, you are a physicist; you know you cannot have motion without friction" (1). A close associate of Doctor MacLaurin's told him he thought he would lose one of these valuable men. MacLaurin answered, "I am afraid I will lose both" (1). Since 1913 Noyes had been aiding in the transformation of Throop College in Pasadena into the California Institute of Technology. In 1920 he resigned as professor at M.I.T. Until his death in 1936, he shared in making Cal Tech the outstanding school it is today.

Walker had come back to M.I.T. briefly after World War I, but left in the early 1920's to devote himself full time to consulting work. One can question whether the departure of one of these two strong men influenced the other also to leave the Institute. As MacLaurin had surmised, two outstanding men were lost to M.I.T. Walker died in an automobile accident preceded by a heart attack in 1934.

In personality, Lewis differed markedly from Walker. He usually reduced his thinking to a quantitative basis, while Walker's thinking was essentially qualitative. Lewis had a remarkable ability to wring information from a few facts and a relatively simple model. Repeatedly, by using only material and energy balances, he was able to develop unexpected relationships. He was bombastic in his nature, and almost explosive in the classroom. On entering a lecture hall he often would take off his jacket, roll it into a bundle, place it on the end of the lecture table, roll up his sleeves, puff out his cheeks, glare at his class, and then often literally blast a question at some unsuspecting student. In his field he was brilliant, and for most of his students he was an excellent and beloved teacher. He enjoyed a lively discussion and got real pleasure out of besting his opponents. He was particularly fond of asking a question to which the obvious answer was the wrong one. He hated to lose an argument and had all sorts of tricks to throw his adversary off the track, often adroitly switching the basis for the argument if he felt he was losing. It took others a long time to realize that he often would read some obscure fact in an encyclopedia and then spring a question that his listeners could not answer.

Though Lewis dressed neatly, he did not particularly care about clothes, and he never smoked or drank. However, for some unexplained reason, many years ago he appeared with a pocket watch, a chain draped across his vest, and at the end of the chain a cutter which was used for clipping the ends of cigars. He also carried a pocket lighter. These adornments lasted about a week, and then disappeared forever.

Lewis was deeply religious and was active in his church. One day while discussing a world problem with an associate, his associate asked, "Doc, don't you trust the Lord?" "Yes," he said, "but sometimes I will be damned if I can understand His methods" (1). He was extremely frugal and never wasted anything. Horace Ford, who so ably served M.I.T. as treasurer for many years, once remarked that he was amazed how much Willy Walker and Warren Lewis could do in chemical engineering with equipment that was primarily pots and pans and pieces of pipe.

Lewis realized that music and the theatre had a place, but never mentioned to his colleagues ever having attended a concert. He did mention having gone to a performance of "Abraham Lincoln," and talked about it many times, but evidently took no active interest in the theatre. Having lunch with a young professor one day, he extolled the cultural advantages of Boston, specifically stressing the art museum. One of his senior associates asked, "But Doc, have you ever visited the art museum?" He answered, "No, but that is no reason why this young whippersnapper should not!" (1).

Doc especially enjoyed telling stories with a moral. "Yarns," he called them. Many of the stories he told have been gathered and bound in a little volume entitled, "A Dollar to a Doughnut." He felt a great debt to Walker and throughout the years he repeatedly referred to him.

This, then, was the man who developed the program in chemical engineering so boldly outlined by Little and Walker. He surrounded himself from the beginning with men of promise in their respective fields. These included William McAdams, Clark Robinson, Robert Haslam, Walter Whitman, Harold Weber, and William Ryan, who unfortunately died in 1933. Robinson died in the late 1940's, Haslam in the early 1960's and Whitman, Lewis, and McAdams in the early 1970's. During the period from the mid-1920's to World War II, Ernst Hauser, Hoyt Hottel, Thomas Sherwood, Edwin Gilliland, Herman Meissner, Glenn Williams, and Edward Vivian would be added. These men formed the framework on which the present staff was built.

World War I

The entrance of the United States into the First World War did much to disrupt operations at the Institute, and temporarily interrupted educational progress, especially in the chemistry and chemical engineering areas. German use of chemical warfare caught the United States quite unprepared. There was immediate need for chemical research and for chemists and chemical engineers to develop an offensive and defensive position for the United States. The staffs in chemistry and chemical engineering responded in numbers. Noyes went to Washington as Chairman of the National Research Council. Walker accepted a commission as colonel in the newly formed Chemical Warfare Service, and Lewis, as a civilian, headed one of the Research and Development divisions of the new service. There were innumerable places where chemical engineers were needed, and relatively few were available. Walker and Lewis turned largely to graduate students in chemical engineering and then to seniors at M.I.T., and finally to members of the junior class. McAdams, then a graduate student in chemical engineering who had transferred from the University of Kentucky, accepted a position as captain, and was sent almost immediately to Cleveland to a top secret unit set up to develop an industrial process for producing mustard gas. It was called the "Mousetrap" because all workers assigned there were, for security reasons, completely isolated from the outside world. Robert E. Wilson, M.I.T. '16, later to be Director of the Research Laboratory of Applied Chemistry, and still later research director at Standard Oil of Indiana and a member of the Atomic Energy Commission, became a major and worked on adsorbent charcoals for gas masks. Weber and Ryan, juniors in chemical engineering and later to become faculty members, served as lieutenants under Walker and Lewis, carrying out work in the construction and operation of gas warfare plants. Walker built and operated Edgewood Arsenal, the first Army Chemical Warfare facility. With the War at an end, M.I.T. once again turned to its main objective—education.

Laying the Foundation

A first undertaking after the close of World War I was the re-establishment of the chemical engineering staff, which had been depleted so thoroughly by chemical warfare demands, and the Practice School which, though started in 1916, had been closed because of the war. Soon McAdams returned as a faculty member. He almost immediately started research on heat transmission, a subject which was to be his life work. Robinson and Whitman, although active in war work, had carried on a semblance of teaching and had not left Cambridge. In 1920, chemical engineering became Course X, a department entirely separate from chemistry, and W. K. Lewis was appointed its first Head.

Although the general concept of the unit operations had been developed by Walker and Little earlier, a quantitative embodiment did not materialize really until about 1920. The actual quantitative expressions by which computations could be carried out were at this time almost exclusively the work of Lewis, Walker, and McAdams. The original edition of *Principles of Chemical Engineering* was followed by a second edition and then by a third in which Gilliland was a fourth author.

One of Lewis' strong interests was in combustion, kilns, and gas production. In the early 1920's, Robert T. Haslam, who had joined the faculty after World War I, chose this part of Lewis' interest as his area of specialization and began with Robert P. Russell of the Research Laboratory of Applied Chemistry to assemble material for a book, *Fuels and their Combustion*, an authoritative volume. This field was so important that Haslam considered the possibility of either a separate chemical engineering division in this area or perhaps a separate course. Fuel and gas engineering did, in fact, maintain a separate identity for a period of several years, but returned to chemical engineering in 1932.

In 1924 the Department awarded its first ScD degrees to John Keats and Charles Herty; in 1925 to Per Frolich, Wayne Rembert, and Ernest Thiele. This was unique among M.I.T.'s engineering departments; the others were not convinced yet that engineers warranted education at the doctoral level.

In the 1920's the demand for gasoline increased to a point where natural gasoline could not be produced in sufficient quantities to satisfy the fuel needs associated with the explosive increase in the number of automobiles. Thermal cracking was the solution, and the quantitative relationships for heat transfer, fluid flow, and distillation—all unit operations—offered to the petroleum industry the possibility of designing cracking units on a quantitative rather than strictly empirical basis. The major oil companies quickly hired large numbers of M.I.T. chemical engineers, and for a number of years these men practically dominated the industry. The chemical engineering senior faculty was almost completely hired as consultants or permanent employees by the oil industry. In addition, research assistants and associates of the 1920's and early 1930's from the department and from its Research Laboratory of Applied Chemistry furnished an impressive list of men who rose to high positions in the oil industry.

Although the unit operations were the basis for early plant designs there was no adequate method for determining the pressure–volume–temperature (PVT) relationships for mixtures of gases at the elevated temperatures and pressures encountered in many industries, e.g. petroleum refining. In the early 1930's Weber and Meissner developed an approximation based on reduced critical properties of gases which solved this problem, not quantitatively from a physical chemical view, but well enough for engineering purposes. Shortly thereafter Lewis, Cope, and Weber published in graphical form the so-called "MU" charts based on the work of Weber and Meissner. For several decades these formed the basis for calculations involving gaseous mixtures at elevated pressures and temperatures.

The cracking-coil furnace was another weak link in thermal cracking to produce gasoline, and the gas radiation charts coming out of Hottel's research became standard basic material for industrial furnace design, particularly in the petroleum industry. It was possibly fortuitous that the explosive growth of that industry, which so badly needed the concepts evolving in Course X at M.I.T., occurred just when these concepts were maturing. Or was it foresight and good judgement on the part of men such as Little, Walter, and Lewis?

It was in this period that a serious attempt was made to duplicate in the amorphous industries area—plastics, rubber, etc.—educational progress similar to that made in the heavy chemicals industries. Ernst Hauser was added to the faculty to aid in the development of educational methods in industrial colloid chemistry, and books by Hauser and by Lewis, Squires, and Broughton were published.

In the 1930's, the fluidized-bed method of handling reacting mixtures of solids and gases originally developed in Germany was brought to a state of industrial practice by Lewis and Gilliland working on grants from industry. This process is used widely by the petroleum industry for producing gasoline by catalytic cracking.

When Lewis resigned as Department Head in 1929, he was succeeded by William P. Ryan. Ryan died suddenly in 1933, and after a brief period of stewardship under Lewis, Walter Whitman was brought back from industry as Head of the Department. His tenure was the longest in the Department's history.

In addition to the classic *Principles of Chemical Engineering* and to *Fuels and their Combustion*, the texts and reference books written by the Department staff were significant and numerous. A text on distillation by Robinson, later to be published in a second, third, and fourth edition by Robinson and Gilliland, gained wide acclaim among not only the engineering profession but also the bootleggers during the prohibition period. Robinson also published, with Hitchcock in M.I.T.'s Mathematics Department, a book, *Differential Equations*, with applications in the field of chemical engineering. Lewis and Radasch published their widely used book, *Industrial Stoichiometry*. McAdams published his authoritative treatise, *Heat Transmission*, to be followed later by second and third editions, translated into five languages. Sherwood brought out his text, *Absorption and Extraction*. Sherwood and Charles E. Reed published a widely used book, *Applied Mathematics in Chemical Engineering*. Weber, and later in a second edition with Meissner, published *Thermodynamics for Chemical Engineers*, a single text covering, for perhaps the first time, the three areas of physical, chemical, and engineering thermodynamics. Ernst Hauser published *Colloidal Phenomena*, and Hottel developed Mollier diagrams for combustion calculations and with Williams and Satterfield published *Thermodynamic Charts for Combustion Processes*. To complete the record—even though it covers a greater time span than the rest of this chapter—a complete list of the 50 or so books (not counting second and third editions) published by the M.I.T. Chemical Engineering Faculty up to 1977 appears in the Appendix. Despite this significant publication list, the Department has continued to rely, to a considerable extent, on dittoed notes changed from year to year.

The Practice School

A young chemical engineer already mentioned, Robert T. Haslam, had graduated from M.I.T. in 1911 and joined the National Carbon Company. He had impressed Lewis during his war activities. With the war ended, Lewis brought Haslam back to Cambridge as General Director of the soon-to-be re-established School of Chemical Engineering Practice. Lewis also brought first William P. Ryan and then Harold C. Weber in as Practice School Directors. Ryan was in charge of the Bangor Station, located at the Eastern Manufacturing Company and at the Penobscot Chemical Fiber Company, both companies being paper and pulp manufacturers. Whitman was made Director of the Boston Station located at the Merrimac Chemical Company in Woburn, Massachusetts and at the Revere Sugar Refinery in Charlestown. D. W. Wilson was Director at the Lackawanna Plant of the Bethlehem Steel Company in Buffalo, New York. The directorships shortly changed, Ralph Price and then Frederick Adams, M.I.T., '21, being placed in charge at Bangor, Weber at Boston, and Ryan at Buffalo. Whitman, Ryan, and Weber after a few years as Practice School Directors returned to the staff at Cambridge.

Practice School students stayed at each station eight weeks and then moved on to the other stations. Registration was limited to three groups of 12 students each. The course was called X–A. Since attendance at the three stations covered a total of only 24 weeks, eight at each station (for which total a student received one full term of M.I.T. credit) the students returned for the rest of the year to take special courses given by the Practice School staff and open only to Practice School students. It is interesting to note that during this period of the early 1920's both Thomas K. Sherwood and Hoyt Hottel were Practice School students and that the following year both became Department Assistants, Sherwood to McAdams at Cambridge and Hottel to Ryan in the Practice School.

The success of the Practice School was so great and the demand for its graduates so large that, although the Practice School had up to this time been a graduate school only, leading to the Master's degree, it was decided to open the course to undergraduates. Beginning in 1921, an undergraduate Practice School course was established, designated as X–B.

The Practice School, an innovation in engineering instruction, has continued to the present day and its graduates are still in great demand by industry. The Practice School is quite different from cooperative courses offered by many universities, where the students work for the companies involved. By the M.I.T. chemical engineering internship method, students study and perform tests on plant equipment continuously under the guidance of a professor and his assistants. Few, if any, other institutions have adopted this method, possibly because of its expense. The Practice School has of recent years been under the direction of Professor J. Edward Vivian.

The Research Laboratory of Applied Chemistry

Simultaneously with the re-establishment of the Practice School, the Research Laboratory of Applied Chemistry, which had been established in 1908 by Walker, was reactivated under R. E. Wilson. The contracts it had with industry rapidly increased, making necessary a continuously increasing staff. Throughout its existence in the Chemical Engineering Department this laboratory directed a percentage of any profit to so-called pro bono publico work. (In the period 1925–1935, the standard contract with industry was for $300 per man per month if the research

was of a publishable nature, $600 if not; the larger figure permitted
support of a second man on pro bono publico research.) When Wilson
left in 1921 to become Director of Research at Standard Oil Company of
Indiana, Haslam was made Director of the Research Laboratory of Applied Chemistry in addition to his Practice School responsibility. One or
two years later Whitman became Assistant Director, but left in 1925 to
follow Wilson in research activities at Standard Oil Company of Indiana.
Haslam, too, left in about 1926 to become Manager of Standard Oil
Development Company. Many of the members of the Research Laboratory of Applied Chemistry were later to become well known in chemical
engineering circles. They were sought eagerly by industry and other
universities because of their unusual training.

The success of the Research Laboratory of Applied Chemistry caused
M.I.T. to expand the concept to cover an Institute-wide organization,
known originally as the Division of Industrial Cooperation, which played
the same role for industry in general that the Research Laboratory of
Applied Chemistry played for the chemical industry. Walker, who
founded this laboratory, could not have foreseen how important the
Research Laboratory for Applied Chemistry would become in the structure of M.I.T.

World War II

The entrance of the United States into the Second World War once
more caused an upheaval in Course X. Since the First War, the army
had adopted the practice of sending numerous chemical officers to M.I.T.
for graduate work in chemical engineering. At the beginning of the war,
research work in chemical warfare was centralized at Edgewood Arsenal.
Bradley Dewey, who had worked under Walker in World War I and who
had been a close friend of Course X throughout the years, convinced the
Defense Department that this was dangerous to national security and that
Edgewood's research effort could be strengthened by association with an
organization like M.I.T. He suggested that the Institute build a new
building for chemical engineering—which the the department needed—
and that the Chemical Corps use this building as a second laboratory, to
be leased by M.I.T. to the Army for the duration of the war. At the end
of the war Course X would occupy the building. This was done and a
highly restricted laboratory employing over 150 employees—half military–
half civil service—quickly occupied the finished building. In this way
Building 12 came into existence. Colonel J. H. Rothschild, who had just
completed work for a Master's degree in Chemical Engineering at
M.I.T., was made officer in charge. Weber was assigned as technical
advisor to the laboratory by M.I.T. at the Army's request. He immediately had two junior staff members of Course X, Roy Whitney and Scott

Walker, assigned to him as technical assistants. All three served full time in these capacities until the war ended.

Meanwhile, the rest of the senior staff was employed on war assignments. Lewis was a member of the National Defense Research Committee (NDRC) and also a member of the Senior Advisory Committee for the Manhattan Project. McAdams conducted war-related research on life rafts and on pressure generation of hydrogen. Sherwood was an early organizer of chemical engineering manpower resources over the nation, for use in building up the NDRC. He later became Section Chief for miscellaneous chemical engineering problems, in Division 11. Hottel and Williams directed research on gas turbine combustors for the Navy. In addition, Hottel became Section Chief for fire warfare in Division 11 of NDRC, and a member of NACA's Gas Turbine Committee; Williams directed a naval torpedo research project on which Mickley and Satterfield served as research assistants while working on their doctoral theses. Gilliland, as Bradley Dewey's assistant on the Rubber Reserve Board in Washington, had a primary responsibility for getting the United States into adequate synthetic rubber production. He later became Assistant Division Chief of Division 11 of NDRC. Whitman was with the War Production Board in Washington, responsible for various chemicals production. Whitman later headed a committee to assess the status of jet propulsion, with Sherwood and Hottel serving as members. The next year a more comprehensive assessment of jet propulsion was made under the direction of Gilliland; Williams served on that study group. In 1943–1944 Hottel went to England on liaison work in the areas of fire warfare and oxygen supply, and in the spring of 1945 Sherwood went to the Continent to report technical findings behind the advancing allied armies.

When the war ended, Course X moved into its new Building 12 and was soon again functioning normally as an education group.

After World War II

During the Second World War the atomic bomb had been developed and it was quickly realized that there would be peacetime uses for nuclear energy. With this in mind, the department established a division of nuclear engineering. Manson Benedict, a Course V M.I.T. graduate then with the Kellogg Company, was appointed Professor of Nuclear Engineering in 1951. The following year Thomas Pigford, who had completed his ScD in Course X, joined Benedict and the first subjects in Nuclear Engineering were offered in the Fall of 1952 in the Chemical Engineering Department. Later Benedict and Pigford would be joined by Edward Mason, another Course X man, ScD 1950, and Theos J. Thompson (both of whom later became Atomic Energy Commissioners).

At Whitman's urging, a nuclear reactor designed by Thompson and directed to the peacetime uses of nuclear energy was constructed on the M.I.T. campus. This was a unique undertaking and represents the first full-scale research reactor directed exclusively to peacetime studies and operated on a university campus. This facility has been used in cooperation with Boston hospitals for medical research and has also served a most useful purpose in the training of engineers in nuclear engineering. The staff has gained an international reputation and has repeatedly been called to aid in government problems. Nuclear Engineering was separated from Course X and became an independent department of the Institute in 1958.

Whitman's career in the post-war period was highly significant. At M.I.T. he directed the first study, in 1948, of the use of nuclear power for airplanes. On leave from M.I.T., he was on the General Advisory Committee of the Atomic Energy Commission and, successively, Chairman of the Research and Development Board of the Joint Chiefs of Staff of the Department of Defense, organized under United Nations auspices of the First International Conference on the Peaceful Uses of Atomic Energy, and Science Advisor to the Department of State.

Although this chapter is chiefly about the early days of the Department, the close association with the first generation of those who had joined the Department by 1950 justifies a few brief comments about the early work of the latter. Sherwood became internationally known for his pioneering and continuing work on mass transfer. Hottel and Williams' work on high-output combustion research attracted widespread attention. Mickley's belief in the need for strengthening chemical engineering science had a profound effect on Department and Institute educational policy. Mickley later became head of M.I.T.'s Center for Advanced Engineering Study. Satterfield turned from early post-war interest in hydrogen peroxide to industrial catalysis and, particularly, the effect of mass transfer thereon. Meissner, whose work on thermodynamics and equations of state has been mentioned, turned to the problem of bringing order into the approach to industrial chemistry. Vivian conducted significant research on mass transfer with chemical reactions, and devoted a major effort to maintaining the quality of the Practice School. Merrill's work on blood flow and dialysis in cooperation with Boston's hospitals presaged the Department's enormous later growth in chemical engineering applied to medical problems.

In the late 1950's, McAdams and Whitman retired. Weber took a leave of absence to become Chief Scientific Advisor for the Army. A decade later Hottel, Sherwood, and Meissner retired, each before or shortly after publishing books in their chief fields of interest; Hottel (with Sarofim) on radiative transfer, Meissner on processes and systems in

Table I. Chemical Engineering Faculty at M.I.T.
(including in the early years only those whose stay was of significant duration)[a]

1921	1926	1931	1936	1941	1946
LEWIS	*LEWIS*	*RYAN*	*WHITMAN*	*WHITMAN*	*WHITMAN*
Walker	Walker	Lewis	Lewis	Lewis	Lewis
Wilson	Haslam	McAdams	McAdams	McAdams	McAdams
Haslam	McAdams	Robinson	Robinson	Weber	Weber
McAdams	Robinson	Weber	Weber	Hottel	Hottel
Robinson	Ryan	Hauser	Hauser	Sherwood	Sherwood
Whitman	Weber	Hottel[b]	Hottel	Robinson	Gilliland
		Sherwood	Sherwood	Hauser	Robinson
			Gilliland	Gilliland	Hauser
				Meissner	Meissner
				Williams	Williams
				Vivian	Vivian
					Mickley

1951	1956	1961	1966	1971	1976
WHITMAN	*WHITMAN*	*WHITMAN*	*GILLILAND*	*BADDOUR*	*SMITH*
Lewis (Em)	Lewis (Em)	Lewis (Em)	Lewis (Em)	Lewis (Em)	—
McAdams	McAdams (Em)	McAdams (Em)	McAdams (Em)	McAdams (Em)	—
Weber	Weber	Weber (Em)	Whitman (Em)	Whitman (Em)	—
Hottel	Hottel	Hottel	Weber (Em)	Weber (Em)	Weber (Em)

Sherwood[c]	Sherwood	Sherwood	Drew (Em)	Hottel (Em)	Hottel (Em)
Gilliland	Gilliland	Gilliland	Hottel	Sherwood (Em)	—
Hauser	Benedict	Meissner	Sherwood	Drew (Em)	—
Benedict	Meissner	Williams	Meissner	Gilliland	Meissner (Em)
Robinson	Williams	Vivian	Williams	Meissner	Drew (Em)
Meissner	Vivian	Mickley	Vivian	Williams	Williams
Williams	Mickley	Satterfield	Mickley	Vivian	Vivian
Vivian	Satterfield	Michaels	Satterfield	Mickley[d]	Satterfield
Mickley	Michaels	Merrill	Merrill	Satterfield	Merrill
Satterfield	Pigford	Baddour	Baddour	Merrill	Baddour
Michaels	Thompson	Reid	Reid	Reid	Reid
Merrill	Merrill	Brian	Brian	Brian	Sarofim
Mason	Mason	Mohr	Mohr	Smith	Evans
Baddour	Baddour	Smith	Smith	Sarofim	Howard
	Reid		Sarofim	Evans	Colton
	Brian		Evans	Bodman	Modell
			Bodman	Mohr	Hites
			Howard	Howard	Clomburg
			Modell	Modell	Donnelly
				Colton	Georgakis
				Virk	Putnam
					Manning

[a] Department heads in capitals.
[b] In Fuel and Gas Engineering, later moved back into Chemical Engineering.
[c] Dean of Engineering.
[d] Director, Center for Advanced Engineering Study.

industrial chemistry, and Sherwood (with Pigford and Wilke) on mass transfer.   Within a period of about two years Gilliland, Whitman, Lewis, McAdams, and Sherwood died.

It is appropriate to record here the names of the distinguished faculty members of the third generation.   To this end Table I indicates faculty members from the founding of the Department to the present (but including only those whose stay was of significant duration).

A Few Notes on the Recent Scene

In the late 1960's, Gilliland resigned as Department Head to concentrate more on teaching.   Raymond Baddour had come to the Department from Notre Dame and earned his ScD degree in Chemical Engineering from M.I.T. in 1951.   His research and publications on ion exchange, heterogeneous catalysis, chromatographic separation, and high-temperature reactions have made outstanding contributions to theory and practice in these fields.   While Department Head he initiated several interdepartmental programs in enzyme technology, catalysis, ion exchange, and high-temperature reactions, and he organized M.I.T.'s Environmental Laboratory.   Baddour had the courage, youth, and endurance to concentrate on developing the strong industrial and financial resources which had their start years before in the department and to bring these together in the magnificent Landau building.

In 1976, after the Department moved into the new Landau building, Baddour relinquished his administrative responsibility.   Kenneth A. Smith, ScD, M.I.T., '62, was appointed Acting Head.   Smith had been a student of Mickley, had spent a postdoctoral year in Cambridge, England under G. I. Taylor and Alan Townsend in the Cavendish Laboratory, and on return began a career of research and teaching in the general area of transport processes.   His nearly 70 publications in the areas of non-Newtonian fluid mechanics, heat and mass transfer, biomedical engineering, desalination, and polymer characterization include many which are outstanding contributions to chemical engineering science.   During his short year as temporary Head of the Department, he made effective contributions to its strength in teaching and research and acquired a reputation for fair and thoughtful consideration of the Department's problems, and decisive actions on them.

In July, 1977, M.I.T. announced the appointment of Dr. James Wei, ScD, '55, as Warren K. Lewis Professor of Chemical Engineering and Head of the Department.   Dr. Wei had conducted distinguished industrial research on catalysis and reaction engineering at Mobil Oil Corp., followed by appointment as the Allan P. Colburn Professor at the University of Delaware.   He also had been accorded visiting professorships at Princeton and the California Institute of Technology.   For the first six months of his appointment, Dr. Wei was on leave and Professor Clark K. Colton served effectively as Deputy Head. Under Dr. Wei's leadership, the Department is confident of further growth in its reputation for the effective education of chemical engineers.

The Future

The Department's progress in the last two decades has been impressive. It is true that M.I.T. does not stand alone in chemical engineering as it did in the 1920's. Today it has strong competition from many universities, but it has advantages others do not have. It is practically free from political interference; it continues to have an administration completely in sympathy with long-range research and development. It has a strong core of young and middle-aged faculty well trained in industrial-type research. It has, as always, the high respect of industry and close contact with it, both through its Practice School and the widespread consulting practice of its faculty. It offers instruction and research opportunities in many areas of excellence. It has great strength in the core subjects of chemical engineering: fluid mechanics, transport processes, and thermodynamics. Its fuels laboratory has an outstanding staff, its work in biomedical engineering is well known, and its range of interest in the industrial chemical area—catalysis, surface chemistry, process engineering, and amorphous materials—is greater than that of many chemical engineering centers.

Enviromental restrictions coupled with decreasing reserves of natural resources now make necessary the redesign of most of our chemical and energy-producing processes. This opens to the chemical engineer a whole new spectrum of research and development opportunities, and no group is better fitted for solving these problems than the chemical engineer with his multidisciplinary training. Certainly, M.I.T.'s Course X will continue the brilliant leadership it has shown always.

Literature Cited

1. Personal communications.

Appendix. Books Published by M.I.T. Chemical Engineering Faculty (During M.I.T. Service)

1922 Elements of Fractional Distillation, C. S. Robinson (McGraw–Hill)

1922 The Recovery of Volatile Solvents, C. S. Robinson (Chem. Catalogue Co., N.Y.) (French Translation, 1937)

1923 Principles of Chemical Engineering, W. H. Walker, W. K. Lewis and W. H. McAdams (McGraw–Hill)

1923   Differential Equations in Applied Chemistry, F. L. Hitchcock and C. S. Robinson (Wiley)

1926   Industrial Stoichiometry, W. K. Lewis and A. H. Radasch (McGraw–Hill)

1926   Evaporation, C. S. Robinson (Chem. Cat. Co., N.Y.)

1927   Fuels and their Combustion, R. T. Haslam and R. P. Russell (McGraw–Hill)

1927   Principles of Chemical Engineering, 2nd Ed., W. H. Walker, W. K. Lewis and W. H. McAdams (McGraw–Hill)

1930   Elements of Fractional Distillation, 2nd Ed., C. S. Robinson (McGraw–Hill)

1933   Heat Transmission, W. H. McAdams (McGraw–Hill)

1936   Differential Equations in Applied Chemistry, 2nd Ed., F. L. Hitchcock and C. S. Robinson (Wiley)

1937   Absorption and Extraction, T. K. Sherwood (McGraw–Hill)

1937   Principles of Chemical Engineering, 3rd Ed., W. H. Walker, W. K. Lewis, W. H. McAdams and E. R. Gilliland (McGraw–Hill)

1938   The Elements of Fractional Distillation, 3rd Ed., C. S. Robinson and E. R. Gilliland (McGraw–Hill)

1939   Colloidal Phenomena, E. A. Hauser (McGraw–Hill)

1939   Thermodynamics for Chemical Engineers, H. C. Weber (Wiley)

1939   Applied Mathematics in Chemical Engineering, T. K. Sherwood and C. E. Reed (McGraw–Hill)

1940   Experiments in Colloid Chemistry, 1st Ed., E.A. Hauser and J. E. Lynn (McGraw–Hill)

1942   Heat Transmission, 2nd Ed., W. H. McAdams (assisted by G. C. Williams) (McGraw–Hill)

1942   Industrial Chemistry of Colloidal and Amorphous Material, W. K. Lewis, L. Squires and G. Broughton (Macmillan)

1942   The Recovery of Vapors, with special Reference to Volatile Solvents (1st publ. in 1922 as Recovery of Volatile Solvents), C. S. Robinson (Reinhold)

1943   The Thermodynamics of Firearms, C. S. Robinson (McGraw–Hill)

1944   Explosives, their Anatomy and Destructiveness, C. S. Robinson (McGraw–Hill).

1949   Thermodynamic Charts for Combustion Processes, H. D. Hottel, G. C. Williams and C. N. Satterfield (Wiley)

1950   Elements of Fractional Distillation, 4th Ed., C. S. Robinson and E. R. Gilliland (McGraw–Hill)

1952   Absorption and Extraction, 2nd Ed., T. K. Sherwook and R. L. Pigford (McGraw–Hill)

1954   Industrial Stoichiometry, 2nd Ed., W. K. Lewis, A. H. Radash and H. C. Lewis (McGraw–Hill)

1954 Heat Transmission, 3rd Ed., W. H. McAdams (McGraw–Hill) (also translations into French, Russian, Yugoslavian, Japanese)

1955 Hydrogen Peroxide, W. C. Schumb, C. N. Satterfield and R. L. Wentworth (Reinhold) (Russian translation, 1957)

1957 Applied Mathematics in Chemical Engineering, 2nd Ed., H. S. Mickley, T. K. Sherwood, and C. E. Reed (McGraw–Hill)

1958 The Properties of Gases and Liquids, R. C. Reid and T. K. Sherwood (McGraw–Hill)

1959 Thermodynamics for Chemical Engineers, 2nd Ed., H. C. Weber and H. P. Meissner (Wiley) (also a Polish translation)

1961 Thermal Regimes of Combustion, translated from Russian edition, L. A. Vulis, Author; G. C. Williams, Editor (McGraw–Hill)

1963 The Role of Diffusion in Catalysis, C. N. Satterfield and T. K. Sherwoood (Addison–Wesley)

1963 A Course in Process Design, T. K. Sherwood (M.I.T. Press)

1964 Probability Theory, J. R. McCord and R. M. Moroney, Jr. (Macmillan)

1966 The Properties of Gases and Liquids, 2nd Ed., R. C. Reid and T. K. Sherwood (McGraw–Hill)

1967 The Application of Plasmas to Chemical Processing, Ed. by R. F. Baddour and R. S. Timmins (M.I.T. Press)

1967 Radiative Transfer, H. C. Hottel and A. F. Sarofim (McGraw–Hill)

1968 Industrial Practice of Chemical Processes, S. W. Bodman (M.I.T. Press)

1970 Colloid and Surface Chemistry, Part 1; Surface Chemistry, S. T. Mayr and R. G. Donnelly (M.I.T. Press)

1971 Processes and Systems in Industrial Chemistry, H. P. Meissner (Prentice–Hall)

1971 New Energy Technology, Some Facts and Assessments, H. C. Hottel and J. B. Howard (M.I.T. Press)

1972 Staged Cascades in Chemical Processing, P.L.T. Brian (Prentice–Hall)

1972 Colloid and Surface Chemistry, Part 2; Lyophobic Colloids, A. Virj and R. G. Donnelly (M.I.T. Press)

1973 Process Synthesis, D. F. Rudd, G. J. Powers and J. J. Siirola (Prentice–Hall)

1973 Colloid and Surface Chemistry, Part 3; Electrokinetics, J. T. G. Overbeek and R. G. Donnelly (M.I.T. Press)

1973 Model Crystal Growth Rates from Solution, M. Ohara and R. C. Reid (Prentice–Hall)

1974 Thermodynamics and its Application, M. Modell and R. C. Reid (Prentice–Hall)

1974   Colloid and Surface Chemistry, Part 4; Lyophilic Colloids, J. T. G.
       Overbeek and R. G. Donnelly (M.I.T. Press)
1975   Mass Transfer (Major revision of Absorption and Extraction), T. K.
       Sherwood, R. L. Pigford and C. R. Wilke (McGraw–Hill)
1976   Properties of Gases and Liquids, 3rd Ed., R. C. Reid, J. Prausnitz
       and T. K. Sherwood (McGraw–Hill)
1977   Dynamics of Polymeric Liquids, Vol. 1, Fluid Mechanics, R. B.
       Bird, R. C. Armstrong and O. Hassager (Wiley)
1977   Dynamics of Polymeric Liquids, Vol. 2, Kinetic Theory, R. B.
       Bird, O. Hassager, R. C. Armstrong, and C. F. Curtiss (Wiley)

RECEIVED May 7, 1979

George E. Davis, Norman Swindin, and the Empirical Tradition in Chemical Engineering

D. C. FRESHWATER

Department of Chemical Engineering, Loughborough University, Loughborough, Leicestershire, United Kingdom

George E. Davis invented the essential unit operation concept and wrote the first textbook on chemical engineering in 1901. Norman Swindin was his only pupil and contributed many new ideas to the practice. Both were empiricists rather than theorists and believed in the absolute need to develop plant and processes through experiment and experience. Davis attempted to found the first Society of Chemical Engineers but died before his ideas came to fruition. Swindin largely rewrote the second edition of Davis' book in 1904 and went on to write pioneer texts such as the Flow of Liquid Chemicals in Pipes *in 1922. In a long and successful life, Swindin showed himself to be near to Davis' "ideal chemical engineer."*

C hemical Engineering has become the fourth great technology through the efforts of many workers who are both practicing engineers and teachers. It is regarded, perhaps a little arrogantly by its practitioners, as being more science based and orientated than its sister professions of civil, mechanical, and electrical engineering. Indeed in recent years there has been a vogue for so-called engineering science—a vogue in which some eminent chemical engineers may be said to have been leaders of fashion.

However there always has been a strong thread of empiricism in process technology throughout its history from Agricola to the present day. There are those who attempt to play down or even deny the role of empiricism and who would argue that chemical engineering is fundamentally different from (and somehow superior to) process technology. But this is to ignore the facts and widen even further the unfortunate gulf that

0-8412-4/80/33-190-097$05.00/1

Figure 1.   George E. Davis

exists between the theoreticians (largely academics) and practitioners (mainly engineers in industry).   Perhaps this damaging dichotomy may be reversed partially therefore by observing that the man who founded the subject as an ordered study was a clear empiricist.

George E. Davis (shown in Figure 1) was born at Eton in 1850 and studied chemistry first at Slough Mechanics Institute and later at the Royal School of Mines (now part of Imperial College).   His formal studies ended in 1870 when he joined Messrs Bealey's Bleach Works at Manchester as Works Chemist.   He moved about fairly often in the next 10 years gaining experience in the chemical industry, mostly in Northwest England.   Then in 1880 he became a consultant with an office in Manchester.

Just over a year later he was invited by Dr. Angus Smith to join the Alkali Inspectorate formed to administer the new Alkali Act.   This was the first legislation to try to control environmental pollution.   Dr. Smith died in 1883 whereupon Davis resigned and returned to his private practice, immensely enriched in experience and knowledge from his work as an alkali inspector. Then, apart from a short break as Manager of Barnsley Gas Works, he continued to practice as a consultant building a very substantial business until his untimely death in 1906.

Throughout his working life, Davis exhibited two characteristics which are notable in the present context. The first was an indefatigable persistence in collecting facts he considered relevant to his past, present, or future professional activities. The smallest detail was noted assiduously for possible future reference. The second was a passion for ordering and classifying this information.

His experience, which was far wider than common for chemists (industrial or otherwise) at that time, together with the two characteristics just outlined, undoubtedly combined to produce his profound concept of the unit operation. It was this idea which codified an immense quantity of previously unsystematized knowledge and laid the foundation for chemical engineering as a major subject in its own right.

In seeing how Davis collected his facts over many years and then ordered them in a new and extremely significant way, one thinks of the similarity between this and the process by which Darwin reached his evolutionary theory. The painstaking collection of data was a necessary but insufficient prelude to the ordering of the facts in a new way—a way which showed a unifying principle which was to have a profound effect on the development of technology.

Before Davis and indeed long after him in some countries, process technologists became specialists in particular chemicals or groups of chemicals. A firm would not think of employing a man specialized in alkali manufacture for the design or management of, say, a nitric acid plant. We still can see the remains of this industry-oriented specialization in the older-fashioned East European institutes—most of which were modeled on the German pattern which itself has changed so dramatically in the last five years.

We do not know for certain when Davis had the flash of inspiration that led him to codify his factual information and experience, not in terms of process or manufactures but under the broad headings of what came to be called unit operations. Certainly it was before 1888 in which year he gave a series of lectures at Manchester Technical School which was to form the basis of his famous handbook. *(1)*

It seems strange that so little is known about so important a discovery. This is no doubt due to the fact that Davis did not appear to seek recognition for the novelty and importance of his discovery, partly because of his early demise. The relative unimportance of the processing industries in general and the chemical industry in particular in Britain in his time also played a part in this neglect. Davis seemed to have been a man who was modest in public while being arrogant in private. He certainly was convinced of the importance of chemical engineering as a concept and was good at persuading other people of the importance of spreading the knowledge of the subject. But like many such persuaders he was prepared to forego pressing his own notions or taking particular credit for them in order to pursue what he saw as the greater good of the spread of knowledge. This can be seen clearly in the part he played in the formation of the Society of Chemical Industry. The Faraday Club formed by Davis for industrial chemists in and around St. Helens in 1875, led eventually to the Society of Chemical Industry of which Davis was the first Secretary (2). It very nearly became, when it was formed in 1880, the Society of Chemical Engineering. However, this title proved too much for all of the founding members to swallow and it was Davis himself who proposed the present name which received general consent. But Davis, while not assertive of his own views when he conceived some greater good, was nonetheless an active and continuous publicist for these views. Thus he founded with his brother in 1887, the *Chemical Trade Journal*, where he styled himself as editor, chemical engineer, and consulting chemist. In the editorial to the first number he wrote, "Chemical engineering, a science so neglected in England during the past, will have its due share of attention." It was in the same year that he gave his now famous course of lectures at Manchester Technical School. What made these lectures significant and the cause of considerable interest was the fact that the subject matter was presented in the new way which Davis had conceived. All of the processes of contemporary chemical technology were analyzed in terms of a series of basic operations. It does not seem that Davis himself used the term "unit operations" nor does he explicitly state his conscious ordering of information in the way that it appears. However, no one can see the structure of these lectures as subsequently reprinted in the *Chemical Trade Journal* without appreciating that Davis indeed had invented the concept of the unit operation and furnished a unifying basis for systematic education in process technology. Limitations of space prevented proper presentation of the lectures in the *Chemical Trade Journal* as anyone who has seen his *Handbook of Chemical Engineering* will appreciate. Due no doubt to the pressure of work, it was not until 1901 that this appeared. It was a massive work, truly encyclopedic in nature as well as size running to some 900 pages (3). It was succeeded in 1904 by a second, enlarged edition in two volumes running over 1000 pages. The list of contents shown in Table I gives some idea of the immense coverage. The many illustrations of plants in the Handbook are often a record of apparatus which now belongs to history, but sometimes the equipment he describes seems relatively modern (*see* Figures 2 and 3).

Table I. List of Contents of the *Chemical Engineering Handbook*[a]

Chapter

   Subject Matter

1 Introduction
General Remarks—Definition of a Chemical Engineer—Difference between Applied Chemistry, Chemical Engineering, and Chemical Technology—Effect of Time, Space, and Surface on Chemical Operations

2 The Fitting of a Technical Laboratory

3 Materials used in Plant Construction—Timber, Stone, Slate, Brick, Clayware, and Stoneware—Enamelledware, Glass, and Porcelain—Mortars, Cements, and Concrete—Asbestos, Rubber, Ebonite—The Metals

4 Weighing and Measuring
English and Metric Standards—Types of Weighing Machines—Measuring Flow of Liquids and Gases—

5 Production and Supply of Steam

6 Power and its Application
Wind, Water, Steam, Gas, Oil, and Electricity

7 Moving Solids, Liquids, and Gases

8 Treating and Preparing Solids

9 Application of Heat and Cold

10 Separating Solubles from Insolubles

11 Dissolving, Condensing, and Compressing Gases

12 Evaporation and Distillation

13 Crystallization and Dialysis

14 Electrolysis and Electrosmelting

15 Packaging

16 Organization and Building (includes a section on safety)

[a]Subheadings have been included in a few instances where some particular significance seemed to be revealed.

The introductory chapter clearly sets out Davis' philosophy. To quote:

"Chemical engineering must not be confounded with either applied chemistry or with chemical technology. Chemical engineering runs through the whole range of manufacturing chemistry whilst applied chemistry simply touches the fringe of it and does not deal with the engineering difficulties in even the slightest degree, while chemical technology results from the fusion of the studies of applied chemistry and chemical engineering and becomes specialized as the history and details of certain manufactured products."

Davis goes on to explain that a book on chemical technology would necessarily describe in detail the chemistry and construction of a plant for each industry. Such a work would not only be exceedingly bulky but would be beyond the competence of one man. On the other hand,

Figure 2.   Timbers for a sulfuric acid plant

chemical engineering deals with the construction of plants necessary for
the utilization of chemical reactions on the large scale without in any way
specifying the industry in which such a plant is to be used.   Davis here
has hit clearly on the idea of generalization—of the approach which marks
the emergence of chemical engineering as a new and significant discipline.   He goes on in his introduction to develop this idea.   "Solids,
liquids and gases have to be moved and measured, they have to be mixed
and otherwise treated with heat or with cold often under pressure, and
the process of lixivation, extraction, of evaporation and distillation, often
exercise the talents of the chemical engineer to the uttermost." There
hardly could be a clearer statement of the unit operations approach.   A
few sentences later he advocates the use of pilot-scale operations for the

development of a new process. After extolling the advantages of experimentation on the kilogram scale (with several examples) Davis goes on: "Still there is a way of going about one's work in chemical engineering, more certain and less expensive than the time-honored process of trial and error, and it is to be hoped that those who essay to become chemical engineers, will discover the fact that science and practice will work together on all occasions where the conditions have been properly studied."

There can be little doubt when considering the handbook that George E. Davis made a major contribution to establishing chemical engineering. Yet it is important to recognize that he was above all an empiricist. The

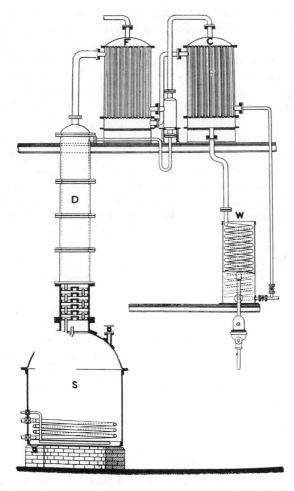

Figure 3.   Complete apparatus for systematic distillation

handbook for all its size is an immensely practical volume, always refer-
ring not only to known technical experience but to costs too.   Theory
intrudes hardly at all and for discussion of theoretical principles the
student is often referred to an elementary textbook.   Time and again
Davis refers to the need to ally theory to practical information and
economic information for developing successful processing plants.   Furthermore, the concept of the unit operation itself is empirical, being a
convenient and revealing classification of subject matter rather than having any basis in either physics or chemistry.   But although it has to some
extent been superseded in recent years by concepts such as transport
phenomena, it nevertheless remains the most usual way of introducing
the student to chemical engineering and the basic pattern for much of our
instruction.

The second (1904) edition of Davis' handbook was not only illustrated
but also largely extended by Norman Swindin to whom we are indebted
for most of our information about Davis other than the bare facts of his
life (3).   Davis said in the preface to his book, "If a student's credentials
included the fact that he knew by heart the whole of the handbook of
chemical engineering and possessed the ability to apply it, he would not
have much difficulty in finding a post."   If this was Davis' ideal chemical
engineer, Norman Swindin must have come close to it.

Norman Swindin (shown in Figure 4) was born in 1880 and after the
early death of his father he started to work as a clerk at the age of
14.   He became interested in engineering and engines at the works
where he was employed and by dint of his own efforts at night school and
by private tuition, got a reasonably good grounding in mechanical engineering and chemistry.   In 1901 he went to work for Davis and thereafter considered himself a chemical engineer.   He was perhaps the only
man to have been trained in his profession by the master and in six years
with George Davis gained more experience than do most young men in
twice that period today.   Nor was this confined simply to technical
experience.   During George Davis' prolonged illness, Swindin took over
as Editor of the *Chemical Trade Journal*.   The practice he thus gained in
writing was to be put to good use in later years.   On the death of George
Davis, Swindin left the firm and began a period of working as a chemical
engineer in a number of somewhat strange enterprises.   First there was
Ashcroft with his obsession with non oxygen, nonwater chemistry; then
there was Elmore whose process demanded the handling of concentrated
salt solutions containing 10% HCl.

Here Swindin's experience with Davis allied to his own native ability
really started to pay off.   His early work on corrosion was thorough
largely because of the importance Davis placed on construction materials.
Swindin realized that the Elmore process represented one of the most
searching tests of materials that could be.   Not only was there a handling

Figure 4. Norman Swindin

problem, these highly corrosive liquors had to be evaporated and the crystals and mother liquor transported. Swindin solved these problems by a combination of experience, insight, and inventiveness. Although he was not the originator of rubber lining, submerged combustion, or the airlift, he was the first to make them work on an industrial scale and in situations where no other solution was feasible (4).

His notebooks are available for this period; they are meticulously ordered and contain reports, usually handwritten, on his daily work (5). One cannot help being struck by the ingenuity of the man as well as his tremendous appetite for work when reading these. They range from a laconic description of what must be perhaps the most esoteric gas lift experiment (reproduced in Figure 5) through detailed accounts of painstaking experiments in the rubber lining of all kinds of vessels. Swindin did not know the expansion coefficient of rubber nor even if it was positive or negative. The experiment he reported on to test this was typical. He filled a small, soft rubber balloon with cold water and

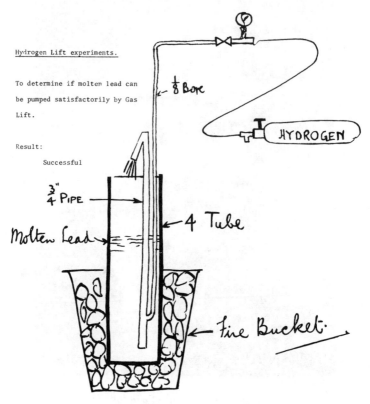

Figure 5.   Hydrogen gas lift

pierced it with a needle, observing the height to which the fountain rose.
He then refilled it with an equal quantity of hot water and observed that
the fountain was higher.   He concludes that "rubber contracts when
hot—the exact amount must be determined so that due allowance can be
made for this." But it is his early work on submerged combustion in these
notebooks that gives us one of the most interesting pictures of the way he
developed the idea.

The first experiment used an ordinary household gas mantle surrounded by a glass beaker which in turn was immersed to most of its
depth in a pot containing water (*see* Figure 6).

This was on May 4, 1902. Further experiments quickly followed on
May 16th and 22nd—the first scaling up and the second dispensing with
the mantle.   This was then but a step from true submerged combustion
(*see* Figure 7).   It was a step that took just over a year due to other
pressures.   In the end, faced with a disbelieving employer Swindin
plunged a lighted gas torch into a large beaker of water and held it there

until the liquid boiled.   (The first proper working model is recorded in his notebook for June, 1923.)

It might be thought that the solution of new and unusual technical problems—what one might call the hallmark of the chemical engineer—would have been enough to have kept a man occupied.   But Swindin was no ordinary man.   Ever since 1916 he had, on and off, been attending evening classes in chemical engineering given at Battersea Polytechnic by Hugh Griffiths.

Griffiths now invited Swindin to combine his practical knowledge with the theory he had been taught and to write a book on pumping as part of a series for which Griffiths was editor.   So in 1922 *The Flow of*

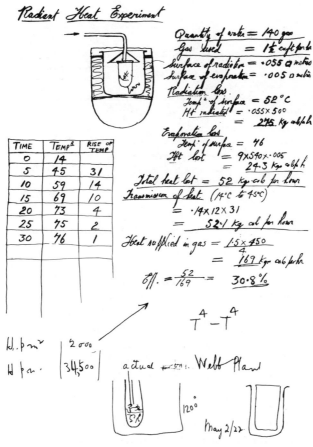

Figure 6.   Early experiment in submerged combustion

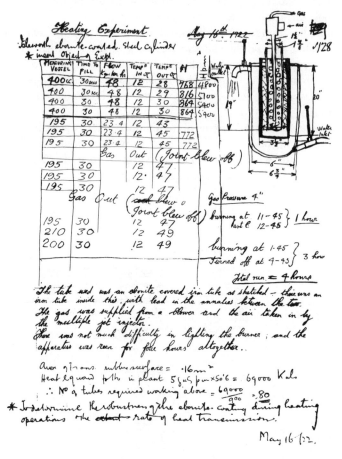

Figure 7.   Later experiment in submerged combustion

Liquid Chemicals in Pipes (reissued a year later under the title of *The Modern Theory and Practice of Pumping*) was published *(6)*.

Swindin had long been impressed with the reputation of Osborne Reynolds—the more so since he actually had heard the great man lecture.   However, as he says, he truly did not understand the implications of Reynolds' work on fluid flow until it was explained to him by Griffiths in his Battersea lectures.   Once he realized the power of the flow criterion particularly to characterize the multifarious fluids with which he had to deal, he was quick to exploit it.   His book well may be the first published practical application of the Reynolds' criterion to processing problems.   Certainly no mention of the Reynolds' number or its application to fluid flow appears in Liddell's Handbook, published the same year *(7)*.

That the Elmore process never worked commercially was due to
inept management and financial inadequacies rather than to Swindin who
solved essentially all of the technical problems.   In doing so he acquired
the know-how to set up a business designing and fabricating equipment
for processes operating with highly corrosive materials.

Much of his time in the next few years was taken up in the nuts and
bolts of starting a new business and putting it on its feet.

It is a tribute to his persistence that at 43 years of age he should start
a business and successfully develop it during the worst period of depres-

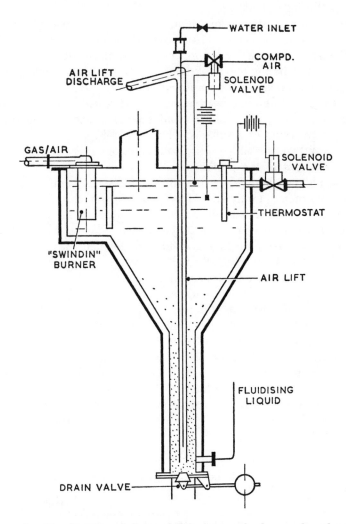

*Figure 8.   Combination of slurry gas lift pump and submerged combustion
in rubber-lined vessel*

sion that the United Kingdom had known. The company which he started eventually flourished and became known worldwide for its expertise in the handling of corrosive materials. It was based firmly on Swindin's three material contributions to chemical engineering—rubber lining, submerged combustion, and the gas lift pump. All three are illustrated in the apparatus shown in Figure 8. But Swindin brought much more to the development of chemical engineering than these three techniques. He was a founder member of the Institution of Chemical Engineers and for many years played an important role in its growth. He was an indefatigable proselyte for the profession through his writing and his work throughout his life.

Just as his master before him, Swindin was above all an experimenter, ready to be guided in the right direction by theory but always recognizing the essential need to find out by trial, the operating conditions. The predicting power of theory can seduce professional engineers to neglect or even abhor the empirical approach. However to do so is to ignore the true difference between engineering which is concerned with the solution of real-life problems and science which is concerned with solving abstractions of problems. Davis and Swindin clearly saw this difference and were enabled thus, each to make a singular contribution to the development of the profession.

It would not be proper to end without some mention of Norman Swindin as a man. I only knew Davis secondhand, through his book and from Swindin, but I was fortunate to know Swindin himself and to become his friend over the last 20 years of his life—from age 76 to 96.

Up to his death he had a vivid and detailed memory and was a natural-born raconteur who talked about the great men of the early 1900's as if they had just left the room for a while.

His memories extended far beyond professional horizons and encompassed his lifelong love of music, art, and literature. Perhaps part of the secret of his longevity was that he was perpetually young in the mind. The twinkle in his eye belied his age and he was never so happy as when he came to Loughborough every year to give his annual lecture to the students—an event much anticipated and always crowded. He truly was an engineer extraordinary. To have known him makes one feel one has a real link with the past going right back to the founder of our profession.

Literature Cited

1. Davis, George E. "A Handbook of Chemical Engineering"; Davis Bros.: Manchester, 1901.
2. Hardie, D.W.F. "The Empirical Tradition in Chemical Technology," The first Davis-Swindin Memorial Lecture, Loughborough University, 1962.
3. Swindin, N. *Trans Inst. Chem. Eng.* 1953, *31*, 187.

4. Swindin, N. "Engineering Without Wheels"; Wiedenfeld & Nicholson: London, 1964.
5. Swindin, N. Laboratory Notebooks 1922 and 1923, Loughborough University Library, Leicestershire, U.K.
6. Swindin, N. "The Pumping of Chemical Liquids in Pipes"; Benn Bros.: London, 1922.
7. "Handbook of Chemical Engineering"; Liddell, D.M., Ed.; McGraw-Hill: New York, 1922.

RECEIVED May 7, 1979.

Pioneers in Chemical Engineering at M.I.T.

GLENN C. WILLIAMS and J. EDWARD VIVIAN

Department of Chemical Engineering, Massachusetts Institute of Technology, Cambridge, MA 02139

In the nearly nine decades since the first bachelor's degree in chemical engineering was conferred on William P. Bryant in 1891, M.I.T. has awarded well over 4000 bachelor's degrees and a nearly equal number of advanced degrees to students of chemical engineering. For over four of these decades it has been the authors' privilege to have been successively students and teachers at M.I.T. and to have been colleagues of all but the first three of the pioneers whose biographies follow. There were, of course, many more faculty who made contributions, both large and small, to the development of chemical enigneering at the institute but we have elected, save the first three, to vignette the careers of those who were full-time faculty prior to 1934 and who completed their careers at M.I.T. during our tenure.

Lewis Mills Norton, 1855–1893

Professor Norton was born in Athol, MA, the son of a protestant minister. He studied chemistry at M.I.T. as a special student for three years, served there as an assistant for two more, and completed the PhD in Chemistry at the University of Göttingen in 1879. He then spent two years as a chemist for the Amoskeag Manufacturing Company in Manchester, NH before returning to M.I.T. in 1881 to serve as a faculty member in organic and industrial chemistry. His research was oriented largely toward practical applications, and in 1888 the faculty approved his proposal to initiate the first course in chemical engineering in the United States. This program combined a rather thorough curriculum in mechanical engineering with a fair background in general, theoretical, and applied chemistry. He fortunately saw the new program established on a firm foundation before his untimely death at the age of 38.

0-8412-0512-4/80/33-190-113$05.00/1

Figure 1.   Lewis Mill Norton

Frank Hall Thorp, 1864–1932

A native of Bloomington, IL, Doctor Thorp was graduated with a degree in chemistry from M.I.T. in 1889.   After serving some time as an assistant in chemistry at M.I.T., he studied chemistry at the University of Heidelberg, receiving the PhD in 1893.   He then returned to M.I.T., replacing the recently deceased Professor Norton as instructor in industrial chemistry and continued in that post for 22 years.   In 1898 Dr. Thorp's text, *Outlines of Industrial Chemistry*, was the first American publication in its field and was used widely for over 15 years.   Doctor Thorp, building on the foundation founded by Professor Norton, was the first to develop a rigorous course of instruction in industrial chemistry in

American technical schools.   Dr. Thorp taught until his retirement because of ill health in 1916 and was once Chairman of the Northeastern Section of the American Chemical Society.

William Hurtz Walker, 1869–1934

Professor Walker, a farm boy from the hill country near Pittsburgh, received his bachelor's degree from Pennsylvania State University in 1890 and his PhD under Otto Wallach at Göttingen in 1892.   After two years as an instructor at Penn State College he came to M.I.T. as an instructor of analytical chemistry.   In 1900 he resigned to join Arthur D. Little in Little and Walker, consulting chemists.   Two years later he was

Figure 2.   Frank Hall Thorp

Figure 3.   William Hurtz Walker

recalled by President Pritchett of M.I.T. to initiate a radical change in the
laboratory instruction in chemical engineering.   In 1905 he described his
ideal chemical engineer as, ". . . first of all, a thoroughly trained chemist,
able to apply knowledge of chemical forces to operate industrial plants
based on chemical reactions."   In that year he also was appointed a
lecturer on industrial chemistry at Harvard University, free to develop a
course which later (in 1923) was the basis for *Principles of Chemical
Engineering.*

In 1907 Dr. Walker completely revised the curriculum in chemical
engineering from a program which was largely mechanical engineering in
nature to one which emphasized chemical training and embodied the
fundamentals of engineering.   Recognizing the need for a closer relation-

ship between academia and industry and the need to train people for industrial research, he founded the Research Laboratory of Applied Chemistry in 1908.   This latter organization grew into M.I.T's Division of Industrial Cooperation, which contributed notably to governmental research during World War II.

In 1916, Arthur D. Little proposed a new method of training which involved work and study in industrial plants under the constant supervision and instruction of Institute staff after students had completed their fundamental training.   George Eastman was persuaded of the merits of this program and contributed generously to its support.   Practice school stations were set up in firms widely representative of the chemical industry, and the success of this unique method of training is attested to by

Figure 4.   Warren Kendall Lewis

the fact that the program, now at the graduate level, has continued to this day, being interrupted only by the two world wars.

With the advent of World War I, Walker was called to form the Chemical Service Section of the army, building Edgewood Arsenal from grass roots. He was awarded the Distinguished Service Medal for his efforts. Returning to the Institute he ran the Division of Industrial Cooperation until he resigned in 1921 to return to private consulting. He died in an auto accident in 1934.

Warren Kendall Lewis, 1882–1975

"For establishing the modern concept of chemical engineering, developing generations of leaders in engineering and service, and pioneering industrial problems that have contributed immeasurably to the progress of mankind"—thus, the citation in 1966 on the award of the John Fritz Medal to this man who came from an ancestral farm in Laurel, DE to Newton, MA to top off his local three-year high school in preparation for M.I.T. After his 1905 Bachelor's degree in Chemical Engineering, he assisted Dr. Walker, who was starting the Research Laboratory of Applied Chemistry, and then went to the University of Breslau for his PhD in Chemistry in 1908. He then returned to M.I.T. as a research associate, took a year off to work in a tannery in New Hampshire, and returned as a faculty member. During World War I he worked with the Bureau of Mines and the Chemical Warfare Service on gas defense.

After the war he worked with Dr. Walker and Professor McAdams on *Principles of Chemical Engineering*, which was used in mimeographed form at M.I.T. prior to its publication in 1923.

On the formation of the Department of Chemical Engineering in 1920 he served as its first head until 1929, when he returned to teaching and research. Just prior to World War II, Dr. Lewis and Dr. Gilliland invented the fluidized-bed method of cracking for petroleum which made possible the enormous production of aviation gasoline so vital in that conflict. Dr. Lewis was Vice Chairman of the Chemistry Division of the National Defense Research Committee, served in the Offices of Scientific and Production R&D and on chemical engineering problems in the development of the atomic bomb.

Combining his broad interests, exemplified here only partially, with his love of teaching and his own unique adversary method, made "Doc" a fantastic teacher, feared by some, until they learned of his soft heart, and remembered with affection by all. Rash young instructors attempted to imitate his style in vain.

In addition to some one hundred papers in broadly diverse areas of chemical engineering, he was coauthor of *Industrial Stoichiometry* in 1926 and *Industrial Chemistry of Colloidal and Amorphous Materials* in

1942.   Dr. Lewis retired in 1948, but was regularly in his office into his mid-eighties. He died at the age of 92 in 1975.

Clark Shove Robinson, 1888–1947

Professor Robinson, a native of Massachusetts, received his BS degree in Chemical Engineering from M.I.T. in 1909 and, after five years in industry, returned to complete a master's degree, joining the staff as an instructor in industrial chemistry in 1915.   His *Elements of Fractional Distillation*, published in 1922, one of the first U.S. texts in this area, was revised through 1950 by Professor Gilliland.   Another 1922 text in the field was *The Recovery of Volatile Solvents*.   In 1923 he collaborated

Figure 5.   Clark Shove Robinson

with Professor Lauren Hitchcock of the Mathematics Department to produce *Differential Equations in Applied Chemistry*. This was followed in 1926 by *Evaporation*. During World War II, while serving as a Colonel of Ordnance, and in large part responsible for the remarkable safety record of the munitions production, he published *Thermodynamics of Firearms, Explosives Chemistry for Safety Engineers*, and *Explosives— Their Anatomy and Destructiveness*.

Professor Robinson's joy was teaching undergraduates in chemical engineering, and he is remembered well by his students for his often risqué historical anecdotes concerning some famed pioneers in chemistry and thermodynamics, as well as for his insistence on numerical accuracy on home problems and quizzes. He was an avid mountain climber and an accomplished solo and choral singer.

William Henry McAdams, 1892–1975

A Kentuckian, Professor McAdams studied at Transylvania College and then the University of Kentucky, where he received BS and MS degrees in Chemistry in 1913 and 1914, and in 1917 a MS in Chemical Engineering at M.I.T. After a year with Goodyear Tire and Rubber Company, he served in the Chemical Warfare Service as a Captain, assistant chief of the development division. He returned to M.I.T.'s faculty in 1919 where he developed and taught graduate subjects in heat transfer, absorption, distillation, and economic balance. From 1925 to 1953 he was a lecturer in chemical engineering at Harvard University.

Together with W. H. Walker and W. K. Lewis he developed *Principles of Chemical Engineering*, and in 1933 produced his classic *Heat Transmission*, which in its third edition was translated into French, Russian, Yugoslavian, and Japanese, and is still a valuable reference.

During World War II Professor McAdams was active in research projects for the National Advisory Committee for Aeronautics, the National Defense Research Committee, and the Office of Naval Research, for which he received the Presidential Certificate of Merit. In 1957 the first National Heat Transfer Conference was dedicated to him for "distinguished services to the engineering profession, and particularly to the art and sciences of heat transmission."

His classroom manner, seemingly very serious, often was interspersed with whimsy such as, "I don't like the hissing sound I make when I say $d\Psi$. From now on it's d (pitchfork)!" Many of these were collected by amused teaching assistants into a private collection called *McAdams Monologues*.

Deeply serious in technical work and often amazing his colleagues with his ability to spot so readily errors in logic or faults in experimental techniques, he could be the life of the party when relaxing; his repertoire

Figure 6.   William Henry McAdams

of southern ballads, whimsical spirituals, and old favorites, while accompanying himself on the piano, guitar, or mandolin seemed never to be exhausted.

Due to ill health, Professor McAdams became emeritus in 1957. He died in 1975 at the age of 83.

Walter Gordon Whitman, 1885–1974

Professor Whitman received BS and MS degrees from M.I.T. in Chemical Engineering in 1917 and 1920. He joined the Institute staff as an instructor in industrial chemistry in 1918 and served as Assistant Professor of Chemical Engineering at the Bangor and Boston stations of

Figure 7.   Walter Gordon Whitman

M.I.T.'s School of Chemical Engineering Practice, and as Assistant Director of the Research Laboratory of Applied Chemistry until 1925. After nine years with the Standard Oil Company of Indiana, he left his position as Associate Director of Research to take the reins as head of the Chemical Engineering Department at M.I.T., a post he held until 1961.

Professor Whitman is probably best known in technical circles for his development of the two-film theory for gas absorption and for contributions in the corrosion field. His later years were devoted largely to administration of the department and to public services such as the War Production Board, the National Advisory Committee for Aeronautics, Science Advisor to the Department of State, and Secretary General of the U.N. Conference on Peaceful Uses of the Atom. His citation on election to the National Academy of Engineering read in part ". . . his contributions to chemical engineering and leadership in providing technical advice to the U.S. Government and to the United Nations." His headship of the department was characterized by a paternal insistence on quality and an open mind to innovation. He was prone to remind his staff if he felt they needed a haircut, and he always wore bow ties. Proficient on the guitar, he thoroughly enjoyed leading a convivial group in a session of song.

Harold Christian Weber, 1895–

Professor Weber, born in Roxbury, MA, received his BS degree in Chemical Engineering from M.I.T. in 1918 and the doctorate from

Figure 8.   Harold Christian Weber

Eidgennosiche Technische Hochschule in Zurich in 1935. He served in the Chemical Warfare Service in World War I and in 1919 joined Walker, Lewis, McAdams, and Knowland as a consulting engineer. He was appointed to the faculty in 1922 and directed the Boston station of the School of Chemical Engineering Practice from 1923–28. In 1939 he published the first text of its kind, *Thermodynamics for Chemical Engineers*, revised with Professor H. P. Meissner in 1957.

During World War II he served as technical adviser to the Commandant of the Chemical Warfare Service Development Laboratory at M.I.T. This laboratory, designed for use by the Chemical Engineering Department, served as its quarters from 1946 to 1976. He was appointed Chief Scientific Adviser to the Chief of Research and Development in 1958 and has lived in retirement in Arizona since 1960. His services as a consultant in the fields of petroleum, textiles, paper, and mechanical and electronic equipment were of great value in introducing the realities of the chemical engineering profession into the classroom, thus making his lectures come alive for the students.

Thomas Kilgore Sherwood, 1903–1976

Professor Sherwood was born in Columbus, OH, received his BS degree from McGill University in 1923 and his MS and ScD degrees from M.I.T. in 1924 and 1929. In 1928 he was appointed Assistant Professor of Chemical Engineering at Worcester Polytechnic Institute, returning to M.I.T. in 1930 as Assistant Professor of Chemical Engineering. Starting in 1940 he was progressively technical aid, section chief, and division member of the National Defense Research Committee.

His text, *Absorption and Extraction*, issued in 1937, was revised with Professor R. L. Pigford of the University of Delaware and again also as *Mass Transfer* with Pigford and Professor C. R. Wilke of the University of California. In 1939 he and Professor C. E. Reed published the first *Applied Mathematics in Chemical Engineering*, revised in 1957 by Professor H. S. Mickley. He coauthored *Properties of Gases and Liquids* with Professor R. C. Reid, revising it in 1966 and again in 1976 (the latter with Professor J. Prausnitz of the University of California). *The Role of Diffusion in Catalysis* with Professor C. N. Satterfield and *A Course in Process Design* were published in 1963. His research and teaching encompassed the areas of drying, heat transfer, absorption and extraction, and eddy diffusion. His colleagues named the dimensionless number used in mass transfer correlations as the "Sherwood number" in his honor. On his retirement in 1969 he became a visiting professor at the University of California, where he continued until his passing.

Figure 9.   Thomas Kilgore Sherwood

Professor Sherwood was active in societal affairs: among other things, he was a founding member of the National Academy of Engineering.   He also served as Dean of Engineering at M.I.T. from 1946 to 1952.

He was an ardent mountain climber and skier, fond of jokes and tricky technical problems.   His students enjoyed his teaching as well as his solicitous concern for their personal problems when he was asked for help.   His many honors, including a doctorate from the University of Denmark and honorary life membership in the Canadian Institute of Chemistry are overshadowed by the esteem in which his former students held him.

Figure 10.   Hoyt Clarke Hottel

Hoyt Clarke Hottel, 1903–

Professor Hottel was born in Salem, IN, and took his BS degree in Chemistry at Indiana University in 1922.   He came to M.I.T. intending to do graduate study and research in rubber chemistry, and instead, became one of the leading authorities in the fields of combustion and radiative transfer.   His pioneering research put the design of large furnaces, particularly cracking coils, on a scientifically sound engineering foundation.   After receiving the MS degree in Chemical Engineering in 1924, he was appointed to the staff as a fellow in fuel and gas engineering and later on to the faculty of the department.   He was Director of the Fuels Research Laboratory from 1934 until his retirement in 1968 and Chairman of the M.I.T. Solar Energy Research Committee, in which latter activity he was also in charge of building three experimental solar-heated houses.

He was awarded the U.S. Medal for Merit and Great Britain's King's Medal for his service during World War II as Chief of the National Defense Research Committee's Fire Warfare Section and on the Gas Turbine Committee of the National Advisory Committee for Aeronautics. After the war he was Chairman of the National Academy of Sciences' Fire Research Committee for 10 years.   He helped initiate and served for many years as the Chairman of the American Committee on Flame Radiation, a supporter of research activities of the International Flame Foundation.

He coauthored with Professors G. C. Williams and C. N. Satterfield *Thermodynamic Charts for Combustion Processes* in 1949, and with Professor A. F. Sarofim *Radiative Transfer* in 1967, both of which texts have had extensive use over the years.

He is known by his students and his colleagues for his meticulous and sound analyses of complex problems and for his untiring efforts in any endeavor he tackles.   Professor Hottel continued to teach his courses in combustion and radiative transfer for some years after becoming emeritus and continues today as a prized consultant to peers, students, and government.

Edwin Richard Gilliland, 1909–1973

Professor Gilliland, a native of El Reno, OK, received a BS degree from the University of Illinois in 1930, an MS from Pennsylvania State University in 1931, and in 1933 the ScD degree from M.I.T.   He served as instructor and professor in chemical engineering, deputy dean, Associate Director of the Laboratory for Nuclear Science and Engineering, Chairman of the faculty, and from 1961 to 1969 as Head of Chemical Engineering.

During World War II he directed an NDRC project on oxygen production, was Assistant Research Director of the Rubber Administration, Chief of the Navy Jet Propulsion Panel, Deputy Chairman of the Chiefs of Staff Guided-Missiles Committee, and Chief of the Office of Field Service.   He later was a member of the President's Science Advisory Committee and the Department of the Interior's Saline Water Committee.

Together with Professor W. K. Lewis, he researched the fluidized catalytic cracking of petroleum and with Professor W. H. McAdams revised *Principles of Chemical Engineering,* and coauthored the revision of *Elements of Fractional Distillation* with Professor C. S. Robinson.   He was a leading authority on distillation, the production of synthetic rubber, and the demineralization of salt water.

Dr. Gilliland was as keenly interested in the education of undergraduates as in that of graduate students and in his researches and those of his colleagues.   He appeared to have boundless energy (always at his desk before most people were up in the morning) and was omnivorous in his appetite for exploring technical problems.   He was noted for his rapid-fire thought processes and for his interjection of the phrase "on the one . . . " as he apparently sorted out the many ideas flashing through his mind to find the most appropriate one for the case at hand.   In spite of

Figure 11.   Edwin Richard Gilliland

his quick thought, which sometimes intimidated new students, he was most considerate and painstaking in elucidating phenomena to his students, and the large delegation of those attending his memorial service gave clear evidence of the high esteem in which he was held by his many very successful alumni.

RECEIVED May 7, 1979.

W. K. Lewis, Teacher

H. C. LEWIS

Georgia Institute of Technology, Atlanta, GA 30332

In educating a person to become an inspiring teacher at the college and university level tactics must vary with time, circumstance, and personality. On the other hand, there may be some permanent principles. The education of W. K. Lewis, a notably influential teacher of chemical engineering, suggests the following principles: (1) developing a strongly rooted interest for each student as an individual; (2) nurturing a lifelong habit of seeking encounters with greatness in areas of human concern in addition to the subject of instruction; (3) including significant achievement in some realm quite different from academic teaching and research; and (4) involving students and colleagues in the teacher training process in more than a perfunctory way.

To many readers the person I am writing about must seem to be a distant figure. There is, however, something only distance can give—perspective. By sketching the origins of W. K. Lewis' skill as a teacher I hope I can suggest some perspectives of interest and permanent value. Figure 1 is a picture of W. K. Lewis in his 90th year. The others in the scene are all former students of his. From left to right they are Cherry L. Emerson, Jr. from Georgia, W. K. Lewis, Jr. and James Donovan from Massachusetts, and Edwin R. Gilliland from Arkansas.

The generation to which W. K. Lewis belonged produced a number of inspiring college and university teachers. That he was one of this group is documented well (1–6). Even when he was in his 90's, long after retirement, a host of former students in every part of the United States and many countries around the globe still remembered him with respect and affection.

How did this come about? In particular, who were the key influences? Answers to these questions may throw light on the problem of how to identify, recruit, and develop inspiring teachers of college and university caliber.

0-8412-0512-4/80/33-190-129$05.00/1

Figure 1. Doc at 90

Concern for Each Student

In searching for key influences I have come across four clues. The first is in an article that appeared some years ago in a popular magazine *(2)*. The article presented the results of an attempt to identify the best college and university teachers in the United States at that time. No eight persons more different in personality and style can be imagined. Nonetheless, the writer of the article observed that they all had one characteristic in common. All taught with their hearts as well as their minds. Evidently, when it comes to great teaching, style is a highly individual affair. It depends on the teacher's background and personality, the subject, the facilities available, and the backgrounds and personalities of the students. However, to teach with heart as well as mind is a must.

As it stands the exact meaning of this observation is not clear. On the other hand, I believe the writer of the article was trying to state a crucial point. When a student learns something of lasting importance, almost always a vital factor is the teacher's concern for him or her as an individual. The student may not realize this concern at the time or even until years later. When the realization comes, though, it implants a teacher's contribution as nothing else can do.

In Lewis, known to all his friends as Doc, concern for each student as an individual was strong. In fact, it was at the heart of his personal-

ity.   Who gave this concern to Doc?   Psychologists tell us that in regard
to emotional attitudes toward other people the prime influence is the
atmosphere surrounding a child during early years.   Hence, I think it of
utmost significance that Doc's parents and the other members of the
extended family in the household where he grew up were all dominated
in thought and action by two of a rare company of individuals.   This
company—a few members well known to history, many unknown—consists of persons who persistently, day after day, year after year, try to be
of service to others.   In the case of Doc's parents and relatives the two
persons who dominated their lives were Jesus of Nazareth and John
Wesley.

When Doc married, he joined his wife's church, Congregational, and
was no longer under the daily influence of John Wesley.   The influence
of Jesus continued.   Throughout their long life together Doc and Rosalind Kenway, the woman he was lucky enough to marry, thought of
themselves as persons trying humbly to be of service to others as did
Jesus.

This suggests a topic historians may wish to explore.   Much has
been written about the conflict between science and religion during the
latter part of the 19th century and the early part of the 20th.   During
this same period what were the relations between engineering and religion?   Were they affected by the conflict between religion and science?

As for Doc, like so many of us he had in mind many unanswerable
questions.   However, on the particular question of religion and science
he came to believe there is really no conflict.   For example, he shared
Harry Emerson Fosdick's view that science has not undermined the
Bible.   Instead, research establishing the approximate time at which
each book of the Bible was written has enhanced greatly our understanding and appreciation of its spiritual insights *(7)*.   Doc agreed with
Albert Schweitzer that critical analysis of the historical evidence concerning Jesus has emphasized, not deemphasized, His relevance to our
times.   In fact, the last paragraph of Schweitzer's work, known in its
English translation as *The Quest of the Historical Jesus (8, 9)*, gives the
basis for Doc's allegiance to the Nazarene.   In religion as in science, Doc
believed, the fruitful approach is the same.   To live effectively one has to
proceed on the basis of hypotheses.   He liked Fosdick's definition of
faith.   "Faith is living by the highest you know" *(10)*.   The highest that
Doc knew was intense respect and concern for each human being.

Encounters with Greatness

In searching for persons of key influence in developing Doc's abilities
as a teacher I found my second clue in a statement by Alfred North
Whitehead.   According to Whitehead, "Education is impossible apart

from the habitual vision of greatness" *(11)*.   He was talking, of course, about education at its best.   He was saying that the most effective teachers succeed in arranging, in various ways, for students to have encounters with greatness.

This can be hard to do.   It can be especially hard in professional courses for scientists and engineers.   The fields are advancing so fast that a teacher of physics, for example, often cannot confront students with such profundity and subtlety of thought as one finds in the writings of Newton or Planck.   Too much time must be devoted to writings on the latest material, authored by persons competent, to be sure, but not of genius.

If teachers in chemical engineering are to confront students with greatness, most of it they must draw from other fields and weave into the teaching of up-to-date chemical engineering material.   Much of the weaving must be subtle, by means of inflection, style, behavior in informal contacts, and the intangible impact of a rich mind.   Some of the weaving can be simple.   Think, for instance, of a perfect gas, that marvelous figment of Maxwell's imagination, and a compressibility plot.   It takes but a minute to point out that, considered together, Maxwell's idea and a compressibility plot illustrate the use of a powerful intellectual tool, exemplified by Plato's *Republic (12, 13, 14, 15)*, and applicable in innumerable areas of human endeavor.   Plato's idea is this.   When one faces a set of complex situations, it is often an immense aid to clarity of thought if one will imagine an idealized model and then consider the differences between the model and the actual situations. When one does this, often a pattern emerges.   As a result, instead of having to think about a welter of actual situations, one can think about two simple ideas—an idealized model and a pattern of deviations.

To do a lot of this kind of weaving one has to have a strong, lifelong habit of regularly seeking encounters with greatness.   In Doc this habit was above all a habit of reading.   Throughout his life he was constantly reading the works of great scientists and engineers.   He also was forever exploring the masterpieces of essayists, short-story writers, novelists, biographers, historians, and philosophers.

The person who planted and cultivated this habit was a first cousin, Mary Witherbee.   Twenty years older than Doc, she taught English at Lasell Seminary, now Lasell Junior College, in Newton, Massachusetts, but spent each summer at the Delaware farm where Doc was born and grew up.   Of the thousand imaginative ways in which she nurtured the habit I will mention only one.   Her wedding present to Rosalind Kenway and Doc was the Temple edition of Shakespeare *(16)*.   Each play was in a separate pocket-size volume.   For years, whenever Doc went on a business trip, he slipped one of these volumes in his pocket to read on the train. No wonder his teaching began to reflect something of Shakespeare's insight into the human heart.

In the weaving process what strands to use of course will depend on a teacher's individual interests, the subject, the particular group of students, and the general cultural climate. On the other hand, Doc felt one strand so important that provision should be made always for at least one teacher of a technical subject to weave it into the professional education of a chemical engineer at some point in the enterprise. This strand is history.

Doc's conviction that history is essential was based on the following observations. For one thing most engineering students have little interest in problems of the past, but for future use they need some kind of a feel for how the greatest scientific and engineering minds work when they run into perplexing problems. Also for future use they need to realize one of the great lessons of history. Major advances rarely are achieved by logical persuasion alone. They demand the living of a life. Finally, most engineering students think in concrete terms. To try to communicate the above ideas merely by talking of abstract principles is hopeless. What most students will remember, and remember for a lifetime, is a dramatic story from history. From history, then, and from many other areas of life Doc continually was confronting students with glimpses of greatness.

Recruiting an Achiever from the Outside World

My third clue is this. Unlike many good teachers, Doc had to be recruited. How he was recruited illustrates a principle. Before I state it, let me sketch the circumstances in which it applies. For many students the college years include periods of deep thought about life. For both students and society it is important that they do this thinking with keen awareness of the wide range of possibilities in life. Not infrequently the persons with greatest potential for communicating such awareness are young graduates who already are making significant achievements in the outside world, quite removed from academic teaching and research. To secure their services those who recruit college teachers must bid against nonacademic employers.

Now for the principle. Attracting some of these young achievers into teaching is important. In recruiting them psychology is more important than money. To show how this operated in Doc's case I must give some background. As mentioned earlier, he was born and grew up on a farm in Delaware. He loved it. When he entered M.I.T. as a freshman, it was not because he had any interest in a career in science or

Figure 2.   Doc as an undergraduate

engineering.   He thought an agricultural school of that day would only
teach him current practices in farming.   At M.I.T. he could learn the
principles of science and engineering on which farming would be based in
the years ahead.   When he graduated, Doc still took it for granted that
his life's work would be as manager and principal laborer on the Delaware
farm.   Although he applied for and won a two-year scholarship for study
in Germany, he thought of this as simply a generous opportunity for
encounters with greatness before he began his career on the farm.

To help earn expenses at M.I.T. Doc spent each summer on the
farm.   Figure 2 is a picture of him there.   For the benefit of those who
never have seen one of the breed, the other character in this picture is a
mule.   Mules are strong, intelligent, and stubborn.   Mules have very
definite opinions, which they express in no uncertain terms.   Mules
express their opinions in ways that make an indelible imprint.   Perhaps I
should not mention this, but persons familiar with both Doc and a mule
have been known to comment, sometimes rather heatedly, on resem-
blances between the two.

At any rate, the influence of the farm was strong.   When Doc
returned from Germany, he switched to engineering and entered industry, saying that self-analysis had convinced him that he made a better
contribution to others when competition was lively.   He realized competition would be stiffer in engineering than on the farm, but even then
he kept up with farm activities and dreamed of returning there after
retirement.

Few people know how much of an achiever Doc was in industry
when M.I.T. recruited him.   The value the company he worked for
placed on him is indicated by the action of the headquarters official to
whom he reported.   I learned about his many years later, not from Doc
but from the official.   According to him, as soon as he heard M.I.T. was
after Doc, he called Doc in, emphasized the company's confidence in
him, found out what salary M.I.T. was offering, and offered twice as much.
Doc said, "No."   The official offered three times as much.   Doc said,
"No."   Finally the official observed, "I suppose there is no amount I can
offer that will keep you from going to M.I.T."   Doc said, "I guess so."   The
official gave up.   He recognized there must be an attraction stronger than
any counterattraction he could muster.

What was the attraction?   There is a story about this.   Although
Doc told me it was a myth, it conveys historical fact.   It rings so true to
the personalities of both individuals involved.   According to the story,
when Walker approached Doc about a job at M.I.T., Doc indicated he
was happy where he was.   But Walker knew his man.   He knew that
sometimes one had to be blunt with him; he knew his sense of humor; he
knew his deep desire to be of service to others; he knew about the
farm.   Walker blew up.   "Warren," he shouted, "I hear you've recently

invented a new kind of manure spreader you hope to use on that farm in Delaware. You completely fail to appreciate the opportunities you'll have for spreading manure if you'll come to teach at M.I.T."

That did it. The young achiever appreciated the opportunities. He came. How he used the opportunities is described gloriously in the booklet in which I found my fourth clue.

Involving Students and Colleagues

My fourth clue I found in that labor of love by Tom Sherwood, that gem, that classic, A Dollar to a Doughnut (3). Compiled at the time of Doc's retirement, it is a collection of favorite stories about him, which Tom solicited from former students. The stories and the illustrations, drawn by a master cartoonist, are priceless.

My clue is this. As every teacher knows, teachers learn from students. A Dollar to a Doughnut reveals that in Doc's case the influence of students in developing his effectiveness as a teacher was abnormally high.

How it happened is intriguing. For one thing it illustrates the successful operation of an educational strategy often overlooked, a strategy that sometimes can bring rich returns. Especially intriguing is the role of serendipity.

The problem requiring strategy is common enough. No teacher's methods are perfect. A student often has a constructive criticism to make. In communicating it to the teacher what strategy is most likely to be effective? In chemical engineering—at least in the United States today—perhaps the most usual practice is for an administrator or student organization to distribute teacher evaluation forms to students at the end of a course. Then the completed, unsigned forms or summarized results are given to the teacher.

In A Dollar to a Doughnut several stories reveal the exhilarating impact of another approach. As instruction proceeded Doc lured students into an amusing activity. Somehow they found themselves vying with each other. Who could be most ingenious in teaching the teacher how to teach more effectively? The result? Much profit and fun!

I turn now to the serendipity in this affair. Serendipity, you recall, is "an apparent aptitude for making fortunate discoveries accidentally" (17). Doc's concentration on teacher–student interaction was on issues other than teaching. One of his great concentrations, for example, was on teaching students to think for themselves. I do not believe that he ever consciously worked for abnormally high input from students on the art of teaching. It was spontaneous combustion.

The reaction mechanism depended on two features of his personality. One exerted a push, the other a pull. The first feature was intensity. A combination of energy, exuberance, a deep desire to help students, a great power of concentration, and the belief that intensity is essential to education—all this made him tend to drive ahead without being as sensitive to each student's individual reaction as one might wish.   The following incident related by an alumnus is typical.   Until Doc read it in *A Dollar to a Doughnut* (3) he was completely unaware that his classroom procedure on a certain day had had a humidifying effect on a certain student in the class.

"It happened in 1930.   It was my first day at Tech, and my first experience with Doc Lewis . . .   The classroom as I remember was quite large with Doc suitably mounted way up on a very high platform.   All very awesome to a newcomer.

'Since you fellows are all graduate students and therefore real scientific experts, I'd like to make a quick rundown of the basic laws of Nature which serve as the background of this course.   Can anyone here name a single infallible law of Nature?'

"Silence (complete).

'You there—the first man in the first row—can't you name one?'

"Answer: 'The law of Conservation of Matter.'

'Humph!   Did you ever hear of cosmic rays?   Those things blow your law to Kingdom Come.   Next man.'

"Answer: 'I had intended to mention the law of Conservation of Matter, but . . .'

"No good.   Next man.'

"And so it went on down the line until Doc reached me.   Each answer in turn had been demolished roundly.   By this time Doc had dismounted from his high perch and was standing about three feet in front of me.   In answer to his pointing finger, I blurted out, . . . 'The Law of Constant Proportion.'

'Did you ever hear of isotopes?'   (By this time he was fairly shouting).

"Yes!   Isotopes are . . .'

'Let me tell you.   Isotopes are things (and here he leaned over to within about 12 inches of me)—are things that spit in the face of the Law of Constant Proportion.'

"He didn't miss, either.   How could he?   We were practically rubbing noses by then" (3).

In fairness to Doc I must note that I have not yet given the final sentence of the alumnus' story.   "But for that shower," he concluded, "I most likely would have remembered neither the incident nor the valuable lesson it contained."   However, this sentence was the calm and reasoned verdict of a mature engineer, looking back on the occasion many years later.   At the time of the event, one can be sure, his feeling was radically different.

Another *Dollar to a Doughnut* story illustrates the second feature of Doc's personality.

"Doc had a student doing a lab experiment which involved a titration depending on getting a blue starch color with free iodine. The man was having no success in checking results. It got so bad that Doc went into the lab to watch. His unerring eye spied the starch, and he demanded to know where it had come from. On being informed that it had come off the shelf, Doc exploded and said, among other things, that he wouldn't let anyone make up his starch indicator. Whereupon Doc mixed some fresh starch, ran off two titrations himself, which checked, and then demanded that the student do it while he watched. The first move of the student was to pour the fresh starch solution down the sink and, when Doc exploded again and wanted to know how come, the student said, "You just told me never to let anyone mix up my starch solution." "Wonderful!" roared Doc. "You know, old horse, sometimes you show faint signs of intelligence!" When the student's report on the experiment came back, the grade was an "H", the highest mark in the grading system at that time. It stood for "Honor" (3).

Invariably, in reacting to a spirited response to his browbeating tactics, Doc showed an open mind, a keen sense of humor, and instant readiness to change direction with good cheer. An amazing variety of imaginative responses turned up with much hilarity. Since the push and pull were integral parts of Doc's nature, they appeared in full force not only in class and lab but also in conversations with his department colleagues. Need I add that his colleagues, a remarkable team of high-spirited individuals, enthusiastically engaged with him in his game of vigorous give and take? In all seriousness, words cannot express the value to Doc of their patience and helpfulness through the years.

Yes, some sort of active, continuing involvement of students and colleagues in teaching a professor how to teach can be a powerful device. On the other hand, "A Dollar to a Doughnut" also reflects the well-known fact that different people like different activities. What about those individuals who did not enjoy the form of exercise involved in Doc's game? Consciously or unconsciously, they responded to his tactics with different kinds of signals. Some of these he learned to recognize. Whenever he noticed one, he reacted with thoughtful consideration and grace. Unfortunately, in many an instance he did not catch a student's signal at the time.

About instances of this latter sort there is something few people know. Often, years later, Doc would remember such an incident, understand the signal, and learn the lesson. The impact would be profound because he would remember the individual involved. The thought that he had hurt this individual made him grieve. As I wrote in discussing my first clue, concern for each individual person was at the heart of his personality. After the handicaps of old age compelled him to stop teaching and he had more spare time, this tendency to remember and learn increased.   Lessons kept coming to mind, kept making him a more sensitive human being.   Truly, students and colleagues can influence teachers more than they ever will realize.

Reflections

Look again at Figure 1.   In his last years this fine old man had some interesting reflections on life and teaching.   Here are a few.

"We live in a universe, not a multiverse.   Among other things this means that science, engineering, the search for justice, the search for beauty, and the search for holiness are interrelated."

"So far as humans are concerned the aim of the universe seems to be the development of character."

"Life is more important than religion."

"The teacher who fails to communicate some sense of the wonder of the universe in which we live is remiss in duty."

"Surgeons operate on bodies.   Teachers operate on minds and spirits."

Conclusion

"Surgeons operate on bodies.   Teachers operate on minds and spirits." Of course Doc knew that good surgeons consider whole persons, not just bodies.   He felt deeply that teachers have a special responsibility for minds and spirits.   In meeting it they need to use a surgeon's exactness and care.

There is something else.   All of us are teachers at times, whenever our words or actions affect the mind and soul of another individual.   So the surgeon is a wonderful symbol of what at times each one of us must be.

I close by quoting another voice from Massachusetts.   One hundred and twenty years ago Emily Dickinson expressed the idea as follows *(18)*:

> "Surgeons must be very careful
> When they take the knife!
> Underneath their fine incisions
> Stirs the Culprit—Life!" *(18)*.

Literature Cited

1. Anon.   *J. Eng. Educ.*   **1947–8,** *38*, 2.
2. Anon.   *Life* **1950,** *29* (16), 109.
3. Anon.   "A Dollar to a Doughnut"; privately printed, 1953.   Reprint, American Institute of Chemical Engineers: New York, no date.
4. Pinck, D. C.   *Technol. Rev.* **1963,** *65* (5), 16.
5. Anon.   *Chem. Eng. Prog.* **1965,** *61* (1), 32–3.
6. Gilliland, E. R.   *Chem. Eng. Educ.* **1970,** *4* (4), 156.

7. Fosdick, H. E. "The Modern Use of the Bible"; Macmillan: New York, 1924.
8. Schweitzer, A. "Von Reimarus zu Wrede"; Mohr: Tuebingen, 1906.
9. Montgomery, W. "The Quest of the Historical Jesus. A Critical Study of its Progress from Reimarus to Wrede"; Black: London, 1910.
10. Fosdick, H. E., personal communication to W. K. Lewis.
11. Whitehead, A. N. "The Aims of Education and Other Essays"; Macmillan: New York, 1929; Chapter 5.
12. πλάτων "HOAITEIA"; 4th century B.C. Mss. of later dates.
13. "Plato's Republic. The Greek Text"; Jowett, B., Campbell, L., Eds.; Clarendon: Oxford, 1894; 3 Vols: I. Text with a Facsimile. II. Essays. III. Notes.
14. Jowett, B. "The Republic of Plato Translated into English with Introduction, Analysis, Marginal Analysis and Index", 3rd ed.; Clarendon: Oxford, 1888.
15. Plato. "The Republic"; Eng. trans. by P. Shorey, "Plato, with an English translation" Heinemann: London, Macmillan: New York, 1914–1955. The Loeb classical library (Greek authors). Greek and English on opposite pages. Contents: 11. Bks. I-V. 12. Bks. VI-X.
16. Shakespeare, W. "The Temple Shakespeare"; Wright, W. A., Ed.; Prefaces, Glossaries, and Notes, I. Gollancz; Dent: London, 1895–97.
17. "Webster's New World Dictionary of the American Language", 2nd college ed.; Guralnik, D. B., Ed.; World: New York, 1970.
18. Dickinson, E. Poem 108. In "The Poems of Emily Dickinson"; Johnson Thomas H., Ed.; The Belknap Press of Harvard University Press: Cambridge, 1951, 1955; reprinted by permission.

RECEIVED May 7, 1979.

8

The Beginnings of Chemical Engineering Education in the USA

J. W. WESTWATER

Department of Chemical Engineering, University of Illinois, Urbana, IL 61801

At the University of Illinois, Urbana, chemical engineering was born in the Department of Chemistry and had its first growth therein. This was common for the early departments such as those located at the Universities of Illinois, Pennsylvania, and Michigan, the Massachusetts Institute of Technology, and Tulane and Case Western Reserve Universities as well as many others. In such cases, the chemical engineering component expanded, and a spin-off was arranged to produce free-standing chemical engineering departments. A rare start is for chemical engineering to emerge from mechanical engineering (Colorado), electrical engineering (Wisconsin), or other nonchemistry units. For young chemical engineering departments a different origin is more usual: the department is established suddenly by fiat. The right to offer the curriculum is cleared, a budget and space are established, and a Head is appointed to select the faculty. The origin has long lasting effects.

The discipline of chemical engineering evolved from chemistry. Most university departments of chemical engineering are spin-offs from prior chemistry departments. This is true for most schools, but not all. This chapter treats the beginnings of chemical engineering in the universities of the United States. In particular it shows that some departments originated in strange and unusual ways.

For centuries, engineering meant just two things: military engineering was concerned with fortifications and other structures for warfare; civil engineering included roads, bridges, waterways, mines, tunnels, and other nonmilitary structures. Other fields of engineering are modern inventions. The first engineering school in the United States, West

0-8412-0512-4/80/33-190-141$05.00/1
© 1980 American Chemical Society

Point Military Academy was established in 1794 at the urging of George Washington *(1)*. Of course chemical engineering was nonexistent at that time.

Some universities were already in existence by that date, including Yale (founded in 1701), Princeton (1746), the University of Pennsylvania (1749), Columbia (1754), Rutgers (1766), Dartmouth (1769), Pittsburgh (1787), and Tennessee (1794). Some of these offered courses in natural philosophy or natural history. A professor of natural philosophy was expected to teach all of science including chemistry, physics, astronomy, zoology, etc. Early technical societies such as the Royal Society, London, included workers in all these fields, and each philosopher was expected to be interested in and knowledgeable in all the others. Benjamin Franklin typified the successful natural philosopher of that era.

Origins in Chemistry

As knowledge accumulated, chemistry began to emerge as a distinct discipline. Its importance grew as commercial applications increased. For example in 1746 the first lead chamber sulfuric acid plant was built in the United States, in 1791 a patent was issued for the Le Blanc process for making alkali from ordinary salt, and in 1842 the Lawes process for making superphosphate from rock phosphate was patented *(2)*. Chemistry courses taught in some early universities included detailed descriptions of these and similar processes. In due time, specific chemistry courses were dedicated to such knowledge, and their titles bore meaningful labels such as Industrial Chemistry, Applied Chemistry, Engineering Chemistry, Technical Chemistry, Practical Chemistry, or Chemical Industries. Many listings like these occur in university catalogs published before 1900.

There is difficulty in pinpointing the exact beginning of chemical engineering education in the United States. Inasmuch as it came into existence by slow evolution and not by a sudden creation, "who is first" becomes a matter of definition. If we argue that old terms such as natural philosophy include chemical engineering, then the beginnings go back to schools existing in the 18th century. If we accept industrial chemistry, applied chemistry, etc. as equivalents to chemical engineering, then some of the early schools include the New Jersey Institute of Technology (a two-year curriculum in industrial chemistry in 1881) and Case Western Reserve (a four-year curriculum in chemical technology in 1884).

It is interesting to examine the early use of the specific words "chemical engineering" in education. These words were coined by Professor George E. Davis of the Manchester Technical School, England

(3). In 1880 he proposed, without success, the creation of a Society of Chemical Engineers. In 1888 he delivered and published a series of 12 lectures on chemical engineering. In 1901 he published A *Handbook of Chemical Engineering.* All of these events were noted in the United States.

By no means was there a rapid, widespread use of the words chemical engineering in the United States. For example, in 1882 Case Western Reserve stated: "The course of study in chemistry will be made as practical as possible" *(4).* There is a clear hint of chemical engineering in this. The next year the catalog states "lectures will be given on theoretical and technical chemistry" *(5).* In 1884 Charles F. Mabery joined the faculty, and the trend accelerated. Individual courses in chemical technology and in gas analysis were added *(6).* The 1904–1905 catalog *(7)* refers to the Department of Chemistry, including Engineering Chemistry. The 1907–1908 catalog *(8)* lists the Department of Chemistry and Engineering Chemistry. In 1909 the degree of Chemical Engineering was offered *(9).* In 1913 the curriculum of Chemical Engineering was offered *(9, 10),* and in 1915 the first BS in Chemical Engineering was awarded *(11).* Finally in 1925 the Department of Chemical Engineering emerged *(11).* Case received accreditation from the AIChE that same year *(9).*

The hesitant adoption of the words "chemical engineering" as shown for Case was common also at numerous other schools. The University of Illinois, Urbana is a second convenient example. Samuel W. Parr joined the faculty in 1885 as a Professor of Natural Science. He became Head of Industrial Chemistry *(12)* in 1891. In 1894 a curriculum in Applied Chemistry with Engineering was offered *(13).* In 1895 the degree of BS in Chemistry with Engineering was offered *(13).* In 1901 a four-year curriculum labeled Chemical Engineering was offered *(14),* apparently the sixth such curriculum in the nation. In 1903 Clarence Bean received the first BS in Chemical Engineering at Illinois *(13).* The unit in which all of this occurred was called the Chemistry Department (1891–1894); Industrial Chemistry Department (1894–1904); Division of Industrial Chemistry in the Chemistry Department (1904–1926); Division of Chemical Engineering in the Department of Chemistry (1926–1953); Division of Chemical Engineering in the Department of Chemistry and Chemical Engineering (1953–1970); and finally the Department of Chemical Engineering in the School of Chemical Sciences (1970–present).

Many other examples exist, showing that the words "chemical engineering" were adopted slowly in universities, particularly among the older departments. At any one school the dates are very different for the first use of these words for an individual course, a curriculum, a degree, the name of a department, and the title of a professor. Frequently several decades elapse at a school before all of these items have simultaneous labels of chemical engineering. This means that any attempt to list schools in the order in which they first taught chemical engineering becomes meaningless unless a precise definition is given of the basis used.

The Massachusetts Institute of Technology is documented well as being the first U.S. school to offer a four-year curriculum labeled Chemical Engineering. The catalog (15) for 1888–1889 describes a BS degree in Chemical Engineering and lists a Course X, Chemical Engineering. The terminology "course" here meant a curriculum. Henceforth in this chapter "course" will be used in the modern context of a one-semester individual offering such as a 3-hr course. Also in this chapter "curriculum" means an organized set of courses, for a group of years, leading to a specific degree. The curriculum in chemical engineering was the tenth curriculum to be offered at M.I.T. and so bore the label Course X. It was described in the catalog as "a general training in mechanical engineering—and a study of the applications of chemistry to the arts, especially to those engineering problems which relate to the use and manufacture of chemical products" (15). The curriculum contained three courses labeled Industrial Chemistry and two labeled Applied Chemistry. None were labeled Chemical Engineering. There were no professors of chemical engineering. Lewis M. Norton was Professor of Organic and Industrial Chemistry. He taught all of the Industrial Chemistry courses, assisted by J. W. Smith, Instructor in Textile Chemistry, and A. J. Conner, Assistant in Industrial Chemistry. This three-man department (or one or two-man department, depending on definition) was called the Department of Chemical Engineering. Clearly its roots were in chemistry. According to Weber (16) it operated as an Applied Chemistry Division in the Department of Chemistry until 1920. At that time the separation from the Department of Chemistry was completed, and W. K. Lewis was named Head.

The University of Pennsylvania in 1892 appears to be the second school to have a four-year curriculum labeled Chemical Engineering (17). It consisted mostly of mechanical engineering (17 courses) and chemistry (six courses). The program had no Professor of Chemical Engineering. No courses were labeled Chemical Engineering, but two of the chemistry courses were titled Applied Chemistry and Selected Method of Industrial Chemistry. The degree offered was a BS in Chemical Engineering. This curriculum operated within the Chemistry Department until 1951 when a free-standing Chemical Engineering Department was established with Melvin Molstad as Chairman.

Tulane University was the first school in the South and apparently the third in the United States to have a four-year curriculum labeled Chemical Engineering. This was in 1894 (18). The degree label was BE in Chemical Engineering and the first recipient was B. P. Caldwell in

1895 *(19).*   There were no courses labeled Chemical Engineering and no Professor of Chemical Engineering.   The most pertinent courses, three in industrial chemistry, were taught by John Ordway.   In 1893 this man bore three titles: (1) Professor of Applied Chemistry and Director of Manual Training School; (2) Professor of Industrial Chemistry; and (3) Professor of Industrial Chemistry and Acting Professor of Civil Engineering.   The courses in industrial chemistry and those in chemistry were taught in the same building.   Thus at Tulane, chemical engineering seems to have its roots in chemistry.

In 1898, two universities became the apparent fourth and fifth to offer four-year curricula labeled Chemical Engineering.   The University of Michigan *(20)* curriculum was an arrangement of existing courses and included no courses labeled Chemical Engineering.   It did contain courses called Chemical Technology, Organic Technology, and Technical Gas Analysis taught by Professor E. D. Campbell and Mr. Alfred H. White. The first BS in Chemical Engineering at Michigan was given to Wareham S. Baldwin in 1901 *(21).*   At Tufts the four-year curriculum in Chemical Engineering also started in 1898 *(22).*   The first student to register in it did so in 1900, and the first BS in Chemical Engineering was awarded in 1905.

All of the chemical engineering curricula mentioned above had their roots in chemistry departments.   Table I lists 55 schools which had similar beginnings.   The listings in the tables in this chapter are incomplete.   The author of this chapter sent letters to 137 department heads asking each about the beginnings of his departments.   All issues of *Chemical Engineering Education* were examined.   A further literature search was carried out, and then correspondence with librarians at 16 universities was initiated.   By all of these means, information was ob-

Table I. Chemical Engineering Departments Which Originated in Chemistry Departments

			Lehigh U.
U. Alabama	Tufts U.	Cornell U.	U. Pennsylvania
Arizona State	U. Detroit	Polytech. of N.Y.	Penn. State U.
U. Arkansas	U. Michigan	Pratt Inst.	U. Pittsburgh
U.C. Berkeley	Michigan State	Rensselaer	U. Rhode Island
Caltech	Michigan Tech.	N. Carolina State	U. S. Carolina
Stanford U.	U. Minnesota	Case W.R. U.	Vanderbilt U.
U. Illinois	U. Mississippi	U. Cincinnati	U. Tennessee
Purdue U.	Montana State	Ohio State U.	Rice University
Tri-State	U. Mo. Rolla	U. Oklahoma	U. Texas, Austin
U. Iowa	Washington U.	Oklahoma State	Texas A.&M. U.
Kansas State	U. Nebraska	Drexel U.	Brigham Young U.
Tulane U.	N.J. Inst. Tech.	Carnegie–	U. Virginia
U. Mass.	Princeton U.	Mellon U.	U. Washington
M.I.T.	N. Mexico State	Lafayette College	

tained for 85 schools. Occasional conflicts were noted between official catalog statements and independent statements of anecdotal nature. For this article, the official catalogs are considered to be correct.

The teaching loads in the early days were prodigious by today's standards. Consider the University of Arkansas which started a four-year curriculum in Chemical Engineering in 1902 (23). The 1903–1904 catalog lists offerings of undergraduate and graduate courses in chemical engineering (24). This program was attached to the Department of Chemistry which had in 1901 a faculty of two people and already a set of 14 courses (25). Small staffs and large teaching loads were the norm everywhere. The faculty who can be identified as chemical engineers numbered one person at Case in 1884; at Tulane in 1894 it was one; at Illinois in 1894 it was one; and at M.I.T. in 1888 it was one to three (depending on definition). As late as 1909 at Michigan it consisted of two chemical engineering professors and one Instructor in Metallurgical Engineering. When the curriculum at Michigan was started 11 years earlier, the curriculum was administered by chemists already there, and the Dean reported to the Regents, "No addition to the teaching force will be needed for this course. It is not expected that the number of students will be large" (21).

From the beginning, nearly every chemical engineering department has engaged in active research. In some schools a highly organized research structure was set up. The first Engineering Experiment Station in the United States was begun at Illinois in 1903. Another was established in 1904 at Iowa State University. Many others soon followed. At the University of Pittsburgh the research establishment became so big and well funded that the administration viewed it as a threat to regular educational activites. In 1927 these research activities were separated from the university and were incorporated as the Mellon Institute of Industrial Research (26). Later, the Mellon Institute merged with Carnegie Institute of Technology a few blocks away to form the present Carnegie–Mellon University. This transfer of a large research group from one university to another appears to be unique.

Nonchemistry Origins

Table II lists 13 universities in which chemical engineering derived from some department other than existing chemistry departments.

Origins in Mechanical Engineering. The University of Colorado produced chemical engineering from mechanical engineering. In 1903–1904, the School of Applied Science offered a four-year curriculum in Mechanical and Chemical Engineering (27). The students had a common program for the first two years, after which they chose either a straight Mechanical Engineering option or a Chemical Engineering op-

Table II. Origins of Chemical Engineering Departments from
Nonchemistry Departments

Department	*School*
Mechanical Engineering	U. Colorado
	U. Rochester
Petroleum Engineering	U. Tulsa
	U. Wyoming
Ceramics & Mining Eng.	Iowa State
Sugar Engineering	Louisiana State
Paper Engineering	U. Lowell
Electrical Engineering	U. Wisconsin
General Engineering	U. Toledo
	McNeese State
Energy Engineering	U. Illinois, Chicago Circle
	U. Wisconsin, Milwaukee
YMCA Extension	Cleveland State

tion. The Chemical Engineering option contained ten mechanical engineering, one physics, two electrical engineering, and ten chemistry courses. The ten chemistry courses were the only items different from the Mechanical Engineering option. These were all taught by one professor in chemistry. The Head of Mechanical Engineering was in charge of both options until 1936. Then a separate Department of Chemical Engineering was established with Henry Coles as Head (28).

Almost two decades after the start of mechanical engineering at Colorado, a similar origin took place at the University of Rochester (29). The union of chemical engineering with mechanical engineering (combined for a time also with chemistry and later electrical engineering) continued until 1959. Today at Rochester the Department of Chemical Engineering is independent.

Origins in Petroleum Engineering. Petroleum engineering gave birth to chemical engineering at several universities. W. L. Nelson started petroleum engineering at the University of Tulsa in 1928. Six years later this split into two parts: petroleum production and petroleum refining. The program of petroleum refining was renamed as the Division of Resources Engineering, Petroleum Engineering, and Chemical Engineering in 1972. A Chairman of Chemical Engineering has existed there since 1954 (30). At the University of Wyoming, a curriculum in Petroleum Engineering was started in 1960. In 1965 this split into two programs, one in Petroleum Engineering and one in Chemical Engineering. At present D. L. Stinson serves as Head of both programs (31).

Origins in Ceramics and Mining Engineering. Chemical engineering at Iowa State University started in 1913 in a Department of Mining

Engineering, Ceramics, and Chemical Engineering. A reorganization in 1916 moved chemical engineering into the Chemistry Department. In 1919, Chemistry and Chemical Engineering separated into two independent department. In 1928, chemical engineering joined in a new Department of Chemical and Mining Engineering. In 1961, chemical engineering again became independent. In 1973 it was combined into a Department of Chemical and Nuclear Engineering. In 1976, chemical engineering once again became independent. In this interesting period of 66 years, chemical engineering had four "marriages" and four "divorces" (32, 33).

Origins in Sugar Engineering. Chemical engineering evolved from sugar engineering at Louisiana State University (34). In 1897 the Audubon Sugar School was started. The content of the curricula in sugar chemistry and sugar engineering gradually shifted to chemical engineering, and in 1905 a BS in Chemical Engineering was granted (35). In the early years the status of chemical engineering concerning its administration seemed uncertain, but in 1919 the program was allied with chemistry. In 1925, chemical engineering became independent in the College of Pure and Applied Science. In 1938 it transferred to the College of Engineering.

Sugar Chemistry was established as a curriculum during the same year 1897 at Tulane just 80 mi from L.S.U. However, this curriculum was not a success, and it was finally abandoned (36).

Origins in Paper Engineering. Chemical engineering originated in paper engineering at the University of Lowell (37). In 1950, a Paper Engineering curriculum was set up at Lowell Technological Institute, and several chemical engineers were on its faculty. The program was heavily based on chemical engineering courses and approach. By 1963, Chemical Engineering degrees were granted as companion degrees to Paper Engineering. In 1975, the University of Lowell was formed by a merger of the Institute with Lowell State College. As part of the reorganization, today's Department of Chemical Engineering emerged.

Origins in Electrical Engineering. The University of Wisconsin is special in that its Department of Chemical Engineering is the only one to have its origins in electrical engineering. The College of Engineering there was established in 1889 with Electrical Engineering as one of the early offerings. In 1895 a Laboratory of Applied Electrochemistry in the Department of Electrical Engineering was set up. It split off as a Department of Applied Electrochemistry in 1898 and so continued until 1905. The catalog (38) for 1904–1905 first uses the words "chemical engineering." The catalog states that a new building was under construction for chemistry and "when this is completed the old laboratory will probably be used for chemical engineering" (38). That catalog lists a BS degree in Applied Electrochemistry, shows a complete four-year curriculum with that label, and lists the names of the three seniors and two faculty members in the Applied Electrochemistry Department.

This small group of faculty and students was the origin of chemical engineering at Wisconsin. Their claim to the vacated chemistry building was of some concern. Hougen *(39)* gives an interesting account of how in 1905 they occupied the territory at night and erected a large sign, "Chemical Engineering," on the building. In 1905, a four-year curriculum in Chemical Engineering was established, and in 1907 the first BS in Chemical Engineering was granted. The leader of these actions was C. F. Burgess (of Burgess Battery fame), who shifted thereby from Associate Professor of Electrical Engineering to Chairman of the Chemical Engineering Department.

Before 1905, Wisconsin had nothing labeled applied chemistry or industrial chemistry. The Chemistry Department did offer individual courses in Electrochemistry (in addition to the Electrical Engineering offerings in this field) and in Chemistry of Gas Manufacture. In essence the time was ripe for chemical engineering, and the vacuum was filled by Burgess from Electrical Engineering.

Origins in General Engineering. Chemical engineering originated in general engineering at the University of Toledo *(40)*. The College of Engineering there was established in 1931, and a curriculum in General Engineering was set up as an interdisciplinary program. In 1946 a chemical engineer was added to the faculty, and a four-year option in Chemical Engineering was offered. The faculty had grown to three by 1963. In 1967 the school became a state university.

Origins in Energy Engineering. The University of Illinois at Chicago Circle was established in 1965. One of the four engineering departments was (and still is) the Department of Energy Engineering. An option in the department is the Chemical Engineering option. The degree is in Energy Engineering. Another school, the University of Wisconsin in Milwaukee, established an Energetics Department in 1965. Chemical engineering is a part of that department. The present faculty contains two chemical engineers, six mechanical engineers, and one civil engineer.

Origins in YMCA Extension. The most unexpected origin for the teaching of chemical engineering is that for Cleveland State University *(41)*. Free evening classes on assorted topics were offered by the YMCA in Cleveland in 1881. The project grew, and engineering (as mechanical drawing) was added in 1890. In 1909 this program became The Technical School. Later it was renamed the YMCA School of Technology. The bulletin for 1926 contains a major heading called Chemical Engineering, under which are listed 29 courses. Twenty-five of these courses are in chemistry, metallurgy, mineralogy, and metallography *(42)*. Two courses are titled Industrial Chemistry, and two are Chemical Engineering. Later this school became Fenn College which today is Cleveland State University.

Free Standing at the Start

Table III lists 17 university Departments of Chemical Engineering which sprang into being suddenly and with no clear connections to prior departments. This usually came about when an administrator perceived the need for a chemical engineering curriculum and proceeded to set it up. The State University of New York at Buffalo is a good example (43, 44). Clifford Furnas, a well-known chemical engineer, was the Chancellor at Buffalo at the time. During 1960–61 he conferred with E. A. Trabant, Dean of Engineering, concerning the lack of chemical engineering at Buffalo. A local consultant, Joseph Bergantz, was hired to examine the situation. Bergantz reported that there was a significant demand for local education in chemical engineering. The dean and chancellor agreed and offered Bergantz the position of first Head of Chemical Engineering. In June 1961, Bergantz was appointed Head, space was assigned, and a budget was allotted for the new department. Bergantz hired two other faculty members in 1961 and three more in 1962. Last year, SUNY Buffalo had a faculty size of ten and awarded 42 BS degrees.

Early Fathers of Chemical Engineering Departments. For any chemical engineering department which started out as a free-standing unit in a single step, it is rather easy to state who the founder is. Three examples are Joseph Bergantz at Buffalo, A. P. Colburn at Delaware, and J. R. Crump at Houston.

For Departments which came into existence slowly by evolution, the "father" may be difficult to identify. For example, Hougen (3) says that W. H. Walker is the Father of Chemical Engineering at M.I.T. A printed announcement of the 1978 Warren K. Lewis Lectureship in Chemical Engineering at M.I.T. states that W. K. Lewis is the Father of Chemical Engineering at M.I.T. However, W. K. Lewis gives credit to L. M. Norton as the Father at M.I.T. (45). There is better agreement

Table III. Chemical Engineering Departments Free-Standing at the Start

U. California, Santa Barbara	U. Kentucky	Oregon State
U. S. California	U. S.W. Louisiana	S. Dakota Mines and Technology
U. Connecticut	Montana State	U. Houston
Yale U.	Rutgers	Lamar University
U. Delaware	SUNY Buffalo	W. Virginia U.
Howard U.	Syracuse U.	

about Samuel W. Parr as the Father of Chemical Engineering at the University of Illinois, Henry K. Benson at the University of Washington, Charles F. Burgess at Wisconsin, Arthur J. Hartsook at Rice, and Samuel T. Sadtler at the University of Pennsylvania. Sadtler became the first president of the AIChE. Many other fine educators undoubtedly deserve credit for important services during the beginnings of chemical engineering education in this country. Unfortunately, many never will be identified because they lived about three generations ago. What little documentation exists may lie buried in university archives.

Conclusions

The words "chemical engineering" were adopted slowly and with great caution by educators in the United States starting in 1888. Approximately 65% of chemical engineering departments in this country had their origins in chemistry departments. About 20% were established suddenly as free-standing units with no prior connection with any department. Roughly 15% had uncommon origins such as departments of mechanical, electrical, petroleum, ceramic, sugar, paper, or general engineering.

Literature Cited

1. Grayson, L. P. "Brief History of Engineering Education in the United States," *Eng. Educ.* **1977**, *68*(3), 246–264.
2. Kobe, K. A. "Inorganic Process Industries"; Macmillan: New York, 1948.
3. Hougen, O. A. "Seven Decades of Chemical Engineering," *Chem. Eng. Prog.* **1977**, *73*(1), 89–104.
4. "Catalogue of the Case School of Applied Science"; Cleveland, Ohio, 1882–83, p. 9.
5. "Catalogue of the Case School of Applied Science"; Cleveland, Ohio, 1883–84, p. 11.
6. "Catalogue of the Case School of Applied Science"; Cleveland, Ohio, 1884–85, p. 16.
7. "Catalogue of the Case School of Applied Science"; Cleveland, Ohio, 1904–05.
8. "Catalogue of the Case School of Applied Science"; Cleveland, Ohio, 1907–08.
9. Angus, John. C., "ChE Department: Case," *Chem. Eng. Educ.* **1977**, *11* (1), 4–9.
10. Anon. "Albert W. Smith: Educator, Chemist, and Engineer"; Case Institute of Technology, Case Western Reserve University: Cleveland, Ohio, 1976; 39 pp.
11. Angus, John C. "Chemical Engineering at Tech," In "Case Alummus"; Case Western Reserve University: Cleveland, Ohio, February 1976; p. 18–20.
12. "The Study of Chemistry at the University of Illinois"; University of Illinois: Urbana, 1907; 31 pp.
13. "Circular of Information on the Department of Chemistry," *Univ. Ill. Bull.* Feb. 21, **1916**, *13*(25), 108 pp.
14. "University of Illinois Department of Chemistry 1941–1951"; University of Illinois: Urbana, 151 pp.
15. "Twenty Fourth Annual Catalog of the Officers and Students with a Statement of the Courses and Instructions and a List of the Alumni 1888–1889"; Massachusetts Inst. of Tech.: Cambridge, MA.

16. Weber, Harold "History of Chemical Engineering at M.I.T.," unpublished data, 1977.
17. "Catalogue and Announcements, 1892–93"; University of Pennsylvania: Philadelphia, 1893; p. 134.
18. "Tulane University of Louisiana, Catalogue 1893–94, Announcement for 1894–95," New Orleans, 1894; 155 pp.
19. "The Tulane University of Louisiana, College of Engineering, Register of Graduate 1889–1921"; New Orleans.
20. "Calendar of the University of Michigan 1898–99"; Ann Arbor; 1899.
21. White, A. H.; Upthegrove, C. "The Department of Chemical and Metallurgical Engineering"; a publication to celebrate a century of engineering at the University of Michigan, Ann Arbor: 1953; p. 1190–1200.
22. Gurnham, C. F. "A Brief History of Chemical Engineering at Tufts College"; section of accreditation application, 1951.
23. "The Chemical Engineering Lecture Series," Sept. 15, 1977 to Nov. 17, 1977, Department of Chemical Engineering, University of Arkansas: Fayetteville.
24. "Catalogue of the University of Arkansas, 1903–1904," 31st Edition; Fayetteville, Arkansas.
25. "Catalogue of the University of Arkansas, 1901–1902," 29th Edition; Fayetteville, Arkansas.
26. "University of Pittsburgh, School of Engineering, Past, Present, Future"; a publication for the dedication of Michael Bendedum Hall, March 18, 1971.
27. "Catalog of the University of Colorado 1903–04"; Boulder, Colorado, p. 134–161.
28. Barrick, P. "Brief History of Chemical Engineering at the University of Colorado"; unpublished data, 1978.
29. Bartlett, J. W.; Miller, S. A.; Perry, R. H.; Wiehe, I. A. "Chemical Engineering at Rochester—1966"; introduction to a report of a Committee on Self-Assessment of a Department.
30. Thompson, R. E. "A Brief History of Chemical Engineering at the University of Tulsa"; unpublished data, 1978.
31. Stinson, D. L., personal communication, Sept. 26, 1978.
32. Arnold, L. K. "History of the Department of Chemical Engineering at Iowa State University"; Iowa State University: Ames, 1970.
33. Larson, M.; Seagrave, R. "ChE Department: Iowa State," *Chem. Eng. Educ.* **1976,** *10*(3), 108–111.
34. Anon. "LSU—A Center of Excellence," *Eng. Educ.* **1979,** *69*(5), 380–2.
35. Harrison, D. P., personal communication, Oct. 3, 1978.
36. Dyer, J. P. "Tulane, The Biography of a University 1834–1965"; Harper and Row: New York, 1966; p. 78.
37. Keeney, N. H., Jr., personal communication, Oct. 3, 1978.
38. "University of Wisconsin Catalogue 1904–1905"; Madison, Wisconsin, 1905.
39. Hougen, O. A. "Beginning of the Chemical Engineering Department"; University of Wisconsin: Madison, unpublished data, August 1976.
40. "Focus on Engineering"; College of Engineering, The University of Toledo: Toledo, Fall 1972.
41. Coulman, G. A. "Department of Chemical Engineering, Fenn College of Engineering, Cleveland State University"; unpublished data, 1978.
42. "The Cleveland YMCA School of Technology Bulletin, Announcement of Courses"; YMCA: Cleveland, Ohio, July 1926; Vol 1, Number 7, p. 88–89.
43. Brutvan, D. R., personal communication, Sept. 18, 1978.
44. Vermeychuk, J. G.; Bergantz, J. A. "SUNY at Buffalo," *Chem. Eng. Educ.* **1973,** *7*(3), 112–116.
45. Lewis, W. K. "Evolution of the Unit Operations," *Chem. Eng. Progr. Symp. Ser.* **1959,** *55* 26, 1–8.

RECEIVED May 7, 1979.

9

The Role of Transport Phenomena in Chemical Engineering Teaching and Research: Past, Present, and Future

R. BYRON BIRD, WARREN E. STEWART, and EDWIN N. LIGHTFOOT

Department of Chemical Engineering, University of Wisconsin, Madison, WI 53706

Experience during and after World War II showed the need for more powerful approaches to the solution of engineering problems. Among these was the unified description of heat, mass, and momentum transfer, referred to as "transport phenomena." The development of this new engineering discipline took place almost simultaneously in many parts of the world. Wisconsin contributions, emphasized in this chapter, took place in the context of a strong engineering science program initiated by Professor O. A. Hougen. This chapter begins with a survey of the role of transport phenomena in chemical engineering. The histories of the basic equations, and of engineering education in transport phenomena, are then reviewed. Finally, some predictions and suggestions are made regarding future directions of activity in this field.

The subject of transport phenomena deals with the transport of mass, momentum, energy, and other entities. These various phenomena can be examined at three different levels: (1) the molecular level, where one describes the viscosity, thermal conductivity, and diffusivities of macroscopic materials in terms of models of the constituent molecules; (2) the continuum level, where one ignores the molecular motions and focuses on the partial differential equations (the "equations of change") which describe the profiles of velocity, temperature, and concentration; and (3) the equipment level, in which one is concerned mainly with relations among input and output quantities for some piece of equipment

0-8412-0512-4/80/33-190-153$05.00/1

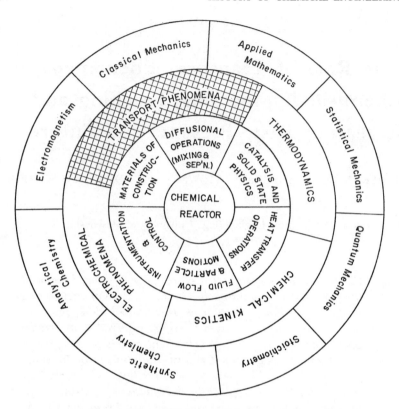

Figure 1. This figure emphasizes that the subject of transport phenomena is one of the engineering sciences which is needed for the understanding and further development of chemical engineering design. The figure is intended to be illustrative and not all-inclusive. Certainly chemical engineering has goals other than that of chemical reactor design. In many biochemical and biomedical areas, such subjects as biology, bacteriology, genetics, and biochemistry are key basic sciences. Other activities in which chemical engineers become involved (e.g., nuclear fuel processing, photographic materials, transistor technology, and lubrication engineering) would have diagrams somewhat different from the above, but transport phenomena would be included as one of the engineering sciences.

or a portion thereof. These three approaches are intertwined inexorably, and the chemical engineer can be called upon to use results from all three.

In order to put the role of transport phenomena in chemical engineering into perspective, consider the diagram in Figure 1. The chemical reactor is shown as a focal point of the field of chemical engineering. Surrounding it are the various engineering subjects which are connected intimately with reactor engineering: separation and mixing processes, heat transfer operations, catalysis, fluid and particle dynamics, instrumentation and control, and materials of construction. These subjects are clearly essential to the design and operation of a chemical reactor.

In the next layer of subjects we list the "engineering sciences" which are needed in various ways for understanding and further developing the core engineering subjects: thermodynamics, chemical kinetics, electrochemical phenomena, and transport phenomena. These engineering sciences, which are themselves interrelated, form the basis for the analytical and numerical description of the chemical reactor and its peripheral equipment. For example, the subject of transport phenomena can be used to analyze diffusion-controlled reactions, separation schemes, transient processes in reactors, thermal processes, flow patterns in reacting systems, corrosion, diffusion in porous media, and other problems connected with reactor engineering.

Transport phenomena occur in many other fields: acoustics, zoology, micrometeorology, plasma physics, combustion, nuclear engineering, fermentation, biomedical engineering, electrochemistry, soil physics, ocean engineering, atmospheric pollution, pharmacology, and polymer processing. In each of these fields the equations of change (i.e., the equations of continuity, motion, and energy) can form the starting point for the description, organization, and systematization of substantial parts of the subject material. The knowledge which a chemical engineering student acquires about the solution of problems in transport phenomena is thus easily transferable to other fields.

It can, of course, be argued that the subject of transport phenomena is really nothing more than a grouping together of three well-known subjects: fluid dynamics, heat transfer, and diffusion. Bringing the subjects together into one subject is, however, advantageous for several reasons: (1) in nature, in biological systems, and in the chemical industry, the three phenomena often occur simultaneously; (2) the mathematical descriptions of the three phenomena are related closely and hence considerable use may be made of analogies among the various phenomena; (3) there are also important differences among the three fields, and these can be emphasized when the subject material is juxtaposed. A person who has studied transport phenomena from this unified viewpoint is in an excellent position to proceed to the study of special treatises and advanced texts on fluid dynamics, heat transfer, and diffusion as well as rheology, electrochemistry, acoustics, combustion, turbulence, boundary-layer theory, and a host of other related fields.

History of Transport Phenomena

To do justice to this topic would require a book. We mention just a few highlights here to give some perspective. The subject of transport phenomena is both old and new. Some of the equations we use have been known for a century or more. Others are of relatively recent vintage. And, of course, there are some parts of the subject which are being developed just now and others which can qualify as terrae incognitae.

The Flux Expressions. We begin with the relations between the fluxes and gradients, which serve to define the transport properties. For viscosity the earliest definition was that of Newton (1) in 1687; however about a century and a half elapsed before the most general linear expression for the stress tensor of a Newtonian fluid was developed as a result of the researches by Navier (2), Cauchy (3), Poisson (4), de St. Venant (5), and Stokes (6). For the thermal conductivity of a pure, isotropic material, the linear relationship between heat flux and temperature gradient was proposed by Fourier (7) in 1822. For the diffusivity in a binary mixture at constant temperature and pressure, the linear relationship between mass flux and concentration gradient was suggested by Fick (8) in 1855, by analogy with thermal conduction. Thus by the mid 1800's the transport properties in simple systems had been defined.

For multicomponent reacting mixtures with simultaneous heat, mass, and momentum transport, it was not until the 1940's that clear definitions of the transport properties emerged as a result of research by Eckart (9), Meixner (10), Prigogine (11), de Groot (12), Kirkwood and Crawford (13), Hirschfelder, Curtiss, and Bird (14), and others. This development had to await the emergence of the thermodynamics of irreversible processes as a new discipline. It must be pointed out, however, that the kinetic theory of gases had provided a great deal of insight for those who developed the thermodynamic codification of the various irreversible phenomena. For an extensive and carefully compiled bibliography on the development of the flux expressions, see the review by Truesdell and Toupin (15).

For structurally complex fluids—notably polymer solutions, molten polymers, suspensions, and emulsions—the linear relation between the stress tensor and the velocity gradients is wholly inapplicable, and in addition these fluids are viscoelastic. The development of an understanding of the nonlinear viscoelastic responses of these complex fluids is the subject of current research programs. From a phenomenological point of view several important suggestions have been made: (1) Oldroyd (16) and others have modified the linear viscoelastic constitutive relation by empirical introduction of nonlinear terms; (2) Rivlin and Ericksen (17) developed a systematic expansion of the stress tensor in terms of the rate-of-strain and higher-order kinematic tensors, leading to the definition of second-, third-, and higher-order fluids; (3) Green and Rivlin (18), Goddard (19), and others have shown how to expand the stress tensor as a memory-integral expansion. Much needs to be learned about the convergence and usefulness of these memory-integral expansions. For an extensive discussion of this aspect of transport phenomena, see the recent book by Bird, Armstrong, and Hassager (20).

Kinetic Theories of the Transport Properties. For dilute gases the kinetic theory has a long history beginning with Bernoulli in the 1700's. Through the centuries increasing refinements in the theories and the modeling of intermolecular forces have occurred. The towering works of Maxwell *(21)* and Boltzmann *(22)* in the 19th century gave tremendous impetus to the study of the interrelation of macroscopic properties and molecular structure. However, it was not until the second decade of the 20th century that the kinetic theory of dilute monatomic gases was developed in a rigorous way by Chapman *(23)* and Enskog *(24)*, independently and almost simultaneously. Their work was generalized to multicomponent mixtures in 1940 by Hellund *(25)* and in 1949 by Curtiss and Hirschfelder *(26)*. Outstanding summaries of modern kinetic theory have been given by Chapman and Cowling *(27)* and by Waldmann *(28)*; applications of the theory have been summarized by Hirschfelder, Curtiss, and Bird *(14)*. A short historical summary of the kinetic theory of gases is given on pp 407–409 of the monograph of Chapman and Cowling *(27)*; this book, which also gives discussions of dense gases, quantum effects, electromagnetic phenomena, and nonspherical molecules, is an extraordinary treatise and a "must" for anyone involved in kinetic theory.

The kinetic theory of liquids has been worked on intensively in this century. The activated-state theories of Eyring *(29)* and co-workers were the first entry into the field. Later Born and Green *(30, 31)*, Kirkwood and his collaborators *(32)*, and others developed rigorous kinetic theories. Because of the complexity of the rigorous theories and the difficulty of getting numerical results from them, attention in recent years has turned to molecular dynamics simulation *(33)*.

For two-phase systems Einstein *(34)* in 1906 was the first to obtain the viscosity for a very dilute suspension of solid spheres; the resulting stress expression is Newtonian. However, it has been only within the past decade that nonlinear viscoelastic expressions for the stress tensor in dilute suspensions have been obtained: deformable spheres *(35)*, ellipsoids *(36)*, emulsions *(37, 38)*. For a survey of activities in this field, *see* the summary by Barthès–Biesel and Acrivos *(39)*.

For polymeric fluids, early kinetic-theory workers *(40)* attempted to calculate the zero-shear-rate viscosity of dilute solutions by modeling the polymer molecules as elastic dumbbells. Later the constants in the Rivlin–Ericksen *(17)* expansion were obtained for dumbbells *(41, 42)* and other more complex models; and only recently have the kernel functions in the memory integral expansions been obtained *(43)*. This rapidly expanding field has been summarized recently in a monograph *(44)*; here, too, molecular dynamics simulation may prove fruitful *(45)*.

The Equations of Change and the Macroscopic Balance. Textbooks and treatises on fluid dynamics, heat transfer, and diffusion seldom give any literature citations to indicate the origins of the equations of change. Truesdell and Toupin *(15)* have ferreted out this information, and we cite a few key references here. The continuity equation for a pure fluid dates back to Euler *(46)* in 1755; and for mixtures the continuity equations were given by Jaumann *(47)* in 1911. The equation of motion for a Newtonian fluid had its origins with Cauchy *(48)* about 150 years ago. The energy equation is virtually as old as the First Law of Thermodynamics. However, very few heat-transfer or fluid dynamics texts give the equation complete with the viscous dissipation term, even though the expression for viscous dissipation was known by Rayleigh *(49)*. It was not until the middle of the 20th century that Prigogine *(11)*, Kirkwood and Crawford *(13)*, and others presented the equation of energy for chemically reacting mixtures.

The macroscopic mass, momentum, and energy balances for pieces of equipment have been used in their steady-state form for many decades, by applying the conservation laws to entire pieces of equipment or parts thereof. The derivation of the unsteady-state macroscopic balances from the equations of change was set forth by Bird *(50)* about 20 years ago; the principal advantage of this derivation is that the mechanical energy balance (or the engineering Bernoulli equation) can be derived more satisfactorily than had been done previously by thermodynamic arguments. A corresponding development for the angular momentum balance was given by Slattery and Gaggioli *(51)* somewhat later *(52)*.

Once the equations of change and the expressions for the fluxes have been established, one has the complete set of starting equations for solving problems. Solving can include analytic solutions, numerical solutions, studies based on time–smoothing of the equations appropriate for turbulent flow, boundary-layer studies, and even dimensional analyses. All of these kinds of problem solving have direct relevance to engineering modeling activities. Because the equations of change form a coupled, nonlinear set of equations, numerical techniques have to be used in most situations. As computing equipment and numerical analysis improve, our capacity for practical problem solving increases. The equations of change thus can be expected to play an ever-increasing role in engineering analysis and design.

Development of Transport Phenomena Teaching at the University of Wisconsin at Madison

During the development of chemical engineering in the first half of the 20th century, chemical engineers quite naturally took over important results from the fields of fluid dynamics, heat transfer, and diffusion as they needed them. This was particularly true in the identification of "unit operations" as a key subject; attention was focused on this subject by the publication of the outstanding textbook by Walker, Lewis, and McAdams *(53)*. This influential book shaped the training of chemical engineers for many decades to come, since subsequent textbooks in the area followed in very much the same spirit.

In the post World War II years chemical engineers began to be called upon to move into a number of new areas in which problem definition and problem solving required a more substantial knowledge of the fundamentals of transport phenomena than that provided in the textbooks on unit operations. As an example, in the polymer-processing industry engineers became involved with fluids of very high viscosity, in which elastic and viscous dissipation heat effects are very important. As another example we can cite the development of new kinds of spray-processing equipment, in which detailed analyses of flow patterns and heat-and-mass transfer rates had to be developed.

By the 1950's many chemical engineering departments were experimenting with teaching courses in transport phenomena—often under titles such as Chemical Engineering Fluid Dynamics, Diffusional Operations, Rate Processes, and Fluid and Particle Mechanics. These experimental courses were offered usually at the graduate level, with the idea of providing research students with an introduction to material needed for their dissertations.

Let us turn now to the events which occurred at the University of Wisconsin. Professors O. A. Hougen and K. M. Watson had been very active in applied chemical kinetics and separations processes. In 1948 W. R. Marshall joined the staff, specializing in drying and spray processing. Because much of the on-going research required an understanding of fluid dynamics and transport phenomena, ca. 1949 Professor W. R. Marshall began to teach classical fluid dynamics, boundary-layer theory, and diffusion in his course on diffusional operations. About the same time Professors J. O. Hirschfelder and C. F. Curtiss of the Chemistry Department were invited to give a special course for the Chemical Engineering department on transport processes; this included the equations of change, shock-wave phenomena, dimensional analysis, dynamics of phase change, and several other topics. By the fall of 1952 Professor Hirschfelder and his two former students, C. F. Curtiss and R. B. Bird, had finished the manuscript of their treatise *(14)*, about one third of which deals with molecular and continuum aspects of transport phenomena. After that R. B. Bird disappeared from the scene to teach one year in the Chemistry Department at Cornell and to spend a very fruitful summer at the Du Pont Experimental Station in Wilmington, Delaware. There he became aware of the large number of problems in the polymer industry which were difficult to solve, partly because of gaps in the traditional chemical engineering training.

In 1953 R. B. Bird and E. N. Lightfoot joined the staff of the
Chemical Engineering Department at Wisconsin; E. N. Lightfoot had
spent several years with the Pfizer Company where he had had extensive
experiences in biochemical processing and fermentation operations. Two
years later W. E. Stewart joined the staff; he had left Wisconsin with an
MS degree in Chemical Engineering in 1947 and completed an ScD
degree at M.I.T. under Professor H. S. Mickley (on simultaneous heat
and mass transfer in boundary layers) and then had six years of experience
at Sinclair Research dealing primarily with catalysis and reactor design.
R. B. Bird's teaching assignments included the development of a new
course in fluid dynamics and taking over Professor W. R. Marshall's
course on diffusional operations.   This gave him an excellent opportunity
to develop his ideas on the teaching of transport phenomena to chemical
engineering students.   At the same time he served on a committee
charged with exploring the idea of establishing a Department of Nuclear
Engineering at Wisconsin.

The Nuclear Engineering Committee made plans for an undergraduate curriculum in nuclear engineering.   The committee decided to incorporate a course in transport phenomena so that the nuclear engineering students would have the required background for coping with
heat-transfer and flow problems, and spent-fuel separations techniques.
When the course outline was sent to the Executive Committee of the
Physical Sciences Division for approval, this committee wanted to know
why the Chemical Engineering Department did not have a course like
this.   On an earlier occasion, the Chemical Engineering Department
had, in fact, rejected a suggestion for instituting a course of this type.
However, because of the request of the Physical Sciences Committee
they were obliged to reconsider the matter.   After lengthy debate it was
decided by a vote of 5–4 to put in the junior year a required three-credit
course on transport phenomena, provided that R. B. Bird would prepare
a   set   of   mimeographed   notes   for   use   in   the   fall   semester   of
1957.   W. E. Stewart and E. N. Lightfoot quickly offered to assist in this
project.   In this way three co-authors with three very different backgrounds came to work together on the preparation of a new textbook.
Most textbooks are prepared after years of teaching experience in a given
field; however, in this instance, the first version of the text was prepared
before any teaching experience with undergraduates had been accumulated.   The detailed plans for the course were presented at a meeting of
Deans of Engineering at Purdue University in September, 1957 (54).

Work on the mimeographed notes began on or about August 1, 1957,
and by the start of the fall term about 13 chapters were ready. During the
fall term the note-writing kept ahead of the teaching activities.   Since
Professor R. A. Ragatz, Chairman of the department, insisted that all of

the professors take part in teaching the course during the first year, R. B. Bird ran a coaching session each week in order to go over the notes, illustrative examples, and suggested homework problems. It was a lot of work for all of us—writing, teaching, coaching, and testing our ideas. During the spring term R. B. Bird taught the course again at the Technische Hogeschool in Delft; there he had the chance to discuss transport phenomena teaching with Professor H. Kramers who had in 1956 published a set of notes entitled *Physische Transportverschijnselen*. Meanwhile, in Madison, E. N. Lightfoot and W. E. Stewart prepared additional illustrative examples and problems to go with the text. The John Wiley and Sons publishers had agreed to put out the notes as a "preliminary edition," a procedure which was being experimented with by a number of publishers at that time. The result was the appearance in the fall of 1958 of *Notes on Transport Phenomena*, by Bird, Stewart, and Lightfoot. Wiley made it possible for a number of chemical engineering departments to use this provisional textbook, and the comments of the teachers and students at these schools were forwarded to the authors. These comments were invaluable in revising the *Notes on Transport Phenomena* for the publication of the final book (55). We will never be able to thank adequately the teachers and students who assisted us in this way.

Throughout the organization and writing of the book, we received a great deal of advice and encouragement from our senior colleague, Professor Olaf A. Hougen. Consequently the book was dedicated to him in an acronym, which may be found by reading the first letter of each sentence in the preface (an additional message may be found by reading the first letter of each paragraph in the postface). The book has appeared in Spanish (1965), Czech (1968), and Italian editions (1970), and an unauthorized Russian edition (1974). (In the Russian edition Chapter 14 was omitted entirely and other unfortunate alterations in the text were made.)

The authors of the book have participated in a number of additional teaching activities in connection with transport phenomena: an NSF three-week short course for engineering teachers in the summer of 1961; numerous industrial- and AIChE-sponsored short courses; R. B. Bird's guest lectures at Kyoto and Nagoya Universities in Japan; W. E. Stewart's guest lectures at LaPlata and Buenos Aires Universities in Argentina; and E. N. Lightfoot's lectures at the Technical University of Denmark, Tunghai University in Taiwan, and the University of Canterbury in New Zealand. In addition, each of us has prepared an advanced text on some special aspect of transport phenomena: E. N. Lightfoot has written a book on biomedical applications (56); R. B. Bird has co-authored a two-volume work on polymeric fluids (20, 44); and W. E. Stewart currently is co-authoring a book emphasizing numerical techniques for solving transport problems (57).

Transport Phenomena Teaching since 1960

In the past two decades some form of transport phenomena teaching has become established in most chemical engineering departments. In some schools it is given as a sort of applied physics course to bridge the gap between elementary physics and unit operations (as is done at the University of Wisconsin). In other schools it is given as a terminal course after unit operations, sometimes as an elective. In still other departments, the materials is taught only at the graduate level. In a few departments the transport phenomena teaching has displaced the subject of unit operations. It was no doubt the specter of such a trend that led Professor T. K. Sherwood of M.I.T. to label our text a "a dangerous book" *(58)*. He rightly cautioned that "if perspective is lost through enthusiasm for scientific and mathematical analyses, the engineer will be less effective in industry" *(58)*. We feel that his words of caution were appropriate and should be heeded.

Since our book was published in 1960 many new books have appeared. The subject of transport phenomena is now available in many languages and flavors. We list here the ones that we know about:

1. Levich, V. G., *Physicochemical Hydrodynamics*, Prentice–Hall, Englewood Cliffs, NJ, 1962. (Actually this is a monograph rather than a textbook; it appeared in Russian in 1952 with a second Russian edition in 1959.)
2. Rohsenow, W. M. and Choi, H. Y., *Heat, Mass, and Momentum Transfer*, Prentice–Hall, Englewood Cliffs, NJ, 1961.
3. Bennett, C. O. and Myers, J. E., *Momentum, Heat, and Mass Transfer*, McGraw–Hill, NY, 1962.
4. Shirotsuka, T., Hirata, A., and Murakami, A., *Kagaku–gijitshusha no tame no Idō-sokudo-ron*, Omu-sha, Tokyo, 1966.
5. Kunii, D., Editor, *Idō-sokudo-ron*, Iwanami, Tokyo, 1968.
6. Koizumi, M., *Idō-sokudo-ron*, Shokodo, Tokyo, 1969.
7. Hanley, H. J., *Transport Phenomena in Fluids*, M. Dekker, NY, 1969.
8. Welty, J. R., Wicks, C. E., and Wilson, R. E., *Fundamentals of Momentum, Heat, and Mass Transfer*, Wiley, NY (1st Ed) 1969, (2nd Ed) 1976.
9. Hiraoka, M. and Tanaka, K., *Idō-genshō-ron*, Asakura-shoten, Tokyo, 1971.
10. Theodore, L., *Transport Phenomena for Engineers*, International, Scranton, 1971.

11. Slattery, J. C., *Momentum, Energy, and Mass Transfer in Continua*, McGraw–Hill, NY, 1972.

12. Eckert, E. R. G. and Drake, R. M., Jr., *Analysis of Heat and Mass Transfer*, McGraw–Hill, NY, 1972.

13. Sissom, L. E. and Pitts, D. R., *Elements of Transport Phenomena*, McGraw–Hill, NY, 1972.

14. Geiger, G. H. and Poirier, D. R., *Transport Phenomena in Metallurgy*, Addison–Wesley, Reading, MA, 1973.

15. Hershey, D., *Transport Analysis*, Plenum Press, NY, 1973.

16. Kobayashi, K., *Idō-ron*, Asakura-shoten, Tokyo, 1973.

17. Foraboschi, F. P., *Principi di Ingegneria Chimica*, U.T.E.T., Torino, 1974.

18. Lih, M. M.–S., *Transport Phenomena in Medicine and Biology*, Wiley, NY, 1975.

19. Beek, W. J. and Muttzall, M. K., *Transport Phenomena*, Wiley–Interscience, NY, 1975.

20. Mikami, H., *Idō-ron*, Asakura-shoten, Tokyo, 1976.

21. Rietema, K., *Fysische transport- en overdrachtsverschijnselen*, Uitgeverij Het Spectrum, Utrecht, 1976.

22. Edwards, D. K., Denny, V. E., and Mills, A. F., *Transfer Processes*, Hemisphere Publ. Co., Wash. D.C., 1978.

23. Ginoux, J. J., *Phénomènes de Transport*, Universite Libre de Bruxelles (3rd Ed), 1978.

Future Activities

With regard to the teaching of transport phenomena, there undoubtedly will continue to be considerable experimentation in the coming years.   The availability of pocket calculators and larger computers will make it possible to apply the principles of transport phenomena in many new ways in engineering design.   These newer applications of the theory inevitably will be mirrored in a changing approach to the teaching of the subject even at the elementary level.

In research there are many fascinating areas for speculation.   At the molecular level there will have to be a small band of devoted scholars who will continue to do the painstaking, detailed derivations which only other kinetic theorists can appreciate fully.   The past few years have seen a renaissance in the kinetic theory of macromolecular fluids, and many new avenues have been opened or suggested.   Because of the development of high-speed computers, there is also considerable interest in developing the molecular dynamics approach to transport properties. It may turn out that transport properties will be predicted more readily by direct calculation of the molecular motions than by doing involved kinetic theory derivations, application of which still requires considerable computational labor.

In connection with the solutions of the equation of change, there will from time to time be found some new and useful analytical solutions. The emphasis surely will be put, however, on the development of more and more efficient numerical techniques, such as collocation and weighted residual methods, implemented with high-speed computers.

Turbulence seems to remain the Mt. Everest of transport phenomena. Many people have attempted to scale its treacherous heights and have been discouraged by the futility of their efforts. On the other hand, the immense amount of information obtained in the past several decades by experimental measurements of turbulence structure and more sophisticated modeling may fall gradually into place and result in useful tools which can be used even by elementary practitioners. Let us hope so.

The chemical engineer will be interested particularly in applying the basic principles of transport phenomena to problems involving separations processes, combustion, polymer processing, interfacial hydrodynamics, multiphase flow, and biomedical engineering. In all of these areas it will be the task of the chemical engineer to utilize the basic theory of transport phenomena innovatively in solving practical problems for the benefit of society.

Literature Cited

1. Newton, I. "Philosophiae Naturalis Principia Mathematica"; London, 1687; Lib II, Chapter IX.
2. Navier, C.-L.-M.-H. *Ann. Chim.* **1821**, *19*, 244–260.
3. Cauchy, A.-L. *Bull. Soc. Philomath. Paris* **1823**, 9–13.
4. Poisson, S.-D. *J. École Polytech.* **1832**, *13*, cahier 20, 1–174.
5. de St. Venant, A.-J.-C.B. *C. R. Acad. Sci.* **1843**, *17*, 1240–1243.
6. Stokes, G. G. *Trans. Cambridge Philos. Soc.* **1845**, *8*, 287–319.
7. Fourier, J. B. J. "Théorie Analytique de la Chaleur"; Gauthier-Villars: Paris, 1822.
8. Fick, A. *Ann. Phys. (Paris)* **1855**, *94*, 59–86.
9. Eckart, C. *Phys. Rev.* **1940**, *58*, 269–275.
10. Meixner, J. *Z. Phys. Chem. Abt. B* **1943**, *53*, 235–263.
11. Prigogine, I. "Étude Thermodynamique des Phénomènes Irréversibles"; Dunod-Desoer: Paris, 1947.
12. de Groot, S. R. "Thermodynamics of Irreversible Processes"; North Holland, Amsterdam, 1951.
13. Kirkwood, J. G.; Crawford, B., Jr. *J. Chem. Phys.* **1952**, *56*, 1048–1051.
14. Hirschfelder, J. O.; Curtiss, C. F.; Bird, R. B. "Molecular Theory of Gases and Liquids"; Wiley: New York, 1954; corrected printing with added notes 1964.
15. Truesdell, C.; Toupin, R. "Handbuch der Physik"; Flügge S., Ed.; Springer: Berlin, 1960; vol. III/1, pp. 226–793.
16. Oldroyd, J. G. *Proc. R. Soc., London, Ser. A* **1950**, *200*, 523–541.
17. Rivlin, R. S.; Ericksen, J. L. *J. Rat. Mech. and Anal.* **1955**, *4*, 323–425.
18. Green, A. E.; Rivlin, R. S. *Arch. Ration. Mech. Anal.* **1957**, *1*, 1–21.
19. Goddard, J. D. *Trans. Soc. Rheol.* **1967**, *11*, 381–399.

20. Bird, R. B.; Armstrong, R. C.; Hassager, O. "Dynamics of Polymeric Liquids: Vol. 1 - Fluid Mechanics"; Wiley: New York, 1977.
21. Maxwell, J. C. "Scientific Papers"; Hermann & Cie: Paris, 1927.
22. Boltzmann, L. "Vorlesungen über Gastheorie"; Barth: Leipzig, 1896.
23. Chapman, S. *Philos. Trans. R. Soc. London, Ser. A* **1916**, *213*, 433.
24. Enskog, D. "Kinetische Theorie der Vorgänge in mässig Verdunnten Gasen"; Almqvist and Wiksell: Uppsala, 1917.
25. Hellund, E. J. *Phys. Rev.* **1940**, *57*, 319.
26. Curtiss, C. F.; Hirschfelder, J. O. *J. Chem. Phys.* **1949**, *17*, 550–555.
27. Chapman, S.; Cowling, T. G. "Mathematical Theory of Non-uniform Gases," 3rd ed.; Cambridge University Press: 1970.
28. Waldmann, L. In "Handbuch der Physik"; Springer: Berlin, 1958; Vol. XII, pp. 295–514.
29. Glasstone, S.; Laidler, K. J.; Eyring, H. "Theory of Rate Processes"; McGraw-Hill: New York, 1941.
30. Born, M.; Green, H. S. "A General Kinetic Theory of Liquids"; Cambridge University Press: 1949.
31. Green, H. S. "Molecular Theory of Fluids"; Interscience: New York, 1952.
32. Kirkwood, J. G. "Selected Topics in Statistical Mechanics"; Gordon and Breach: New York, 1967; a collection of reprints from the years 1946–1958.
33. Hynes, J. T. *Ann. Rev. Phys. Chem.* **1977**, *28*, 301–21.
34. Einstein, A. *Ann. Phys.* **1906**, *19*, 289–306 (erratum: ibid., **1911**, *34*, 591–592).
35. Goddard, J. D.; Miller, C. *Rheol. Acta* **1966**, *5*, 177–184.
36. Leal, L. G.; Hinch, E. J. *J. Fluid Mech.* **1972**, *55*, 745–765.
37. Schowalter, W. R.; Chaffey, C. E.; Brenner, H. *J. Colloid Sci.* **1968**, *26*, 152–160.
38. Frankel, N. A.; Acrivos, A. *J. Fluid Mech.* **1970**, *44*, 65–78.
39. Barthès-Biesel, D.; Acrivos, A. *Int. J. Multiphase Flow* **1973**, *1*, 1–24.
40. Kuhn, W.; Kuhn, H. *Helv. Chim. Acta* **1945**, *28*, 97–127.
41. Giesekus, H. *Kolloid-Zeits.* **1956**, *147–149*, 29–45.
42. Prager, S. *Trans. Soc. Rheol.* **1957**, *1*, 53–62.
43. Armstrong, R. C.; Bird, R. B. *J. Chem. Phys.* **1973**, *58*, 2715–2723.
44. Bird, R. B.; Hassager, O.; Armstrong, R. C.; Curtiss, C. F. "Dynamics of Polymeric Liquids: Vol. 2, Kinetic Theory"; Wiley: New York, 1977.
45. Gottlieb, M. *Computers in Chemistry* **1977**, *1*, 155–160.
46. Euler, L. *Mem. Acad. Sci. Berlin* **1755**, *11*, 274–315.
47. Jaumann, G. *Sitzngsber. Akad. Wiss. Wien, Math.—Naturwiss. Kl., Abt. 2A* **1911**, *120*, 385–530.
48. Cauchy, A.-L. *Ex. de Math.* **1827**, *2*, 108–111.
49. Strutt, J. W. (Lord Rayleigh) "The Theory of Sound," 1st ed. (1877), 2nd ed. (1894); Dover Reprint: New York, 1945.
50. Bird, R. B. *Chem. Eng. Sci.* **1957**, *6*, 123–131.
51. Slattery, J. C.; Gaggioli, R. A. *Chem. Eng. Sci.* **1962**, *17*, 893–895.
52. Bird, R. B. *Chem. Eng. Progr. Symp. Ser.* **1965**, *58*, (61), 1–11.
53. Walker W. H.; Lewis, W. K.; McAdams, W. H. "Principles of Chemical Engineering"; McGraw-Hill: New York, 1923.
54. Bird, R. B. In "Recent Advances in the Engineering Sciences"; McGraw-Hill: New York, 1958.
55. Bird, R. B.; Stewart, W. E.; Lightfoot, E. N. "Transport Phenomena"; Wiley: New York, 1960; 19th printing, 1978.
56. Lightfoot, E. N. "Transport Phenomena and Living Systems"; Wiley: New York, 1974.
57. Stewart, W. E.; Sørensen, J. P. "Collocation and Parameter Estimation in Chemical Reaction Engineering"; Wiley: New York, unpublished data.
58. Sherwood, T. K. *Chem. Eng. Sci.* **1961**, *15*, 332–333.

RECEIVED May 7, 1979.

A Century of Chemical Engineering Education in Canada

L. W. SHEMILT

Professor of Chemical Engineering and Dean of Engineering, McMaster University, Hamilton, Ontario, Canada L8S 4L7

The development of chemical engineering education in Canada can be traced from the applied chemistry degree program at Toronto 100 years ago to the present national scene of 18 accredited schools in seven provinces. Factors and traditions in curriculum development, including the influence of resource development and industrial patterns, are highlighted through the major contributions, as well as the educational genealogy, of a number of pioneering chemical engineering teachers. Postgraduate programs and research developed strongly after 1945, as measured in degrees granted (now about 100 masters and 50 doctorates annually) and research output. International participation, the special role of chemical societies, American and British influences and connections, and the indigenous factor of bilingualism, all have shaped the pattern of Canadian chemical engineering education.

. . . And in the Beginning

A quotation from Emil Fischer appears on the program for the 6th Annual Dinner of the Industrial Chemical Club at the University of Toronto in 1914 and reads, "My way is to begin at the beginning" *(1)*. Quoting from a European chemist typifies one of the significant influences in the development of applied chemistry and chemical engineering in Canada, just as its presence on a piece of University of Toronto memorabilia is not untoward at that University which saw the first graduate of a chemical engineering program. But the true beginning is earlier. Historically, it goes back to Canada's first settler, Louis Hébert, who, like his father at the time of Louis XIV, was an apothe-

0-8412-0512-4/80/33-190-167$08.00/1

cary, and literally Canada's first chemist (2). Since he practiced medicinal herb extraction, lime burning, and the brewing of beer at Quebec where he had settled in 1617, he more accurately might be deemed Canada's first chemical engineer.

Engineering based on chemistry of the new resources of the new land continued from that early effort on through the 18th century, with, for example, the development of the iron ore deposit in the St. Maurice Valley in Quebec and the first commercial metallurgical plant in North America beginning operation in 1736 (3). The 19th century saw, among many others, the invention of the process of kerosene oil by Abraham Gesner, who was distilling the product for home and street lighting in Halifax by 1850 (nine years before Drake's gusher in Pennsylvania). The application of chemistry and the need for practical chemistry grew apace, and the early educational institutions and governments took note. The province of Quebec was the "the birthplace of chemical education in Canada" (4), and in 1833 the first textbook in chemistry published in Canada appeared in Montreal in the form of J. B. Meilleur's *Cours abruge de lecons de chymie*. Laval University had, from 1856, its Professor of Chemistry and Minerology in the person of Dr. T. S. Hunt, who also served as Professor of Applied Chemistry and Minerology at McGill from 1862. Ecole Polytechnique in Montreal traces its beginnings to 1873, and in 1874 McGill University offered instruction in applied science, including that in civil engineering and in practical chemistry and assaying. This latter one was the forerunner of the present course in chemical engineering. Though this third quarter of the 19th century professorships in chemistry, natural science, or natural philosophy, also were being established at several Maritime universities in Nova Scotia and New Brunswick. However, in only one of these institutions, the University of New Brunswick, was a chemical engineering program eventually developed. The practical chemistry aspects in all of the universities mentioned so far were developed either in relation to geology, minerology and assaying, or medicine. While the basis for later chemical engineering programs can be said to have been established, the connecting elements were not present in any truly viable sense. For such a continuity and for the real antecedents of chemical engineering education, we must turn to Ontario and specifically to the University of Toronto.

There Is a First: The Pilot Plant

As early as 1871 the Ontario Legislature passed a Bill to establish a School of Technology in Toronto. Although the scheme never was developed fully, the school began operations in 1872 and one of the three instructors appointed was a W. H. Ellis in Chemistry. Political disagreements were obvious, however, and a new Act was passed in 1873 to establish a School of Practical Science for "instruction in mining, engineering, and the mechanical and manufacturing arts" (5). As one historian–engineer has stated recently "the period in which Canada came to nationhood coincides, therefore, with the realization, in many parts of the continent, that society was in urgent need of the benefits to be derived from organized instruction in technology and applied science" (6). By September, 1878 a new structure, thus named the School of Practical Science, had been built on the University campus. The professors and lecturers in science at University College were available to the students entering the new school, and among the departments formed to offer three-year diploma programs was one in Analytical and Applied Chemistry. The present Department of Chemical Engineering and Applied Chemistry was the direct descendant from that original entity. It has celebrated thusly its centennial in 1978, the basis for the claim, now advanced, of a century of chemical engineering education in Canada. In that first department, in the new School of Practical Science of 1878, H. H. Croft of University College (who had been Professor of Chemistry since 1843) was the first Professor of Chemistry, with Ellis as his assistant. Croft was followed two years later by W. H. Pike. An only too familiar type of academic struggle ensued. The outcome, in 1882, saw Pike remaining as University Professor and Head of the Department of Chemistry, but Ellis appointed Professor of Applied Chemistry and Head of the Department of Analytical and Applied Chemistry. By 1889 the School of Practical Science was affiliated formally with the University and in 1906 became the Faculty of Applied Science and Engineering. Meanwhile in 1893 an optional fourth year became available as a degree program, and after 1909 only the four-year programs were available. Not long after, the department's name was changed to its present format.

Although Ellis served the University for 48 years, the last five as Dean, his contributions to the initiation of chemical engineering education possibly never have been recognized adequately. It was under his guidance that a young demonstrator in Applied Chemistry, J. W. Bain, began his university work in 1899. It was Bain, again with Ellis' unequivocal support, who established what some have claimed (4) to be the first course in chemical engineering in a Canadian university as outlined in the calendar for 1904–05. Although the world's famous first program at M.I.T. had been established in 1888, there were fewer than half a dozen other programs in existence at the time of Toronto's first announcement in 1904.

Dr. Ellis (*see* Figure 1) was born in England, migrated to Canada, and graduated from the University of Toronto in Arts and Medicine. Forsaking medicine for chemistry, he became, "a man of multitudinous

Figure 1. The Dean at work. Dr. W. H. Ellis, Professor of Applied Chemistry 1882–1919, Dean of the Faculty of Applied Science and Engineering at the University of Toronto 1914–1919.

professional activities, Professor of Applied Chemistry in the Faculties of Applied Science, of Arts and of Forestry, and Professor of Toxicology and Medical Jurisprudence in the Faculty of Medicine, public analyst for the Federal Inland Revenue Department, medicolegal expert and one-time president of the Canadian section of the Society of Chemical Industry" *(4)*. (The Society which British conservatism had prevented from becoming the world's first society of Chemical Engineers.)

Professor J. Watson Bain, who carried responsibility for the new program from 1904, became Professor of Chemical Engineering in 1916 and retired in 1946. By the time Bain assumed full direction of chemical engineering and Ellis had moved on to become Dean, the numbers in the chemical engineering program considerably exceeded those in the original Analytical and Applied Chemistry, even though much of the instruction was common. The latter program was discontinued soon, and the department, with a fully integrated single program was led strongly by Bain. Graduating originally in Mining Engineering from the School of Practical Science, Bain also had become a prominent industrial chemical consultant, one of the founders of the Candian section of the Society of Chemical Industry, and, in 1930, the first recipient of its Canada Medal. He in turn was succeeded as Head of the Department of Chemical Engineering and Applied Chemistry by Dr. R. R. McLaughlin, who also obtained his degrees at the University of Toronto. He joined the department in 1931, and had the distinction of becoming the first President of the Chemical Institute of Canada when it was formed out of three original chemical societies in Canada.

The traditions of this department, established through Ellis, Bain, and McLaughlin, were related firmly to applied chemistry, industrial chemical plant, and an engineering environment. Through this department's priority in the field in Canada, the first 25 years of the century under discussion belongs to this pilot plant for chemical engineering education. The same design forces were at work elsewhere—chemical engineering in the universities was ready for expansion.

Development: Full-Scale Design

Development to today's scale, the end of the century, is seen best in two phases, almost equally divided in length (1905–1945 and 1945–1978).

While it has been claimed *(4)* that chronologically the second school of chemical engineering to be established in Canada was at Queen's University in Kingston, the evidence denies it. The 1902–1903 calendar for the School of Mining, affiliated with Queen's University, announces a four-year BS course in chemical engineering with the footnote that it "provides for the education of a class of engineers of growing importance" *(7)*. The University of Toronto was beaten by two years—although its first graduate in chemical engineering was a year earlier than the first at Queen's! The curriculum included two courses in industrial chemistry, and one in Chemical Works and Engineering. Dr. John Waddell was the only lecturer in chemistry in the school, and it depended greatly on the Department of Chemistry in the university. A reorganization of that department in 1909 led to the assignment of Dr. L. F. Goodwin as an Assistant Professor of Physical Chemistry and Industrial Chemistry *(4)*. Goodwin, with a PhD in Physical Chemistry from Heidelberg in 1903, had designed and operated nitric acid plants in England. With this experience and interest, he energetically led the organizing of a full chemical engineering curriculum by 1912, and by 1922 the formation of a separate department of which he became Head.

Essentially in parallel, programs were beginning at U.B.C., Alberta, and Saskatchewan. These were all relatively new universities, but had developed strong chemistry departments, all of which included one or more faculty members with interest in industrial chemistry. A unique aspect pertained, however, to each case. At U.B.C. the first few years of the program in chemical engineering were essentially unintegrated. However in 1921, W. F. Seyer (PhD, McGill) came from Alberta where he had been a Professor of Physical Chemistry. He spent a brief period of time at Columbia studying its chemical engineering program, and immediately adopted and organized a more integrated approach to a chemical engineering curriculum at U.B.C. under the administration of the Department of Chemistry. His interests maintained a strong physio-chemical flavor, with problem applications in a wide range of industrial chemical situations. Until his departure in 1949 to become the first Professor of Chemical Engineering at UCLA, he nonetheless provided the leadership for a distinctive chemical engineering program in the chemistry environment.

At the University of Alberta, Dr. O. J. Walker (BA, Saskatchewan; MA, Harvard; PhD, McGill) of the Department of Chemistry, supported and assisted in the establishment of a chemical engineering course in 1926 although its development was very slow. The first appointment of a lecturer trained in chemical engineering was G. H. Johnston in 1936, succeeded in 1940 by G. W. Govier (BASc, U.B.C.; MSc, Alta.; PhD, Mich.). Dr. E. H. Boomer (BSc, U.B.C.; PhD, McGill; PhD, Cambridge), who had joined the Chemistry Department in 1925, became Chairman of the Committee on Chemical Engineering in 1942, and a year later was appointed the first Professor of Chemical Engineering. General support for chemical engineering, in the area of applied chemistry and industrial chemistry, was given by Dr. J. W. Shipley who had come from the University of Manitoba in 1930 to be Head of the Department of Chemistry. By 1946 a separate Department of Chemical Engineering was created, headed by J. A. Taylor until 1948, and then by Dr. Govier.

At the University of Saskatchewan the key figure was again a Professor of Physical Chemistry, Dr. T. Thorvaldson (BA, Manitoba; PhD, Harvard) who gave forceful leadership as Head of the Chemistry Department. He developed a strong research emphasis in the department, with his own interests related to both practical and theoretical problems in cement. The problem of alkali water corrosion of concrete structures initially had been pressed on him by Dean C. J. Mackenzie of the College of Engineering (8). He became a world authority in this area, and it is not surprising that the first chemical engineering program in Saskatchewan in 1934 was entitled Chemical and Ceramic Engineering. At the University of Saskatchewan, the Chemical Engineering program grew, prospered, and, unique in Canada, has remained within the Department of Chemistry, although in 1964 the department adopted the joint title of Chemistry and Chemical Engineering. No Chemical Engineering program or department ever developed at the University of Manitoba, even though professors such as Shipley, and later A. N. Campbell, long Head of the Department of Chemistry, had a vigorous interest in applied chemistry.

Meanwhile in eastern Canada, McGill University had offered a course in chemical engineering as early as 1914 (4), although it has been claimed that the first degree was awarded in 1911. The Department of Chemical Engineering, however, was not organized until 1947 with Dr. J. B. Phillips (MSc, PhD, McGill) as Chairman. Through those intervening years a strong industrial chemistry influence, again within the Department of Chemistry, was evident. Noteworthy was the emphasis on the pulp and paper industry, exemplified by the establishment of a Professorship of Industrial and Cellulose Chemistry at McGill. The initial development of the Pulp and Paper Research Institute of Canada on the McGill campus provided a medium for both chemistry and chemical engineering professors, with their students, to be directly and jointly involved in industrially oriented research problems.

At L'école Polytechnique, founded in 1873 and affiliated to the Université de Montreal in 1920, a course in industrial chemistry can be traced to 1912 when Laval University was authorized to grant to Ecole Polytechnique graduates a "diploma d'ingenieur-chimiste" for an additional year of study. In 1942 a regular BASc program in industrial chemistry was established, although its change to a full chemical engineering program took place in what we are terming the second phase of development.

The Department of Chemistry at Laval University also provided the first chemical engineering program through the initiative and wisdom of the Head at that time, Dr. Paul–E. Gagnon. The first students were allowed to complete their course at McGill, but with the addition of other staff, degrees were granted by 1944. By the end of 1945 a separate department was established under Dr. M. A. Cholette (M.Eng., Laval; MS, DSc, M.I.T.)—a smooth and rapid transition from an initial course to departmental status. Table I presents a summary of this development phase. Measured by eight programs, just under a score of professors spread from Vancouver to Quebec City, the development was marked by the presence of an able, energetic, and committed group of department Heads as program directors. They are the small band to which chemical engineering education in Canada remains ever grateful.

The myriad impacts of World War II were felt as powerfully in the universities as elsewhere. Certainly in chemical engineering education, little opportunity for change existed beyond what has been noted. The hiatus, however, was but a prelude to the next major stage of expansion, which also cut away from the traditional curricular patterns as well as from the organizational ones typified by the firm connections to the chemistry

Table I. Chemical Engineering Program and Departments: First Phase 1878–1945

University	Program Established	Department Established	First Graduates Bachelors	First Graduates Masters	First Graduates Doctorates
Toronto[a]	1878	1878	(1903)	—	—
Toronto[b]	1904	1908	1905	1922	1933
Queen's	1902	1922	1906	1913	1966
Ecole Polytechnique[c]	1912	1958	(1930)	1940	1941
Ecole Polytechnique[d]	1942	1958	1943	—	—
Ecole Polytechnique[e]	1958	1958	1960	1962	1967
McGill	1914?	1947	1911	1938	1935
U.B.C.	1915	1954	1917	1920	1961
Alberta	1926	1946	1928	1945	1961
Saskatchewan	1934	—	1933	1935	1962
Laval	1940	1945	1944	1956	1963

[a] As the Department of Analytical and Applied Chemistry.
[b] As the Department of Chemical Engineering and Applied Chemistry.
[c] For "Diploma d'ingenieur-chimiste" granted by Laval.
[d] Program in industrial chemistry.
[e] Program in chemical engineering and Department of Chemical Engineering.

departments.   The factors marking these developments in chemical engineering education and leading to rapid growth in the 1950's and 1960's were complex.   Among the Canadian forcing functions were certainly the rush to university of the war veterans, many heading for engineering, the gradual acceptance in engineering schools that chemical engineering was effectively a major discipline area and their responsibility, and the overall curricular recognition that a body of knowledge separate from chemistry was inherent in that discipline area.   American developments in chemical engineering, albeit attenuated and delayed, had arrived in noticeable form.

The second phase of development beginning immediately after the war and continuing for almost two decades, saw over a dozen new programs in chemical engineering.   With one very special national exception, none were affiliated directly with or grew from chemistry departments.   All began within faculties of engineering and were given departmental status; thus unencumbered, all were free to initiate what they considered to be most current and appropriate in chemical engineering content.   Many were begun in newly established faculties of engineering and several within the newer post-war universities.

Chronologically, this new phase (*see* Table II) found its beginning at Nova Scotia Technical College with Dr. M. R. Foran (BSc, MSc, Saskatchewan; PhD, McGill) called to Head a new Department of Chemical Engineering in 1947. The University of Ottawa, with Dr. L. Madonna (BS, Iowa; PhD, Washington) appointed to head a new department in 1954, was followed soon by the University of Windsor with Dr. M. Adelman (BASc, MASc, PhD, Toronto) as its first Head.   At McMaster University, a new Faculty of Engineering began a Chemical Engineering Department under Dr. J. W. Hodgins (BASc, PhD, Toronto) who was also Dean, and soon to be succeeded by Dr. A. I. Johnson (BASc, Toronto; MS, Brooklyn; PhD, Toronto) who came from the University of Toronto.   At the University of New Brunswick, with the strong support of the Head of the Chemistry Department, Dr. F. J. Toole, a new department was established in 1960 with the writer privileged to have the opportunity to help build a new program and a new department in an engineering school of proud and lengthy tradition (the first lectures in engineering in Canada was offered there in 1854).   At the recently established University of Calgary, Dr. R. A. Ritter came from the University of Alberta to initiate a new program in 1962; and at the Université de Sherbrooke another francophone department was established as part of its new Faculty of Engineering in 1965. Meanwhile, growing directly from its long-established applied science venue, the program in industrial chemistry at Ecole Polytechnique developed into a strong Department of Chemical Engineering.

Table II. Chemical Engineering Program and Departments: Second Phase 1945–1978

University	Program Established	Department Established	First Graduates		
			Bachelors	Masters	Doctorates
Nova Scotia Technical College	1947	1947	1949	1958	1965
Ottawa	1954	1954	1956	1957	1963
Western	1956	—	1958	1966	1971
Windsor	1956	1956	1961	1962	1968
Waterloo	1958	1958	1962	1968	1968
McMaster	1958	1958	1961	1963	1966
R.M.C.	1958[a]	—	1962	—	—
U.N.B.	1960	1960	1962	1963	1966
Calgary	1967	1967	1969	1970	1971
Sherbrooke	1968	1972	1973	1971	1971
Lakehead	1972	—	1974	—	—

[a] Degree program from 1948 required one additional year elsewhere.

To this list must be added several special ventures. The University of Western Ontario had founded an integrated school of engineering and established certain disciplinary traditions without departmental organization. Chemical engineering was selected as one of those and by 1968 had been accredited as a program even though the administrative function was not formulated in the departmental style. The University of Waterloo, on its new campus, founded its outstanding Faculty of Engineering on the cooperative work–study principle (later adopted by the Université de Sherbrooke also). Its new Department of Chemical Engineering began in 1958, at first under Dr. T. L. Batke and then briefly under Dr. L. Bodner. Through the 1960's, it grew rapidly under the energetic leadership of Dr. D. S. Scott (B. Eng., Alberta; PhD, Illinois) who had come from U.B.C. Today it vies with Toronto as the largest department in Canada.

The one federally supported higher education institution with a full engineering program is the Royal Military College (R.M.C.) in Kingston. This college's chemical engineering program from its beginning in 1954 under J. W. Hodgins, while four years in length, required its graduates to complete their chemical engineering with one further year at other Canadian universities. In 1962 it began to offer the full program and—in this one unique post-war example—remains administratively with the chemistry department. A very special Canadian (rather Canadien) situation arose when a multicampus Université de Québec was instituted in the 1960's. From that development chemical engineering programs were promoted at Chicoutimi and at Trois Rivières, but today remain unaccredited. Indeed, the program at the latter campus has been put in abeyance. The final lines in the saga have been written at Lakehead University in Thunder Bay, Ontario, where a degree program has been instituted primarily for students who have received a three-year postsecondary diploma in chemical engineering technology or related technologies. Dr. R. Rosehart (B.Eng., PhD, Waterloo) was initially responsible for this program and its evolution to degree status.

Names have been sprinkled liberally in this recitation—not only to give due acknowledgment to those who assumed the leadership roles in this country, but because of our belief in the importance of chemical engineering genealogy and its influence on course, program, and educational philosophy. Space does not permit the addition of all who should be mentioned as one traces the development of this second phase. However, the leadership, both administrative and academic, of Epstein at U.B.C., Robinson at Alberta, Tollefson at Calgary, Shook at Saskatchewan, Rhodes at Waterloo, Graydon and Charles at Toronto, Hoffman at McMaster, Plewes and Clark at Queen's, Furter at R.M.C., Ratcliff and Douglas at McGill, Lu at Ottawa, Bergougnou at Western, Grenier and Cloutier at Laval, Corneille at Ecole Polytechnique, Coupal

at Sherbrooke, and Picot at U.N.B. is acknowledged clearly by their colleagues coast-to-coast. In many instances new chairmen and heads have succeeded ones that have been mentioned, and have made or will be making their contributions to what is truly a common and exciting enterprise. It is now an indigenous enterprise, well developed, securely founded, and bright with the promise that strong leadership has bestowed.

Among obvious indicators that lend themselves to graphical proof of the growth and development of the total chemical engineering education system in Canada are not only the number of programs and departments, but also the number of degrees granted and the total professorial strength. Figures 2, 3, 4, and 5, which also will be referred to below, thus illustrate these manifestations of the century of development.

Curriculum: Major Trends

A century of curriculum development in Canadian chemical engineering pleads for a depth of analysis well beyond the scope of this review. The American influence has dominated for almost three--quarters of that time so it is not surprising that Hougen's "seven decades" (9) can be invoked as a matching scale, usually with some time lag, for

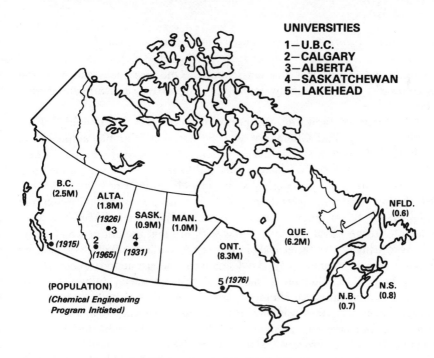

Figure 3. Chemical engineering degree programs I

6—WINDSOR (1956)
7—WESTERN (1956)
8—WATERLOO (1958)
9—McMASTER (1958)
10—TORONTO (1904)
11—QUEEN'S (1902)
12—R.M.C. (1958)
13—OTTAWA (1954)

14—McGILL (1914)
15—ECOLE POLYTECH. (1912)
16—SHERBROOKE (1968)
17—LAVAL (1940)
18—U.N.B. (1960)
19—N.S. TECH (1947)
(Chemical Engineering Program Established)

Figure 4. Chemical engineering degree programs II

changes and trends in Canadian schools.   One also seeks and finds the
benchmarks of a few highly definitive, progressive, and influential textbooks.

As Figure 6 indicates, the period from 1878 to 1920 could well be
termed the Applied Chemistry Era with its considerable emphasis on
analytical chemistry and industrial chemical processes.   F. R. Thorp's
Outlines of Industrial Chemistry first published in 1898 was in common
use, as were other monographs on specific industries.   As a further
interesting example, several editions and translations (both French and
English) of R. Wagner's *New Treatise of Industrial Chemistry* were
available in the Laval library.   Links with mining, metallurgy, and assaying were usually strong, at least in eastern Canadian universities, and the
dependence on teaching by the Chemistry Departments (except at Toronto) helped to maintain a major analytical chemistry influence.   The
transition from industrial chemistry was, in one direction, to material and
energy balances.   This was foreseen through Lewis and Radasch's *Industrial Stoichiometry* in 1926, and accelerated by Hougen and Watson's
famous *Chemical Process Principles* in 1945–1947.   That curriculum
element of material and energy balances remains entrenched.   As a 1972

AIChE survey *(10)* indicated, Canadian schools (16 respondents) were not noticeably different from their U.S. counterparts, except for a slightly higher average number of curricular hours, a higher tutorial effort, and more related laboratory.

The period 1920 to 1945, a "Transition Era" in many respects, encompassed the transition from industrial chemistry in the direction of chemical plant and hence unit operations. The concept of unit operations now has provided the crucial part of the structure of chemical engineering for almost half a century. As outlined so well by Brown *(11)* it was presaged first by Professor George Lunge's address at the World Exposition in Chicago in 1893, developed explicitly by A. D. Little in his famous 1915 report to the President of M.I.T., and given its full classical exposition by Walker and his colleagues in 1923 *(12)*. The impact in Canada was almost immediate and the move into the central part of chemical engineering curriculum came inexorably. The 1923 calendar for the University of Toronto mentions the equipping of a "laboratory for carrying on chemical operations on a small factory scale" *(13)*. At U.B.C., the Walker, Lewis, and McAdams' text was adopted immediately after its publication and continued, through succeeding editions, as the central text until 1949.

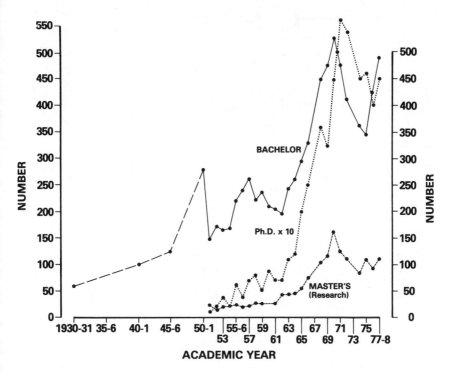

Figure 5. Chemical engineering degrees granted by Canadian universities

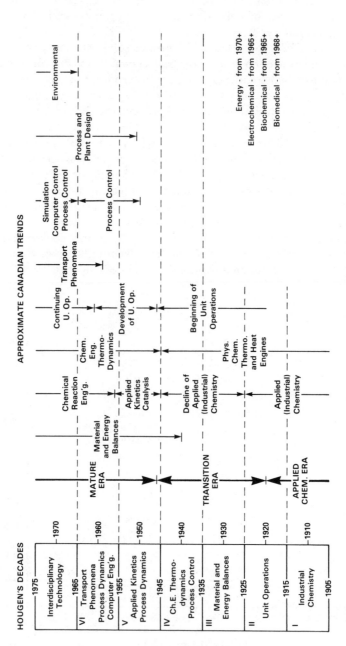

Figure 6. Approximate Canadian curriculum trends

Queen's likewise adopted it, but other universities were slower to follow suit. Lecture and laboratory courses involving unit operations, however, were established firmly at Toronto, McGill, and Queen's by the mid 1930's although it took almost another decade for their inclusion to be complete in other programs existing at that time. The state of development in the chemical industry, and certainly the apparent success of the industrial chemistry graduate combined to slow down the full acceptability of the new unit operations approach.

Process design has been a major component of many chemical engineering curricula for 40 years. In Canada, however, there is little trace of it prior to 1945, and hence it fits best into the description of the "Mature Era" covering the last 30+ years. With appropriate linkages to related professional development courses (e.g. technology, engineering economics, optimization and equipment design, etc.), normally sequential to unit operations, and thus broadly defined, process design has provided a significant emphasis in all Canadian programs. Again, there is wide variation among schools—but no wider than among American departments. A survey in 1965 *(14)*, which included 13 Canadian departments and 40 U.S. schools, verified the breadth of approach, but overall indicated a considerable continuation of the unit process or technology courses (reminiscent of industrial chemistry), and relatively little emphasis on economic analysis. The challenges to curriculum planners and teachers were outlined at the same time *(15)*. The rigorous accreditation approach in the United Kingdom resulting in an almost standard plant design requirement stands in strong distinction to the more laissez-faire attitude here. From 1970, however, the Canadian picture has been highlighted with our own national accreditation process, and a quinquennial examination of the actual design content of our programs. Since this was required to stand at 25% of the total program for some years (though now relaxed to a combined 50% in conjunction with engineering sciences) the response of departments led to a more concentrated effort on design and synthesis.

There has been no national effort to develop a unified approach such as provided for many years in the case study opportunity by the AIChE with its annual design problem competition. Apparently only two Canadian schools (U.B.C. and McGill) participated in those, and captured their share of awards—D. Cianci at U.B.C. standing first in 1956. It is fair to attribute, however, to one Canadian group, some of the praise due the innovators. The pioneering work of A. I. Johnson and his colleagues at McMaster in developing undergraduate plant simulation projects in cooperation with industry *(16)* has been an important educational tool in the process design area. It also led to the first textbook in this field *(17)*.

This mature period saw also the transposition of other major curriculum stems which had been bound previously in physical chemistry.

Chemical engineering thermodynamics, while certainly initiated by H. C. Weber's book in 1935, was launched in the 1940's with the texts by Dodge in 1944, then by Hougen and Watson, and finally by J. M. Smith. Canadian adaptation was rapid (and virtually complete by the early 1950's), much more so than in the area of applied kinetics where it took almost another decade and the successive and increasing impact of the texts by Hougen and Watson, Walas, J. M. Smith, and O. Levenspiel.

Just as these advances gave scientific strength to chemical reaction engineering, so too did the rise of studies in transport phenomena for the now-traditional unit operations. With the famous text by Bird, Stewart, and Lightfoot in 1960 signifying a new contribution, the unit operations curriculum stream broadened, partaking, to varying degrees in different schools in Canada, of the new input.

Beginning in 1957 there has been a series of analytical studies through questionnaire-type surveys of aspects of the undergraduate chemical engineering curriculum carried out by the Education Projects Committee of the AIChE. Since these studies have included the Canadian schools, it is possible to make internal comparisons as well as to see any significant differences that arise on a national basis.

Some already have been referred to earlier (10, 14). Others (18, 19, 20) dealing with chemical engineering thermodynamics, kinetics and reactor design, and with mass transfer, respectively, give evidence that there is no really distinctive difference in a group of Canadian schools of chemical engineering as compared with an American group as far as these subject areas are concerned. Some difference can be discerned, however, in the teaching of process dynamics and control where the 1975 survey indicated a higher percentage of Canadian curricula with compulsory or core courses in this area (21). Several Canadian departments moved early into courses in process control, U.B.C. requiring such a course as early as 1950 (22).

Curriculum: Laboratory and Other Elements

Historically in chemical engineering education there always has been great importance attached to laboratory work. A common claim is that the laboratory really has been the classroom where the student learned engineering. The objectives and consequently the nature of the laboratory programs have changed in discernible patterns through the century. Probably today's laboratory effects are more diffuse and designed to try and meet a wider range of objectives than ever before. In Canadian engineering schools the more evident sequences have no obvious unique features. The first decades concentrated almost entirely in descriptive and analytical chemistry (*see* Figure 7). Industrial chemical analysis with the dual emphasis on process understanding and analytical skills was the

Figure 7. University of Toronto laboratory ca. 1910

dominant part of the laboratory program up to the 1920's and was still significant until the 1940's in most curricula. The development of the unit operations concept accelerated the installation of plant-type laboratory facilities. At the University of Toronto a new laboratory combining both unit operations and some small-scale chemical reactions in industrial processes was instituted in 1923. By 1929 Queen's University had a wide range of unit operations equipment in teaching laboratories. Other than ancillary laboratories in other areas of chemistry and in other engineering subjects, it was noticeable in the interwar period that all of the programs gradually stressed more and more the inclusion of some laboratory exercises in unit operations but essentially for the cause of teaching practical engineering.

An increase in emphasis in instrumentation took hold beginning in the 1940's and led within a decade to the establishment of numerous experiments in process control in the 1950's. Industrial and analytical chemistry gradually faded from the scene, although certain aspects of analytical laboratory courses and industrial chemical analysis laboratory remained in many curricula until the 1950's. With the arrival of such textbooks as Zimmerman and Lavine (1955) and McCormack (1948), a broad pattern of unit operations experiments following established procedures was a common feature of all of the existing curricula.

The benchmark established in 1960 with the textbook on *Transport Phenomena*, supported by Crosby's laboratory text soon thereafter, caused an immediate move to add some of this more fundamental approach to unit operations experiments. It did not, however, completely sweep the board, and in a number of instances led only to the provision of a few additional experiments in a general chemical engineering laboratory program.

The last two decades have seen essentially two major changes, again, not unique in the Canadian scene. First, there has been a more extensive emphasis on process control and, more recently, computer control; and second, there has been a significant evolution towards a project-type laboratory orientation away from the single-session, carefully defined, straight-forward unit operations experiments. Further, there has been in the last two decades the increasing inclusion of laboratory equipment and experimental outlines directly designed to amplify chemical reactor design principles, although there is a paucity of chemical kinetics experiments (19).

A recent review by Woods and Patterson (23) on teaching via the undergraduate chemical engineering laboratory covered responses from over 50 institutions in the United States and Canada. They found a strong trend away from the concept of short (3-hr) experiments toward "teaching the process of experimentation" (23). They have noted also the addition of specialized areas, such as polymers, vacuum science, and ceramics. There was some increase in an emphasis on the introduction of detailed prelab work, including design of experiments and continuing concern for improvement of communication skills in reporting. The 14 Canadian schools responding showed a slightly more uniform total scheduled laboratory time than the U.S. institutions, but with a lower average time.

Some curriculum elements other than chemical engineering subjects overall have shown a decrease as chemical engineering as a discipline area has itself grown and developed. However the contrary occurs in the field of humanities and social sciences. Long neglected in all of engineering, such liberal studies underwent major revisions in the Faculty of Applied Science and Engineering at the University of Toronto in the 1940's thus sparking an increase in the required courses, and influencing other universities. This change was supported further by the Canadian engineering accreditation standards set about 1970 which required 12.5% of the total curriculum content in that area. The same figure holds for the Engineer's Council for Professional Development (ECPD) accreditation in the United States, and a 1976 survey verified the steady increase that had occurred from 1957 (24).

Increased mathematics, especially since 1960, and senior electives in specialized fields (environmental science, polymers, nuclear engineering, biochemical engineering, and biomedical engineering) are to be noted in both Canadian and U.S. department offerings. The areas sacrificed in turn have been in chemistry (particularly from 1940 to 1970) and a rather steady elimination of courses from other fields of engineering. The search for new approaches, in addition to design, that exhibits the integration of chemical engineering subject matter, has had some success in the recognition of problem solving as an academic area *(25)*. Overall for the Canadian schools founded prior to 1940, the Wisconsin pattern *(9)* is generally applicable in its analysis of courses added and eliminated over seven decades. For programs initiated in the last 30 years, the dynamic of curriculum development had less of a change to effect.

The Product Line

Classifying the granting of degrees at both the bachelor's and advanced levels as a production process is but an alternative way of indicating a university response to social, economic, and technological forces. For chemical engineering there were no graduates until after the turn of the century. From then until 1914 only Queen's, Toronto, and McGill had graduates, and although precise data have not been obtained, the total in those years would seem to be less than 50. The University of Toronto graduated, for example, one, three, and six in the years 1913 to 1915, respectively. By 1930 with three more programs underway, the annual total just exceeded 50. The immediate post-war years showed the expected influx of students. At Toronto the total undergraduate population in chemical engineering peaked to almost 300 in 1920, dropping to 75 by 1925 *(5)*. Meanwhile the first graduate in Western Canada received his degree from U.B.C. in 1917 (Dr. C. H. Wright, who also received U.B.C.'s first MSc degree in Chemical Engineering, and later had a distinguished career with Cominco), and apparently the first woman graduate in Canada, Rona Hatt Wallis, graduated in 1922, also from U.B.C. *(26)*. The Toronto data *(5)* for BASc degrees show an oscillatory growth to 1940, similar in nature to that for the total Canadian scene following World War II *(see* Figure 5). As perceived in a recent sophisticated analysis of production of bachelor's degrees in engineering *(27)*, such production curves indicate "two degrees of freedom—the inertia or stored kinetic energy effect of education delayed by military service, and the potential effect of increased demand for holders of engineering degrees" *(27)*.

For both bachelor's and advanced degrees there was roughly exponential growth from 1950 to 1970 followed by an approximately steady-state period to the present *(see* Figure 5). With the predominant provincial nature of higher education, a similar pattern exists in almost all of our institutions. Production of master's degrees, restricted to five universities prior to 1955, became widespread in the 1960's, although Toronto has continued to dominate in numerical terms. While PhD

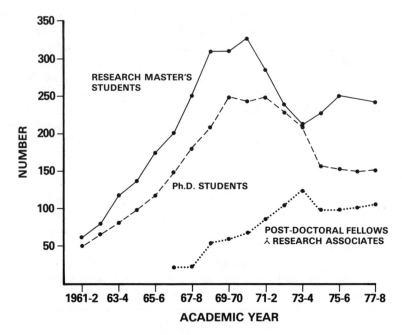

Figure 8. *Total number of full-time graduate students enrolled in chemical engineering programs*

degrees are offered now at 17 universities, it was Toronto that provided the majority of such recipients until 1960. In the doctorate area a few universities predominate—with Toronto, Waterloo, Western, and McGill accounting for at least half of such degrees in the last four years. Through a more limited time span, 1961 to 1978, Figure 8 shows the total number of students registered at each of the advanced degree levels. There is further evidence here of exponential growth from 1960 to 1970 but a distinct downturn after 1970. This is due in large measure to the research funding policies of the National Research Council which in 1969 forbade financial support from their grant funds to foreign students. The excellent employment situation for engineers also contributed to the lower numbers entering graduate study.

    Further analysis, not possible here, of the data base now available could well be carried out using for example, some of the innovative approaches suggested by Hartwig (27), as well as by more traditional methods (28). The current picture as we leave this century under discussion is an annual production of about 500 bachelor's level graduates, 100 at the master's level and 50 doctorates.

    In the sense of marketing the product, it is worthwhile examining the records for the last 20 years for one department, which have been published in an annual series of papers, (see e.g. Ref. 29). While neither definitive nationally nor readily extrapolated, such information gives an interesting and useful picture—and challenges other departments to assemble such historical records.   Table III give a summary of these data in five-year groupings.   The low percentage attracted to graduate study is a well-known phenomenon.   The increase in employment in the petroleum industry was to be expected, and the fairly steady move into pulp and paper bespeaks another regional resource emphasis.   The employment mobility picture is true of many other parts of Canada, but full analysis requires total input–output data *(28)*.   The chemical engineering educational role of provincial universities responds clearly to meet two types of societal demand: that of its secondary school graduates for professional education which is a local or provincial demand, and that of industry and government for highly qualified manpower which is a provincial and national demand.

Accompanying the product line above, and reflecting the research component of the total educational enterprise, are research papers, publications, and theses.   Figure 9 illustrates some of the related indicators, corresponding primarily to the last 20 to 25 years.   The rapid period of growth to about 1970, and the relatively steady period since, are exemplified both in the total number of papers presented to the annual Canadian Chemical Engineering Conferences, and to the number of papers from university sources.   The latter, with some specific exceptions, tend to be about 40% of the total, indicating the strength of support from the academic field to the professional society, the Canadian

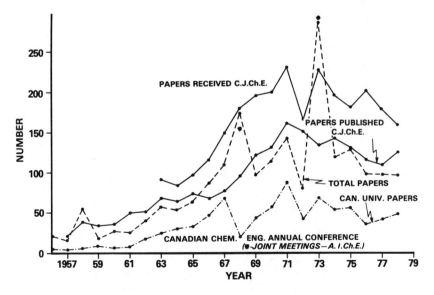

Figure 9. Some research activity indicators

Table III. Employment Profile: U.B.C. Graduating Classes

Years	1959–1963	1964–1968	1969–1973	1974–1978
Total Number	100	129	151	85
First Employment (percent)				
Industry	77	83	66	68
Graduate Study	5	6	3	9
Other	18	11	31	23
Type Industry (percent)				
Chemical	50	28	22	31
Pulp and Paper	21	22	31	25
Petroleum	15	36	26	33
Metallurgy and Chemical	7	11	13	4
Other	7	3	7	7
Location (percent)				
B.C.	36	42	52	51
Other Canada	60	37	39	49
U.S.A. and Other	4	21	9	0

Society for Chemical Engineering (CSChE) (a constituent Society of the Chemical Institute of Canada). The CSChE was derived from the former Chemical Engineering Division of the CIC. It was the latter which assumed, in 1957, responsibility for the *Canadian Journal of Chemical Engineering*. With total university contributions to its pages running at about 90%, there always has been a very major participation from Canadian chemical engineering schools. Its pages, as well as the annual conferences, have benefited also by the periodic joint meetings with the AIChE. The numerical trends, not unexpectedly, correspond closely to the advanced degree trends, and indeed to the numbers of professors.

On the qualitative side of the research output, the topical nature has followed the curriculum changes and emphases in chemical engineering, and, in recent years, the areas of national concern expressed by federal government research granting agencies. As a simple illustration of the change, there were noted in the 1925 University of Toronto Engineering Research Bulletin, four research papers in chemical engineering, all in applied chemistry. These ranged from a colorimetric determination of platinum to an analytical method of separating nickel and copper from iron. The department's research publication list for 1945 to 1949 included 17 papers of which 11 were in applied chemistry and two were in unit operations. By the 1960's the research topics under investigation by graduate students across the country *(30)* could be classified as follows:

Research Topic	*Percent of Graduate Students Doing Research*
Applied Chemistry	30%
Kinetics and Catalysis	8%
Chem. Eng. Thermodynamics	9%
Unit Operations	47%
Reactor Design	3%
Process Control	3%

Detailed analyses of more recent trends have been made *(31, 32, 33, 34)* using later versions of the directory source noted *(29)*. For the years 1971–1972 to 1974–1975 sharp increases in popularity occurred for the areas of interfacial phenomena and properties, process design, and particulate phenomena, while the area of thermodynamics decreased. A significant amount of activity, reflecting both environmental concerns and energy-related problems, was also evident *(34)* as in Table IV.

Appropriate as a further product is the category of monographs and textbooks. It is here that there is real paucity. Certainly in the area of undergraduate textbooks there are few to note beyond that of Crowe and his colleagues *(17)* and D. R. Woods' recent (1975) book on *Financial*

Table IV. Research Areas of Societal Concern and New Interest.

Category	Number of workers 1974–1975
Water treatment	47
Real-time computer applications	13
Solid waste treatment	12
Energy resource models	7
Synthetic fuels	9
Oil sands technology	11
Air pollution	19
Biochemical processes	18

Decision Making in the Process Industry.   Dr. L. A. Munro of Queen's authored *Chemistry in Engineering* in 1964 for nonchemical engineers, while C. E. Wales, R. A. Stager, and T. R. Long wrote *Guided Engineering Design* in 1974.   In the last few years, a number of specialized areas have been covered with such monographs as those by G. W. Govier and K. Aziz on *The Flow of Complex Mixtures in Pipes* (1972), K. B. Mathur and N. Epstein's on *Spouted Beds* (1974), *Heat and Concentration Waves: Analysis and Application* by G. A. Turner, and *Bubbles, Drops, and Particles* by R. Clift, J. R. Grace, and M. E. Weber (1978).   Dr. R. Luus has coauthored *Optimal Control of Engineering Processes* (1967), and E. Rhodes and D. S. Scott co-edited *Co-current Gas Liquid Flow* in 1969. R. B. Anderson has edited two volumes of *Experimental Methods in Catalytic Reseach* and K. F. O'Driscoll co-edited *Structure and Mechanism in Vinyl Polymers* in 1969.   In addition, of course, there has been Canadian authorship of chapters in the various Advances in Chemistry Series and of major review articles in such foreign publications as the *AIChE Journal.*

International Connections: Countercurrent Flow

The external sources for developing science and technology in a developing country were clearly set in the Canadian scene.   A century ago and continuing until almost 1914, there was a strong background influence of the great chemistry and applied chemistry centers in Europe and Great Britain.   The genealogy was clearly from chemistry.   By the 1920's the gradual shift to stronger connections with the growing American chemical engineering effort was apparent.   The European connection then became almost negligible, although it was not until the 1940's that the Toronto curriculum dropped a four-year requirement in German!

Perhaps the key relationships for this last half-century have been both a strong connection with the United Kingdom and an even stronger connection with the United States.   The genealogical form is shown clearly for the most recent decade in Figure 10 which indicates the origin

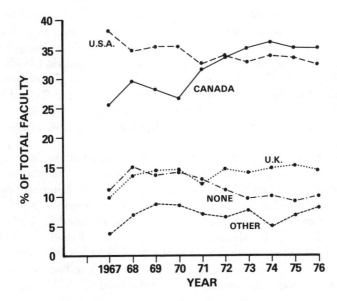

Figure 10. Origin of professorial PhDs

of Canadian chemical engineering professors in terms of their PhD studies. There is a very significant U.S. involvement, decreasing however, in percentage and greatly in numbers. The percentage of doctorates from the United Kingdom has increased from 10% to about 15%, which corresponds to a large numerical increase. In the last 15 years there has been a marked increase in the number of professors who have received their graduate training in Canadian institutions joining the Chemical Engineering Departments in Canada.

A truly significant factor in the sense of establishing the "British connection" was the unique Athlone Fellowship scheme. Over a period of 20 years (1951–1970) some of the most outstanding engineering graduates of Canadian universities were offered two-years fellowships for study or industrial experience in Great Britain. Of the 810 Athlone Fellows, 108 were chemical engineering graduates—and of these, 15 now hold university positions in Canada. The influence of this scheme on chemical engineering education has been a marked one. Perhaps it has been most effective in increasing exchanges, visits, and wider appreciation of knowledge of British chemical engineering, including the educational aspects. While Canadian chemical engineering graduate curricula follow more closely the American pattern of a significant prescribed course situation, the emphasis on comprehensive knowledge, self-paced study, and course freedom epitomized in the United Kingdom curricular developments has had its influence.

Some evidence of the interplay of these two channels of influence can be seen through major textbook adoptions that existed in Canadian chemical engineering departments. In the immediate post-war years widespread use of Brown's *Unit Operations* was displaced to only a very limited degree by the textbook by Foust, but in a number of institutions by considerable use of the British volumes by Coulson and Richardson. In the field of chemical engineering thermodynamics there also has been moderate use of Denbigh's text. In other areas, American textbooks predominate.

Connections are bilateral. It is not unimportant to indicate what well may be said to be an important influence on American chemical engineering education when one notes that among American chemical engineering educators of note who received their education to at least the first degree level in Canadian universities can be included T. K. Sherwood, J. G. Knudsen, A. B. Metzner, and T. W. Fraser Russell. Others now teaching in chemical engineering in the United States who received one or more of their university degrees in chemical engineering in Canada include K. E. Cox, A. Varma, H. Y. Sohn, and C. J. Huang.

The Canadian Identity

One of the most significant factors in development and maintenance of a strong chemical engineering identity in Canada, as centered around its chemical engineering educational establishment, has been the aspects of nonimperialism shown by professional institutions. Neither the British Institution of Chemical Engineers nor the AIChE made any move to establish sections or divisions in Canada. This is in marked contrast to their sister institutions in other branches in engineering—for example, mechanical and electrical engineering. It is true that in the case of the British connection the Society of Chemical Industry (SCI) had a Canadian section. It, however, played a supportive role and put its membership activities in abeyance in the forming of the Chemical Institute of Canada (CIC) in 1945.

The AIChE not only refused to establish operating sections in Canada, but only by special request did it permit the establishment of student chapters. The two primary ones that were established and existed for many years were at U.B.C. and McGill. It was also possible for such a student chapter to be an affiliated student chapter of the Chemical Engineering Division of the CIC and later of the Canadian Society of Chemical Engineering (CSChE) at the same time. It is undoubtedly true that the formation of the CSChE in 1966, which preceded the forming of similar Canadian societies in other engineering disciplines, by almost 10 years, was accelerated by this indirect type of support. Probably the most significant factor though was the continuing close cooperation of the chemical profession in Canada—both chemists and chemical engineers. The university relationships, with a strong tie between chemistry and chemical engineering through teaching as well as through research in applied chemistry, carried on through their graduates as co-workers in industry. The continuing close interdependence and mutual strengthening, in the face of vast geographic differences and sparse population, provides a heartening and indeed unique relationship—a true chemical bond.

In a major way, the Canadian identity too has been fostered by the CSChE role in international joint conferences. These have included the two joint meetings with the AIChE held in Canadian locations in 1958 and 1973, and the Tripartite Conference in Montreal in 1968 with the British Institution of Chemical Engineers and the AIChE. In these, as well as in InterAmerican Federation organization and meetings, Canadian chemical engineers—especially from the universities—have played an important role. W. J. M. Douglas of McGill in InterAmerican activities, and a large group of others in their participation in AIChE committees and national meetings all have helped to ensure a meaningful and mature Canadian presence.

The Canadian presence in international connections has been exemplified in the preceding section as well as directly above. It is an indissoluble part of the Canadian identity when international participation and contributions to other nations are made in such a sense of partnership.

One looks internally too to assert the substance of Canadian identity in chemical engineering as contributed from the educational wing. Several of the presidents of CSChE have been professors of chemical engineering, and at least two of the recent presidents of the CIC; two of the four editors that have spanned the 22 years of the *Canadian Journal of Chemical Engineering* came from academe. The R. S. Jane Memorial Lecture Award for exceptional achievement in the field of chemical engineering or industrial chemistry has been bestowed on such truly eminent chemical engineering professors as W. H. Gauvin, G. W. Govier, W. H. Rapson, and A. Cholette. The Erco Award initiated in 1970 for a distinguished contribution in the field of chemical engineering by a person under 40 has gone to T. W. Hoffman, M. Moo–Young, A. E. Hamielec, B. B. Pruden, and M. E. Charles. These awards reflect in turn the significant, identifiable, and often special contributions being made through the university groups in chemical engineering—contributions marked high on any international scale. Some of the departmental specialities which thus stand out include spouted beds at U.B.C., applied thermodynamics at Alberta, computer control studies at McMaster and Alberta, environmental studies at Calgary, pipeline flow studies at Saskatchewan, pulp and paper studies at Toronto (through the world-famous work of Rapson), fluidization at Western, polymers at Mc-Master and McGill, reactor dynamics at Laval, process innovation at Waterloo, and a score of others. Open to the world of chemical engineering through a myriad of associations and the many facets of communication, possessing programs in chemical engineering in two world languages—French and English, and seen in part through a journal that accepts articles in both languages, Canadian chemical engineering education has indeed a maturity and an identity.

The last few words rest with one of Canada's most remarkable chemical engineers, one whose life span, and indeed career, covers most of the century under discussion. John S. Bates, after receiving a chemistry degree at Acadia University, decided in 1909 on a chemical engineering career, proceeded to Columbia University and received a Chemical Engineering degree in 1913 and the first PhD in Chemical Engineering in 1914. Returning to Canada and to over 50 years of superlative professional activity, he was indubitably Canada's first chemical engineer practicing with that advanced degree. Beginning his studies when, as he says, "chemical engineering was only about five years old in a few universities . . . . Today it is hard to realize how youthful the chemical industry was in North America during the second decade of this century" (35). He then goes on to talk to the "unquenchable spirit of motivation, determination, and calculated risk" (35) of the early chemical engineers and chemists working together at the Shawinigan plant—the forerunner of the world-famous Shawinigan Chemical Company. In tracing such movements as that of professional associations, including the CIC, he concludes that chemical engineers "over the period of some 60 years, deserve applause for dedication of their members, continuation of progress, all-out cooperation, and major accomplishments" (35). These wise observations in his 91st year lead Bates to challenge us all to "apply ourselves with the same vim and vigor" (35). Canadian chemical engineering education through its first century has indeed shown dedication, progress, cooperation, and accomplishment through its professors and its products.

Acknowledgments

The author is indebted greatly to the *History of Chemistry in Canada* by Warrington and Nichols which, while only occasionally quoted and referenced, was the basis for much of the factual information. He is also most grateful to Dr. D. R. Woods of McMaster University and Dr. L. E. Jones of Toronto for source material and helpful discussions. A host of colleagues from all of the chemical engineering departments in Canada were most helpful in providing data and general information. The opinion expressed are, of course, the author's sole responsibility, as are the errors of omission and commission which must occur in a review covering such space and time.

Literature Cited

1. Program, Sixth Annual Dinner, Industrial Chemical Club, University of Toronto, 1914 (Archives, Department of Chemical Engineering and Applied Chemistry, University of Toronto).
2. "A History of Science in Canada", Tory, H. M., Ed.; The Ryerson Press: Toronto, 1939.
3. Habashi, F. "Chemistry and Metallurgy in New France," *Chem. Can.* **1975**, 27 (5), 25–27.
4. Warrington, C.J.S.; Nicholls, R.V.V. "A History of Chemistry in Canada"; Sir Isaac Pitman & Sons (Canada) Ltd.: Toronto, 1949; p. 502.
5. Harris, R. S.; Montagnes, I. "Cold Iron and Lady Godiva—Engineering Education at Toronto 1920–1972"; University of Toronto Press: Toronto, 1973.
6. Jones, L. E. "University of Toronto Engineers Celebrate Their Centenary," *Eng. Digest (Toronto)* **1973**, 19 (4), 35.
7. "Calendar of the School of Mining, Kingston, Ontario, 1902–1903", British Whig Office: Kingston, 1902.
8. "The Department of Chemistry, University of Saskatchewan, 1909-1959"; University of Saskatchewan: Saskatoon, July, 1959.
9. Hougen, O.A. "Seven Decades of Chemical Engineering," *Chem. Eng. Prog.* **1977**, 73 (1), 89–104.
10. Woods, D. R.; Bennett, G. F.; Howard, G. M.; Gluckman, M. J. "Teaching Undergraduate Mass and Energy Balances—1972," *Chem. Eng. Education* **1974**, Spring, 82–88.
11. Brown, G. G.; "Unit Operations"; Wiley: New York, (1950); p. 611.
12. Walker, W. H.; Lewis, W. K.; McAdams, W. H. "Principles of Chemical Engineering"; McGraw-Hill: New York, 1923.
13. Woods, D. R. "Teaching Process Design: A Survey of Approaches Taken," Faculty of Engineering Report #22, McMaster University, October, 1965.
14. "Calendar: Faculty of Applied Science and Engineering, University of Toronto, 1923–1924", University of Toronto: Toronto, 1923.
15. Woods, D. R. "Teaching Process Design in Universities," *Chem. Can.* **1966**, 18, 17–19.
16. Baird, M. H. I. "Things are Humming at McMaster," *Chem. Eng. Education* **1970**, 4 (3), 112–116.
17. Crowe, C. M.; Hoffman, T. W.; Johnson, A. I.; Shannon, P. T.; Woods, D. R. "Chemical Plant Simulation"; Prentice-Hall: 1971.
18. Eisen, E. O. "Teaching of Undergraduate Thermodynamics," presented at the 68th AIChE. Annual Meeting, Philadelphia, PA, Nov. 13, 1973.
19. Eisen, E. O. "Teaching of Undergraduate Kinetics," presented at the 69th AIChE Annual Meeting, Washington, DC. December 4, 1974.
20. Eisen, E. O. "Teaching of Undergraduate Mass Transfer," presented at the 71st AIChE Annual Meeting, Miami Beach, FL, Nov. 16, 1978.
21. Seborg, D. E. "A Survey of Process Control Education in the United States and Canada," presented at the 71st AIChE. Annual Meeting, Miami Beach, FL, Nov. 12–16, 1978.
22. Shemilt, L. W. "Instrumentation as a Course in Chemical Engineering," *Chem. Can.* 3, 195.
23. Woods, D. R.; Patterson, I. "Teaching Via the Undergraduate Laboratory: 1978," *Chem. Eng. Education*, in press.
24. Barker, D. H. "Undergraduate Curricula 1976," *Chem. Eng. Education* **1977**, 60–63, 96.
25. Woods, D. R.; Wright, J. D.; Hoffman, T. W.; Swartman, R. K.; Doig, I. D. "Teaching Problem Solving Skills," *Ann. Eng. Education* **1975**, 1 (1), 278.

26. Cavers, S. D., personal communication.
27. Hartwig, W. H. "An Historical Analysis of Engineering College Research and Degree Programs as Dynamic Systems," *Proc. IEEE*, **1978**, *66* (8), 829–837.
28. Shemilt, L. W. "Research Report on Engineering Education in the Maritimes," Maritime Provinces Higher Education Commission, Fredericton, N. B., 1976, 360.
29. Cavers, S. D. "Employment of Graduating Chemical Engineers—1978," *Chem. Can.* **1979**, *31* (1), 27.
30. "Directory of Chemical Engineering Research in Canadian Universities," 5th ed.; Chem. Inst. of Canada: Ottawa, 1965–6.
31. Burns, C. M., "Postdoctoral Research Grows in Chemical Engineering," *Chem. Can.* **1972**, *24* (9), 16–18.
32. Grace, J. R. "Graduate Programs Shrink in Chemical Engineering," *Chem. Can.* **1973**, *25* (7), 12–14.
33. Grace, J. R. "Increased Social Motivation in Chemical Engineering Research," *Chem. Can.* **1974**, *26* (7), 18–20.
34. Seborg, D. D. "Increased Graduate Student Enrollments in Chemical Engineering," *Chem. Can.* **1975**, *27* (11), 25–27.
35. Bates, John S., personal communication, Feb. 1979.

RECEIVED May 7, 1979.

11

The Association Aspects of Chemical Engineering in Canada

T. H. G. MICHAEL,

The Chemical Institute of Canada, 151 Slater Street, Suite 906, Ottawa, Ontario, Canada K1P 5H3

This chapter outlines the development of the technical, educational, and professional organizations that have culminated in the present associations. The parallel development of bodies regulating the practice of chemical engineering in Canada is sketched. Similarities to and more significantly the differences from the structures and practices in other countries are pointed out. References are made to some of the more significant personalities associated with these developments.

The history of the development of chemical engineering organizations in Canada follows two separate if interrelated paths. These relate to the so-called professional organizations, and the so-called technical or scientific ones. For lack of more precise definitions, we may consider the technical societies as those that are concerned primarily with the transfer of technical and scientific knowledge and information. The professional associations are concerned with the status of the practitioner, his relation to government and the public, and his qualifications to practice.

The same distinction is found in many of the professions, and is not peculiar to engineering or chemical engineering.

The following discussion is perhaps of primary interest to students of the field in countries other than Canada, since most of the general patterns are known to Canadian chemical engineers, and to science historians. However, it is easy to overlook even recent history, and some of the points mentioned may be of more general interest.

Chemical engineering is of course one of the subdisciplines of engineering. As such, it is subject to any regulations governing the engineering profession that may be in force. In Canada the situation is somewhat complex.

0-8412-0512-4/80/33-190-199$05.00/1

The Canadian constitution is known as the British North America Act, passed by the British parliament. It took effect in 1867. One of the main features of this act was that it assigned responsibility for certain areas to the federal government and others to the provincial governments, of which there are now ten. An important area assigned to the provinces was labor relations. This has been interpreted to include responsibility for the regulation and governance of the professions.

As a result of this, each of the ten provinces has passed legislation providing for the regulation of the engineering profession. Generally these acts establish an Association of Professional Engineers, essentially self-governing, and requiring that in order to practice engineering, an individual must be a member of the association. The earliest of these associations, that in New Brunswick, was formed in 1920. The majority of the governing bodies of these associations are elected by the membership, and a minority are appointed by the provincial government.

Certain exemptions to the membership requirement are provided. The most important of these is that persons teaching engineering are not required to be members. But in general, any person who is practicing engineering, who calls himself or herself an engineer, or who signs engineering drawings or reports must be a member of the association.

In la province de Québec, the governing association is known as L'Ordre des Ingénieurs du Québec. It is one of 24 professions that are subject to the Québec Professional Code, and is likewise a member of the Interprofessional Council. It is noteworthy that apart from certain "grandfather" and transient provisions, L'Ordre des Ingénieurs requires proficiency in the French language as a requirement for its members.

The provincial associations and L'Ordre are all members of the Canadian Council of Professional Engineers. This was formed to provide a mechanism for consultation between the provincial organizations, and to enable joint action when necessary. One such important joint action was the formation of the Canadian Accreditation Board. This board establishes and maintains the academic levels required for membership in the provincial bodies. It has been able to maintain uniform standards across Canada and thus has ensured that acceptable graduates from any Canadian university engineering faculty may to able to join any provincial governing body.

The provincial bodies that have been discussed are concerned with professional matters, the control of the profession, and educational requirements insofar as they affect entry into the profession. They are not concerned with the continuing transfer of technical knowledge and information. This is traditionally the function of the technical, scientific, learned, or publishing societies.

In Canada, the Engineering Institute of Canada was founded in 1887 to perform these functions for engineering. Initially formed by civil engineers, its outlook and emphasis were primarily directed to serving civil and mechanical engineers. The engineering institute made attempts to provide technical programming for the emerging subdisciplines of the engineering field. However, in Canada the needs of the practitioners in these fields were met in two ways. The first was by American and British organizations forming Canadian or local chapters. Conspicuously successful in this way were the Institution of Mechanical Engineers and the American Institute of Electrical Engineers, one of the forerunners of the Institute of Electrical and Electronics Engineers (IEEE).

The second way was the formation of indigenous Canadian specialized engineering organizations, often related to resource industries. Perhaps the two most successful of these are the Canadian Institute of Mining and Metallurgy (CIMM) and the Technical Section of the Canadian Pulp and Paper Association (CPPA). Both of these attract a considerable membership of chemical engineers who are working in their respective fields. The CIMM founded in 1898, originally was concerned with the technical and related aspects of the mining industry. Because of the nature of the exploited Canadian geology, it primarily was interested in the Laurentian Shield surrounding Hudson's Bay, and also in the mining conditions and problems in the Western mountain regions. More recently, the extraction of petroleum, principally in Alberta, has led to a great deal of interest in the engineering aspects of well drilling, pipelining, and refining. The CIMM formed a subsidiary organization, the Petroleum Society, to concentrate on this field. Naturally it includes many chemical engineers among its members, and provides programming services in these subjects, but particularly in Alberta.

CPPA is primarily a trade association of that part of the forest products industry producing paper and its intermediates. This is an industry of great concern and value to Canada. Tremendous stands of timber are found in most provinces, particularly in British Columbia, Ontario, Quebec, and three of the four Atlantic provinces. The pulping process and the formation of paper are chemical engineering operations. Probably because of the lack of a suitable medium for the exchange of technical information in its field in Canada, the CPPA formed the Technical Section in 1915. It has maintained an active information transfer program ever since.

Chemistry and chemical engineering always have been closely related, if not inseparable. The first chemical engineers, although not described in that way, were chemists who found themselves in charge of plant processes and who had to obtain satisfactory process equipment, and manage the reactions to be efficient and economical. Gradually specialized education was developed to meet the need, and the chemical engineer was born. As we have seen from other chapters in this volume, chemical engineering came to be a recognized subdiscipline of engineering.

In the late years of the 19th Century, Canadian universities commenced giving courses in applied chemistry, as contrasted to "pure" chemistry. These gradually evolved into the chemical engineering courses that we know today. In an attempt to meet the needs of the practitioners in this budding field, the Engineering Institute soon set up a Chemical Engineering Division, but by and large its activities were lost among the Institute's other areas of activity.

The Society of Chemical Industry (SCI) is an international organization, based in Britain, and devoted to furthering all aspects of the chemical industry. The Society encouraged the formation of overseas sections, and the first one in Canada was established in 1902 in Toronto. This was followed rapidly by the establishment of sections or branches in Montreal and Ottawa, and in Vancouver in 1917. These organizations held their own separate meetings until 1918, at which time a joint meeting was held in Ottawa, being the first Canadian Chemical Conference. These sections provided opportunities for chemical engineers and others to exchange information and ideas, and made the publications of the SCI available to their members.

In 1921 the Canadian Institute of Chemistry (CIC) was incorporated, primarily to meet the professional and scientific needs of chemists. However, from the earliest days many chemical engineers took an active part in it, and its programming reflected this.

By the mid 1940's the SCI sections and the CIC, together with a more amorphous body known as the Canadian Chemical Association, were questioning the need for three independent organizations in the field, and eyeing the probable increased effectiveness and efficiency and the influence of a single larger body. The result of mature consideration and negotiation was the formation of the Chemical Institute of Canada in 1946. This Institute was designed, and enjoined in its Charter to further the professions and the sciences of chemistry and chemical engineering.

It is interesting to note that although the three pre-existing organizations amalgamated their activities into the new Institute, the Society of Chemical Industry in Canada coalesced to form one Canadian Section, in order to maintain and continue its awards program. This it has continued to do to this day.

One of the first actions of the newly formed Institute was to establish the Chemical Engineering Division. This was the beginning of the present organizational activity of chemical engineers as such in Canada. The Division from the first showed great activity. Under the dynamic leadership in its early years of such chemical engineers as Lyle Streight, Adolf Monsaroff, R. R. McLaughlin, I. R. McHaffie, G. W. Govier, and J. D. Leslie, it established an annual Canadian Chemical Engineering Conference. The first, held in 1951, attracted a substantial attendance and set the stage for a continuing successful series. The most recent, the 33rd, was held in Sarnia in 1979. At intervals along the way, the 18th incorporated the well-remembered Tripartite Conference in Montreal in 1968, and the 23rd was the equally well remembered Joint Conference with AIChE in Vancouver in 1973.

It soon was realized that a technical–scientific journal was a requirement for a growing society, in a period of rapid expansion of research, development, and production. The new Institute already had set up a news journal, *Chemistry in Canada*, in 1949, but a specialized chemical engineering journal was required. The result was the establishing of the *Canadian Journal of Chemical Engineering* in 1957. It continued in this new style, the former *Canadian Journal of Technology*, a specialized part of the Canadian Journals of Research series, published by the National Research Council of Canada. While facing the problems common to all such journals the CJChE is prospering, and is producing Volume 57 in 1979. Since its establishment in 1957 the editorship has been held by four eminent Canadian chemical engineers—W. M. Campbell, Albert Cholette, G. L. Osberg, and L. W. Shemilt.

The Chemical Engineering Division did not only look toward meeting the needs of its own members, but it also looked outward. The CIC, on behalf of the Division, was one of the early supporters, and took a very active part in the formation of the Interamerican Chemical Engineering Confederation. This organization is major attempt to organize and meet the needs of chemical engineers in the Latin American countries. Along with the AIChE, the Canadian organization has contributed greatly to the organization, continuity and financial stability of the ICEC since 1962.

It gradually became evident that the status of a division within the CIC did not represent properly the needs, membership, and activities of the chemical engineers. After careful planning, a major evolutionary step was taken in 1966 with establishment of the Canadian Society for Chemical Engineering (CSChE) as a constituent society of the CIC. As a constituent society, it has almost complete autonomy and freedom of action, and at the same time access to all of the programs, activities, and services of the Institute as a whole. The CSChE is represented on the governing bodies of the CIC. As a result of the changes in structure effected in 1978, that representation and influence have been increased significantly.

No apology is needed to conclude this chapter with some thoughts as to possible trends or patterns in the future. The pressures for public accountability will force the provincial professional governing bodies to consider the public interest as increasingly important, compared with the interests of the individual. The growing doctrine of individual accountability for all actions, as opposed to the tradition of corporate accountability, will cause the individual chemical engineer to examine every action that he takes in the light of this doctrine. The growing concern over the environmental effect and impact of chemical operations will cause the practicing chemical engineer to consider these effects in even greater depth. The probable result of these factors will be a growing dependence by the chemical engineers on their technical and professional organizations.

RECEIVED January 28, 1980.

The History of Chemical Engineering in Italy

GIANNI ASTARITA

Istituto di Principi di Ingegneria Chimica, Università di Napoli, 80125 Naples, Italy

Three historical epochs are considered: from antiquity until the birth of Italy as a nation in 1860; from 1860 until 1950, when Italy became an industrial society in the modern sense; and from 1950 up to the present time. A concise survey of chemical technology and of engineering education is given for the first period, while parallel sections are dedicated to industrial and academic development for the next two. The historical account reveals the very strong interference of industrial and academic developments, with closely parallel times of growth and stagnation. Political and cultural influences are discussed.

W hile great works of engineering were realized by civilizations prior to the Romans, theirs was the first one in the history of mankind whose very existence was based on engineering. The centralized government of the Roman Empire was possible because of the extraordinarily perfect system of roads, many of which are still today main thoroughfares of European traffic; the network of aqueducts made survival of large towns, first of all Rome itself, possible; the conquest of the Empire was made possible by the naval and military engineering developed by the Romans. Yet they were not chemical engineers, and did not contribute anything significant to chemical technology. (Pliny's *Natural History* makes this quite clear. Chemical technology in Roman times is discussed thoroughly in Ref. 1).

Whatever chemistry was known in Roman times, in the Middle Ages it went into the blind alley of alchemy, which was universally accepted as a scientific theory if occasionally opposed on moral grounds ("Me per l'alchimia che nel mondo usai/danno' Minos a cui fallar non lece" *(2)*). Yet Dante seems to have had a better understanding of elementary physical

chemistry than justified by only reading Aristoteles: "Ben sai come nell'
aere si raccoglie/quell' umido vapor che in acqua riede/tosto che sale
dove il freddo il coglie" *(3)*). Technology in the Middle Ages was basically the same as in Roman times, if not regressed *(4)*. As in most fields
of human endeavor, the origin of the contemporary culture must be
traced to the Italian Renaissance.

In 1505, Bartolomeo della Valle, Giovanni Pietro Bassi, Lazzaro de'
Pallazzi, and Maffeo de' Glussiani established the "Statuti et Regole per
l'Ingegneri et Agrimensori del Ducato di Milano", the first known by-
laws of a professional engineering society. Between 1537 and 1551, the
mathematician Niccolò Tartaglia wrote treatises on ballistics, which may
be regarded as the first scientific works on military engineering.
(Archimedes' great feats of military engineering, vividly described by
Plutarch in the Life of Marcellus, are not discussed in any of his great
scientific works. Tartaglia is, of course, best known for his "triangle,"
the simple algorithm for calculating the coefficients of the nth power of a
+ b). Leonardo's contributions to engineering are too well known to be
discussed. (The relevance of ff. 3v and 48v of Ms. A, and of the back-
cover of Ms. F, for scale-up theory have been discussed in Ref. 5 ff. 34v
and 44v of Ms. F suggest the use of tracer particles for determining the
kinematics of a flow field. I did not conduct a thorough research of
Leonardo's works, and was unable to trace any other element which
might be related to chemical engineering.) What is less generally
known, however, is that the Italian Renaissance produced the first known
treatise on chemical engineering with the posthumous publication in 1540
of the *Pirotechnia* of Biringuccio (*see* Figures 1 and 2). Georg Bauer of
Joachimsthal published his book in Latin and even changed his name to
the Latin form of Agricola; and the *De Re Metallica* is popularly, but
erroneously referred to as the oldest book on chemical engineering,
although in fact it was published 16 years after the *Pirotechnia*. Incidentally, the very fact of writing in Latin makes Agricola much more of a
Middle Ages man than his predecessor Biringuccio, who wrote in
"Volgare," the language of the "Volgo," or common folk. (The first
edition of Vannuccio (or Vanoccio) Biringuccio's *Pirotechnia* was published in Venice in 1540 (*see* Figure 1). Three more Venetian editions
are dated 1550, 1559, and 1678. I have consulted my own copy of the
1678 Bologna edition published by Gioseffo Longo (*see* Figure
2). There are four French editions (1556, 1572, 1627, 1859) and a Latin
edition published in Köln in 1658. Of the modern editions, a partial
Italian one was published in Bari in 1914; a German one in Brunswick in
1925; and an English translation by C. S. Smith and M. Gnudi was
published in New York in 1943. As for the *De Re Metallica* by Georg
Bauer (Agricola), it is curious that a German translation had to wait, as far
as I was able to ascertain, until 1929 when it was published in Berlin. An

Figure 1. Front page of the first edition of the Pirotechnia

English translation by H. C. Hoover and L. H. Hoover was published in New York in 1940. Agricola studied in Italy at the Universities of Padova and Bologna.)

Reading the *Pirotechnia* is a fascinating experience for a chemical engineer. Mining, melting, and refining metals, separation processes, alloys, foundry technology, distillation, and production of ceramics are the subjects of the first nine chapters. The tenth one deals with explosives, and gives the name to the whole book. Biringuccio chooses to conclude as follows: "I don't want my writing to end in a tragical tone; so that I have decided to also talk to you (after certain fires made of impetuous and horrible matters, which give great and hurting fear to men) of others which are made for gaiety and enjoyment" (6). And he proceeds by first describing the technology of firecrackers, and finally talking of that fire which burns completely and leaves no ashes, the fire of love. The style is colloquial, but the substance is serious; in regard to

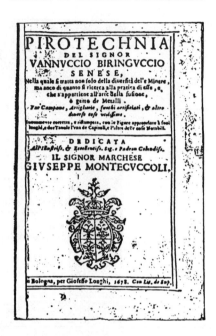

Figure 2. Front page of the 1678 edition of the Pirotechnia. *Quotes in the text are from this edition.*

alchemy, Biringuccio states that "even if it were true. . . . I must not and want not to consider it at all" *(6)*. He was, after all, writing of real, existing chemical technology; his honesty of mind and concreteness of beliefs could be imitated profitably by any contemporary engineer. (The translations of Biringuccio's quotes are my own ones.)

Of course, chemical engineering in a modern sense cannot conceivably develop unless large manufacturing industries come into existence. The Renaissance culture spread all over Europe in the 16th and 17th Century, and most everywhere the seed was planted in the fertile soil of emerging economies of national size. In contrast with this, Italy was divided, up to 1860, in a myriad of small independent nations, the economy of each one having too small a base to make industrial development a promising venture. Consequently, whatever technology developed sporadically in Italy, it did not reach the industrial stage. (An illuminating case history is told in curiously uncritical style in Ref. *(7)*. During the Middle Ages, use of alum almost had disappeared in Europe, though it was still mined, refined, and used in the East. In the 15th Century, the Italian Giovanni di Castro, after living some time in Constantinople, found some mines near Rome, and in 1462 he was granted the rights of mining by Pope Pius II. Roman alum was considered the best in the world down to the end of the 19th Century, and its exports to all of Europe were a major source of income for the Papal State. Yet the technology did not improve over the centuries, and by the end of the 18th Century the French had developed such a good technology for the production of artificial alum that their imports from Rome dwindled from 200,000 francs' worth in 1803 to 90,000 in 1805.) On the scientific side, the great development of the eminently empirical subject of chemistry in the period 1770–1900 had only a minor contribution from Italy, due to the work of Avogadro who in fact was culturally a Frenchman (8).

Turning attention back to earlier times, the academic development is also of interest. The very word "engineer" probably dates back to an Italian document of the 12th Century, a deed stipulated in Genova on April 19, 1195, which refers to a "Rainoldus encignerius." Yet engineering was not regarded as an academic subject for a long time after the first universities came into existence. The first Italian universities formed, between the end of the 12th and the beginning of the 13th Century, by spontaneous clustering of scholars, with the exception of the University of Naples which was founded in 1224 by the Emperor Frederick II, who wanted a school capable of producing the clerks required for the administration of the Empire. (I am here not considering the Medical School in Salerno which was already in operation in the 10th Century. It must have been an exceptionally progressive school, since Moslem, Jewish, and even female students were eligible for admission. An academic institution, later to develop into a university, is recorded in Pavia as early as 825 A.D.) With that notable èxception, universities were not, at the time, supposed to produce professionals useful in the surrounding society, but only academic savants; with such a conception, engineering was excluded by definition. Incidentally, such a conception is not as antiquated as it may appear at first sight; as recently as 1973, it has been claimed (9) that producing academic savants is the role of any university, with any professional ability acquired by graduates being an incidental and largely irrelevant by-product.

The first historical example of an engineering school of academic level is possibly the establishment by Francesco d'Este, in 1690, of the Academy of Military Architecture in Modena (10). (According to Masoni (11), the early engineering schools were: the Royal Military Academy of England, 1741 (i.e., 51 years later than the Modena Academy); Collegium Carolinum of Brunswick, 1745; the French School of Drawing established in 1747 by Louis XV (later to become, in 1760, the famous Ecole de Ponts et Chaussées); a school for engineers established by the Austro-Hungarian Empire in Milan in 1795.) In 1800, during the Napoleonic domination of the region of Lombardia, it was established that a university degree was required in order to become an engineer, and the University of Pavia's Physicomathematical Faculty began holding courses for engineers. The first school of engineering in Italy was established in Naples in 1811 (12). (See also Figure 3. Since the man who theorized the repetitiousness of history, Giovanbattista Vico, was a Neapolitan, it is

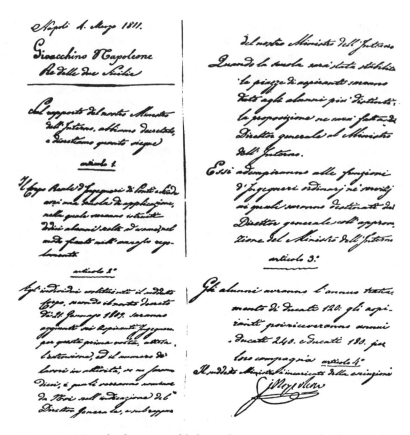

Figure 3. Murat's decree establishing the engineering school in Naples

perhaps appropriate that, 587 years after the University of Naples had been founded by Emperor Frederick the II, the Engineering School of Naples also was founded within the scope of an Imperial policy.)

During the short (1805–1815) Napoleonic domination of the Kingdom of Naples, Napoleon's brother-in-law, Murat, who had been made King of Naples, established the Scuola di Applicazione del Corpo Reale di Ponti e Strade, which was modeled after the French Ecole de Ponts et Chaussées (see Figure 3). The school was willing, in its first year of operation, to accept up to 12 students (see Figure 4). (Today, about 2,000 students per year enroll in the Engineering School of Naples. Which one of the two figures is more reasonable is an interesting question.) Murat was less lucky than his brother-in-law when the restoration came, and he was executed in 1815. The School of Engineering was closed temporarily, but started operations again in 1819 (13).

While chemistry had been taught in Murat's school (see Figure 5), it was not taught after 1819. Only in 1835 there was a Professor of Chemistry, Filippo Cassola, but he was paid only 10 ducats per month (i.e., as much as any student in Murat's school, *see* Figure 3), while the Professor of Mathematics was paid 40, and, interestingly, one of the janitors was paid 12 *(14)*. Cassola was fired in 1851, and his place was taken by Arcangelo Scacchi, whose salary was 15 ducats per month, and who was still in charge when the Kingdom of Naples (by then of the two Sicilies) collapsed in 1860. The relevance that chemical technology had in the Kingdom is easy to understand when one reads the official Annals, where matters even remotely related to industry and/or chemistry were granted only about one third of the number of pages granted to the minute description of the life of members of the Royal Family *(15)*. (I have consulted the copy in the library of the Engineering School. Pages relating to industry had never been edge-cut, so that I was the first one to read them. Pages on the Royal Family's life were worn out.) The only industry related to chemical engineering which had substantial exports was the mining of sulfur in Sicily, which was conducted with methods which had been described by Biringuccio.

A School of Engineering was established in Rome in 1817 by Pope Pius VII. Up to the unification of Italy in 1860, the two schools in Naples and Rome were the only ones where a (supposedly) complete

C O R P O R E A L E
DI PONTI E STRADE.

SCUOLA DI APPLICAZIONE.

CONCORSO PER L'AMMISSIONE DEGLI ALUNNI.

A V V I S O.

CONFORMEMENTE a clocch'è prescritto nel decreto de' 4 marzo 1811, e ne' regolamenti che vi si approvano, gli esami per l'ammissione de'dodici alunni alla scuola di applicazione di ponti e strade saranno aperti presso il Sig.r Direttor generale (o nel locale del soppresso convento di S. Maria di Caravaggio, ove sarà stabilita la nominata scuola) nel giorno 17 del prossimo alle ore otto di Francia del mattino. Le conoscenze che si esigeranno dai candidati sono le seguenti:
1°. Tutte le matematiche pure.
2° La statica applicata all'equilibrio delle macchine le più semplici.
3° La traduzione, in presenza degli esaminatori, di qualche pezzo di un autore latino in prosa, ed in seguito l'analisi grammaticale di qualche frase italiana della stessa traduzione.
4° Lo stesso per la lingua francese.
5° Che scrivino correttamente e con chiarezza l'italiano.
6° Che copiino una testa di un disegno, che sarà presentato loro dagli esaminatori.
I candidati dovranno farsi iscrivere al burò di ponti e strade nel convento di Caravaggio in Napoli, ove presenteranno la loro dimanda, indicando nome, cognome, patria, età, e domicilio.
L'iscrizione sarà chiusa nel giorno precedente quello dell'apertura dell'esame. Napoli 1 maggio 1811.

NELLA TIPOGRAFIA DI ANGELO TRANI

Figure 4. The announcement of the beginning of operation of the Naples Engineering School

Figure 5. Decree of appointment of a Professor of Chemistry at the Engineering School of Naples in 1811

engineering curriculum was offered. This was not, however, in any sense typical of Italy; what was typical was the very low level of industrial development. The only (partial) exception was the region, including Milan, which was part of the Austro-Hungarian Empire; that region is still today the most heavily industrialized one in Italy.

Early Industrial Development: 1860–1950

When Italy became a nation in 1860, its economy was still essentially an agricultural one; industrial development was far behind that of other European countries. In 1880, the seven largest producers of iron were as follows *(16)*:

Great Britain, 8,600,000 tons;
United States, 4,700,000 tons;
Germany, 3,400,000 tons;
France, 2,000,000 tons;
Belgium, 700,000 tons; and
Austro-Hungarian Empire, 570,000 tons.

Italy produced only 17,000 tons in 1880, i.e., exactly as much as had been produced in Great Britain a hundred and forty years earlier in 1740. Raw materials are extremely scarce in Italy, and therefore mining always has been minor. There were few manufacturing industries; the main production was the textile industry in the Po Valley.

While the unification of Italy had created a sufficiently large internal market to make industrial development possible, the beginning was very slow, and the agricultural crisis of 1887–1888 had serious effects on whatever industry had developed so far. Only at the end of the 19th Century did industry become a non-negligible factor in the Italian economy, with about 1,000,000 people working in it. Yet the gap with other countries had in fact become wider; in the period 1860–1900, industrial production had grown 328% in Germany, 146% in France, 123% in Great Britain, and only 94% in Italy (17).

A tariff law was established in 1887, and was enforced until 1921. After 1900, industry developed rather rapidly, and, in spite of the recession during the war years, production more than doubled in the first 25 years of this century (18). Production kept rising until 1929 when the great depression hit also Italy; it decreased to a low in 1932 and then started growing again to the high of 1935, when it was three times as large as it had been in 1900 (19).

In the early 1920's, the modern chemical industry came into existence in Italy. Guido Donegani transformed the Montecatini firm from a mining venture into a verticalized chemical firm, having as a base the know-how acquired with the development of the process for ammonia synthesis by G. Fauser; SNIA entered the field of artificial fibers. Separate statistics for the chemical industry have been available since 1922; its growth in the 1922–1935 period paralleled that of industry as a whole (20).

In spite of the industrial growth, the Government had been unable to keep the balance of payments in check, and during the whole 1922–1935 period Italian imports always exceeded exports; gold and currency reserves of the Central Bank had decreased to a dangerously low level by 1935. Furthermore, the Government entertained dreams of acquiring a colonial Empire (although it was at least 100 years late for such an enterprise, and in fact only twenty years before the end of colonialism). In 1935, an isolationist policy was started, which indeed did decrease Italian imports, but even more Italian exports. Therefore, with no advantage whatsoever for the balance of payments, the only concrete results were a slowdown of industrial development and a tendency for industry to neglect the requirement of being competitive on the international market. When Italy entered World War II, industrial production was still essentially at the 1935 level, and the technological quality of it was in fact rather low.

During the first two years of war, industrial production slowed down; after that, the situation degenerated to the point of total collapse. Industries in Italy were bombed by the Allied Forces until 1943 (and until 1945 in the North), by the Germans after that; a bitter and cruel war was fought in Italy from July 1943 to April 1945. During their slow northbound retreat, the German troops destroyed systematically whatever they could. When the war ended and the Italians started licking their wounds, Italy was a land of ruins with no industry to speak of; in 1946, industrial production was practically zero. (A single example is illuminating. Production of gasoline had been, in 1939, 520,000 tons. In 1944, only 2,000 tons were produced, and in 1946 less than 1,000 tons.)

Italy had, however, obtained one advantage from the war: it had gotten rid of the Fascist Government. The post-war years were heroic times, which required heroic efforts; and Italians are good at those. Politicians in power had been underground during the Fascist period, and they were, as politicians go, honest men. The general atmosphere was one of hope, of belief in future possibilities, and the ensuing economic growth took place at an extraordinarily fast rate. In only three years (1948–1950) industrial production rose from zero to the pre-war level.

The growth was going to continue for several years; the quality of industry, starting in about 1950, was going to change, and Italy was going to enter the restricted circle of modern industrialized societies. Up to 1950, Italian industry remains essentially in its infancy. In particular, the chemical industry was not very diversified, and its production was concentrated on few basic processes the technology of which did not really require competence in modern chemical engineering. It was, however, growing so fast that the future need for such a competence was being felt at least by the most intelligent individuals in the field.

Early Academic Development: 1860–1950

The Kingdom of Sardinia, with its capital in Torino, was a constitutional monarchy which essentially annexed the rest of Italy and transformed it into the Kingdom of Italy in 1860. Some rudimentary engineering had been taught in Torino since 1845 in the Istituto Tecnico run by Carlo Giulio, but up to 1859 citizens of the Kingdom of Sardinia had to go abroad, say e.g., to Rome or Naples, in order to get a complete engineering education. Shortly after the industrial region of Lombardia was annexed, university-level engineering education was established with the Law of November 13, 1859, which made the Istituto Tecnico part of the University of Torino. The same law established that an engineering school was to be established in Milano; the latter started operation in 1863 (21).

With the unification of Italy, the whole system of higher education was reformed. The Napoleonic concept of centralized lawmaking and administration was adopted in Italy, and the university regulations which had been enforced one year before in the Kingdom of Sardinia were applied to all Italian universities and other institutions of higher education, and thus in particular also to the Engineering School in Naples (Rome was not yet part of Italy, and was annexed only in 1870.) The latter became in fact less autonomous than it had been under the absolute monarchy of the Kingdom of the two Sicilies (*22*). The essential features of the Law of 1859, quoted earlier remain even today the basis of Italian universities.

Universities are—with very few exceptions, and none in engineering schools—run directly by the State; all members of the faculty are civil servants with a rigid pay structure. (Up to World War II, assistants and nonteaching staff were on the university budget, but even these had been made civil servants immediately afterwards. The Director of the Engineering School of Milano, F. Cassinis, said on December 16, 1948:

> ". . . the less active individuals . . . believe (and perhaps
> rightly so) that true happiness consists in being civil servants,
> since no matter how little one may work, and whichever may
> be individual deficiencies—the salary keeps going and the job
> is guaranteed for the whole life" (21).

In spite of a myriad of warnings such as this one, the career structure in Italian universities has become more and more rigid with passing time.) Universities are divided into faculties (Medicine, Engineering, Science, and so on) which confer only one degree; there is no graduate school system. Faculties in turn are divided into administrative units called institutes, and up to the late 1960's there was a one-to-one correspondence between full professors and Institutes.

During the 19th Century, engineering was, above all, civil engineering; after all, it had started as a School of Bridges and Roads. The word "civil" was used to distinguish it from military engineering. The development of industry, however, imposed the requirement of training industrial engineers, with a cultural background different from that of civil engineers. Teaching of industrial engineering in Italy started later than in other countries, which in view of the lower level of industrial development is not surprising. The situation at the turn of the century was as follows (*11*).

The engineering schools in Naples, Rome, Torino, and Milano had been supplemented by three more in Bologna (started in 1874), Padua (1880), and Palermo (1886). The school in Milano was an entirely independent institution; Bologna, Naples, and Torino were independent schools who admitted students after they had taken two years of mathematics, physics, and chemistry at the local university; the other schools were part of the local university; all schools were regulated by the law of October 8, 1876. In Torino there was also the Royal Industrial Museum, which granted degrees in industrial engineering since the law of July 3, 1879. The school in Milano had granted degrees in both civil and mechanical engineering since the beginning of its operation in 1863; in 1868, a Chair of Chemical Technology was established. At the same time, an Institute of Industrial Chemistry was established, and Angelo Pavesi was appointed as Head. He kept the position until 1872, and was followed by Luigi Gabba, 1872–1914; Ettore Molinari, 1916–1926; M. G. Levi, 1927–1939; and Giulio Natta, 1939–1974 *(23)*. All other schools had only civil engineering curricula, and no degree in chemical engineering was offered anywhere in Italy.

A chemical subdivision of the industrial engineering curriculum was started in Milan in 1900; industrial engineering was established in Naples in 1901, with two subdivisions—electrical and electrochemical engineering. (In 1905, an Institute of Electrochemistry was established in Naples, and Oscar Scarpa was appointed as Head; he kept the position until 1924. He was followed by Francesco Giordani, 1925–1942; Mario Jacopetti, 1942–1963; and myself, 1966–1976. I was the first one not holding a Chair of Electrochemistry, but one of Principles of Chemical Engineering, and the name of the institute was changed accordingly. In the 1977 edition of the Annuary of the University of Naples, however, it is listed, in supposedly alphabetical order, immediately before the Institute of Electrotechnics. A School of Electrochemistry was established also in Milan on September 25, 1902, and the first Head was Giacomo Corrara, 1904–1925; he was followed by Oscar Scarpa, 1927–1948 and Roberto Piontelli, 1948–1975.) However, it was stated clearly in the original decree that electrochemical engineering was to be understood as industrial chemistry, i.e., in the same sense as chemical engineering was understood in Milano, where an Institute of Industrial Chemistry was in existence since 1868. (An Institute of Industrial Chemistry was established also in Naples in 1937, when Marussia Bakunin decided to give a technological slant to her work in organic chemistry. She was the Head until 1943, and was followed by Giovanni Malquori, 1943–1967 and Leopoldo Massimilla, 1967–1976. Institutes of Industrial Chemistry were established at different times also at Bologna, Genova, Palermo, Pisa, Roma, and Torino. Another subject which traditionally has been very important is applied chemistry, in many senses undistinguishable from industrial chemistry. Institutes of Applied Chemistry have been established at Cagliari, Naples, and Trieste.)

This was the beginning of a cultural qui–pro–quo which lasted unchallenged until 1960, and is not overcome completely even today. Italian academic chemical engineering was born as industrial chemistry, and the cultural formation of the industrial chemist became the predominant one when, first in the United States in the early 1920's, and then in the rest of the world (with perhaps the exception of Germany), chemical engineering transformed from the process-by-process approach of industrial chemistry to the unit operations approach, nothing of the same kind happened in Italy, where the academic world of chemical engineering was dominated by chemists.   Biringuccio had been an industrial chemist, not a chemical engineer; up to 1960, Italian academic chemical engineering had improved quantitatively, but had not progressed qualitatively, over Biringuccio's viewpoint.   (I became acquainted with Biringuccio's *Pirotechnia* when I was a student in the early 1950's; a copy was kept in the Library of the Institute of Industrial Chemistry of Naples. It was not in any special antiquarian section; it was just one, if the oldest, of the available books on industrial chemistry.) Since the beginning of chemical industry in Italy, and up to about 1950, it was still in an early stage of development, and no pressure developed for a change in academic policy.

Of course, the situation just described was not, in the first 20 years of this century, different from the one in other countries.   Hougen's description (24) of the situation at the University of Wisconsin in 1911 would apply to Italy as well:

> *"Instruction in mathematics, physics, and other branches of engineering was excellent, but there was no integration between chemistry and engineering. . . .   A required course in surveying served no purpose in chemical engineering"* (24).

The difference lies in the fact that such a situation, in Italy, was implanted into a rigid university system which resisted stubbornly, particularly under the Fascist Goverment, any cultural change: surveying was a required course in the chemical engineering curriculum as late as 1960.   (Academic curricula are established by an Act of Congress in Italy.   This implies that a required course easily will survive 30 years after its cultural obsolescence.)   While this cultural immobility was not too serious a problem in fields where no basic cultural evolution took place in the world in the period 1920–1950, it was lethal in the case of chemical engineering.

Chemical engineering curricula—all of the general engineering plus chemistry type—were established in other Italian universities during the first 30 years of this century.   (In Torino, where industrial engineering had been in existence since 1879, the chemical subsection was established in 1908.)   As said before, however, cultural evolution came almost to a standstill with World War I, and even worse with the characteristically anticultural philosophy of the Fascist Government.   Indeed, university professors were required to swear allegiance to the Fascist party; it is sad to say that only 13 had the courage to refuse and be fired.   (One of them was Luigi Einaudi, Professor of Economy at Torino, who was later to become the first President of the Republic of Italy). Many others, in a typically Italian attitude, did swear but did not take seriously the whole exercise. The growth of industry in 1922–1929 was a growth of quantity, not of quality; from 1930 onwards, the Fascist Government was established strongly enough to allow itself an orgy of stupidity, and Italian culture went rapidly downhill. (Continuum mechanics was one subject in engineering where Italians were particularly strong; Truesdell and Noll (25) state that "Knowledge of the true principles of the general theory seems to have diminished except in Italy, where it was kept alive by the teaching and writing of Signorini" (25). Yet the same authors relate that later on "Signorini, however, had meanwhile stopped reading . . ." (25).) When Italy entered World War II, its academic chemical engineering was, at best, at the cultural level of the beginning of the 20th Century, having failed to undergo any of the cultural changes which American chemical engineering had undergone, which have been reviewed by Hougen (24).

At the end of World War II, Italian universities were in as bad a shape as Italian industry. German troops, before leaving the towns in their slow retreat, destroyed whatever they could, including universities. (The University of Naples went through the ordeal on September 12, 1943; the story has been told passionately by G. Malquori (26). German troops entered the university and systematically threw hand grenades into laboratories and set fire to libraries. During the ensuing 18 days, war was fought in the region of town near the harbor where the university is located, and German troops would capture any men in sight and send them to extermination camps in Germany. If anything was saved in the university, the merit is of the women who braved the war and did what they could. On October 3, 1943, the Allied Forces held the town; American troops were stationed at the engineering school. Fortunately enough, the Officer in charge seemed to be afraid of whatever had been left by the Germans of the mysterious gadgets in the chemical laboratories, and the Institute of Industrial Chemistry was spared military occupation.) In the period 1943–1946, the entire school system in Italy, and of course any research activity, came to a complete stop. Indeed, very few libraries have the 1943–1946 issues of scientific journals.

The same heroic spirit which sparked the industrial reconstruction was, however, operative also in universities, and perhaps even more so; and also Italian universities rebuilt rapidly from the ashes of war, and by 1950 they had reached—physically—the pre-war level. Furthermore, cultural ferments were active in the new democratic life of Italy, and a major evolution was in the making. In 1950, Italian academic chemical engineering was at least 30 years behind the United States in its cultural evolution; but there were the seeds for a spectacular growth, which in the next 20 years would in fact take place and succeed in filling the gap.

Recent Industrial Development: 1950–Today

The growth of Italian industry in the immediate post-war period, which had brought it back to pre-war level by 1950, continued spectacularly for the next 20 years. The chemical industry did even better than industry as a whole; in the first five years (1950–1955) it increased production by 120% (*27*), and it began diversifying into the whole range of chemical products. In the ensuing 13 years (1955–1968) it grew at an average rate of 12.25%, topped only by the USSR chemical industry which grew at an average rate of 12.5%. In 1968, the production of the Italian chemical industry was ten times as large as it had been in 1950, and it represented 4.45% of the total world production. Italy was the seventh largest producer of chemicals in the world, after the United States, USSR, Germany, Japan, France, and the U.K. Montedison, the result of the merging of the Montecatini and Edison firms, was the tenth largest chemical industry in the world, its production having reached almost one half of the largest one, E. I. Du Pont (*28*).

The growth of the Italian chemical industry in the 1950–1970 period was not only a growth of the amounts, but also of the number of chemicals produced. Correspondingly, a large variety of technologies came into existence in industry, and the required chemical engineer became one who could cope with a variety of different processes, rather than one who would know in detail a few specific ones. This had important effects on academic chemical engineering, which transformed during the same years from the industrial chemistry stage to the modern chemical engineering one.

Beautiful as the statistics appeared in the late 1960's, there were hidden problems; and some intelligent people pointed out very explicitly where the problems were. Foraboschi, in 1971 (*17*), emphasized the following ones:

1. While the Italian industry produced 4.4% of the total world production of chemicals, the percentage of high-priced ones was much lower: e.g., for pharmaceutical products it was only 2.8%. Italian industry was producing much more low-priced than high-priced chemicals, which is a very poor policy for a country like Italy which has practically no raw materials.

2. A large part of the production increase had been obtained by increasing the size of plants, so as to achieve economy of scale. This implies low levels of employment per unit of capital invested, again a very poor policy for a country with chronic unemployment problems. In 1970, the chemical industry produced 14% of the total production of manufacturing industries, but employed only 6.7% of the people.

3. Foreign capital was heavily present in the Italian chemical industry, particularly in that area producing high-priced chemicals: 100% of the the Italian photographic industry was owned by foreign firms.

4. While most large chemical firms in the world invested about 5% of their total income into research, Montedison invested only 2.1%. Practically no fundamental research was done in Italian industry; people working on research in industry had very poor career opportunities. By 1970, the growth rate of production was slowing down, and the qualitative growth of technology had come to a stop.

Warnings such as given by Foraboschi were left largely unanswered; the Italian chemical industry did essentially nothing in this regard. Times are too close to allow historical perspective, but at least a try at understanding the reasons of the crisis which took place in the 1970's can be made.

The post-war reconstruction had required such a large investment that the State had to intervene and furnish the necessary capital. With this, the State had acquired direct or indirect control of a very large fraction of Italian industry, and the fraction controlled by the State kept increasing during the post-war years. The progressive deterioration of the political climate, possibly related to the progressive corruption of the party which has been in power since 1946 (the Christian Democrats), reverberated through industry. Decisions in industry have not been taken on technical grounds, but as part of a complex political game. Political action in turn has not been geared to any long-range political goal, but more and more it has been related only to the short-term goal of not losing power. People in key positions in industry have been appointed for political reasons, rather than for technical competence. The "black funds" (money going through underground channels from industry to political parties and/or individual politicians) have become a larger and larger fraction of the true, if not the official, industrial budgets. The very bad economical crisis of the 1970's was, in such a situation, inevitable.

Recent Academic Developments: 1950–Today

The more sensitive people in academic circles were aware of the evolution that chemical industry was undergoing from 1950 onwards, and they realized that the type of chemical engineering education which was being offered in Italian universities was out of phase. Some of the people in charge were too old, or too lazy, or too unimaginative to do anything about it. They knew, or hoped, that during their active lifetime they were personally unlikely to become visibly obsolete, and they didn't have enough interest in Italian culture to do anything for the next generation. Other ones, however, were either younger, or more imaginative, or more interested in the university, and they came to the realization that the next generation of academic chemical engineers would need an entirely new cultural background. They understood that the German tradition of industrial chemistry, with its attention to specific examples of chemical technology, just couldn't possibly do for the modern chemical industry. Only the most honest ones openly admitted that they really didn't know what would do, except for the vague knowledge that the necessary cultural background was available in the English-speaking world.

Two men were the most active: Giulio Natta in Milan and Giovanni Malquori in Naples. Natta, the only chemical engineer who has ever been awarded the Nobel prize, recognized that no matter how advanced the understanding of the chemistry of a process might be, engineering know-how is required in order to make it economically attractive. He developed a school of chemical engineering along two lines—chemical reaction engineering and polymer physical chemistry. His personality, however, was so overwhelming that he developed his school without feeling the need for substantial cultural input from abroad (over and above what was a complement of his own cultural input); and even today chemical engineering at Milan is rooted deeply in the tradition of industrial chemistry, although interpreted in a modern sense. (The viewpoints of the Milan and Naples schools have been discussed at length, *see* Ref. *29–34.*)

Malquori at Naples was bolder in his approach. He came to the conclusion that the best policy was to hire young graduates and send them abroad to acquire the necessary cultural background. They would need afterwards to develop on their own, and introduce into Italian academic chemical engineering, the Anglo-American viewpoint. He began such a policy in the early 1950's, and it was remarkably successful. Malquori was a fascinating personality, and he was interested in young people; so he ended up having a profound influence also on the young people who were working in those Italian universities where the old generation had just given up the exercise of thinking.

Italian chemical engineers who made contact with chemical engineering abroad in the 1950's found themselves exposed to the cultural evolution which was taking place at the time, namely, the transition from unit operations to transport phenomena as the key cultural nucleus of chemical engineering. (H. Kramers had been teaching transport phenomena at Delft for several years before he published his book *Fysische Transport verschijnselen* in 1961; the University of Wisconsin had started courses in transport phenomena in 1957 *(35)*). Italian chemical engineers who were exposed to this influence brought back to Italy the seeds of this cultural evolution, and academic chemical engineering in Italy evolved directly from the industrial chemistry to the transport phenomena stage, without having to go through the Unit Operations one. (Of course, such broad generalizations are liable to confutation by counter example. In some Italian universities, even today chemical engineering cannot be said to have outgrown the unit operations stage; yet these are the ones where the older generation in the 1950's was lazy enough to do nothing, yet active enough to resist younger people doing something.)

Cultural influence from abroad was made available also to those chemical engineers who actually did not leave Italy. The Donegani foundation of the Accademia dei Lincei had been holding summer schools of chemistry for several years, and in 1960, the school was dedicated to transport phenomena (36). The best European chemical engineers were invited to give review lectures. Probably awed by the cultural tradition of the Accademia they presented extremely lucid and well-organized lectures, some of which have become classics in the field. (The Accademia dei Lincei was established in 1603 by the 18-year-old Federico Cesi, together with three other gentlemen. None less than Galileo joined the Accademia in 1611, and the first publications were the three letters of Galileo on the sunspots, published in 1613. Works of Caratheodory, Hadamard, Landau, Levi–Civita, Truesdell, and Volterra (to quote only a few in the field of exact sciences) have been published by the Accademia. In spite of frequent attempts of whoever has been in power in Rome since 1603 to gain political influence on the Accademia or destroy it altogether, it has survived to the present day as the most prestigious cultural institution in Italy.) Chemical engineers in the 25–35 age group were present in Varese, and they absorbed eagerly the cultural message. A sense of being responsible for the realization of a badly needed cultural change was brought back to almost all of the Italian universities where chemical engineering was being taught.

On January 31, 1960, a new law regulating engineering schools was passed by the Congress. Chemical engineering became a degree standing on its own feet, rather than a section of industrial engineering as it had been before. Correspondingly, compulsory requirements of credits in mechanical, electrical, and civil engineering were relaxed, while the number of credits required in chemical engineering subjects was increased. (The fact that curricula are established by an Act of Congress is responsible for the slowness of change in Italian universities. In 1910, A. Sayno, Vice Director of the Engineering School in Milan, had written, "Possibly not too far away in the future, also in Italy, as is already the case in foreign engineering schools, differences between curricula will be emphasized, and general subject requirements will be decreased in favor of specific ones" (37). It took the Italian Government exactly 50 years to do what Sayno had indicated in 1910.) The freshly acquired competence in modern chemical engineering therefore could be transmitted directly into the teaching pattern, and chemical engineers graduating from Italian universities from about 1965 onwards had received an education comparable with that of their colleagues in the best schools of the Western World.

The people who had been young in the 1950's were not so young anymore in the 1960's, but they had the opportunity offered by a fresh crop of competent graduates, and a second generation came into being. In the early 1970's, good schools of chemical engineering were consolidated, or being established, in all ten Italian universities offering a chemical engineering curriculum. Research papers of good quality were being published by Italians in international journals more than occasionally; an Italian journal dedicated to chemical engineering which had started publication in 1965, published good papers monthly (*Quaderni dell Ingegnere Chimico Italiano,* monthly supplement to *La Chimica e L'Industria.* (The latter journal is the official one of the Italian Chemical Society, and has been published since 1919.)) Books on several aspects of chemical engineering had been published by Italians. (Some are listed as Refs. *38–42.* These books are of varying quality, but none of them was culturally obsolete at the time of writing. The book by Foraboschi is a significant step in the cultural evolution of chemical engineering (*43*).) It is fair to state that in 1970 chemical engineering in Italian universities was as good as anywhere else in the world, which is very good indeed, since it had been lagging by 30 years in 1950.

But, again, there were hidden problems. The rate of growth of industry was slowing down, and the technological quality was in fact going downhill. Graduates from Italian universities were culturally overqualified for the existing job market. Academic chemical engineering was becoming divorced from industrial reality, as witnessed by the very scarce industrial support of academic research. This was particularly true for universities in the South since the headquarters of Italian industry are in the North; and indeed research at, say, Napoli, Cagliari, and Palermo was perhaps culturally more advanced, and hence in fact farther away from industrial reality, than it was in say Torino or Milan, located in the hub of Italian industry.

The law of 1960 had not instituted a system of graduate education. In the 1960's, young graduates were hired as assistants, and, in their first years of work they got essentially the type of training that a PhD candidate gets in the United States; as long as the whole educational system was expanding, after a certain length of time they would get a teaching position. The system was, in a sense, equivalent to a graduate school.

In the late 1960's the expansion of the educational system essentially stopped. Since even graduates were overqualified for the industrial job market, there was practically no industrial opportunity for people with qualifications equivalent to a PhD. Young graduates hired by universities from 1970 onwards had no future. Indeed, hiring itself has come to a stop, since previously hired people cling to the available positions.

But the real cause of disaster was the uncontrolled increase of the number of students. The chronic unemployment problem of Italy had become very bad in the late 1960's, and the government found a way out, with typical political shortsightedness, by making access to universities easier and easier. After all, an 18-year-old person does not consider himself or herself unemployed if enrolled in the university; neither is a student registered in the unemployment statistics. If the university is free (as it always has been in Italy) and the student gets a small allowance from the State (as is the case in Italy), unemployment is swept neatly under the rug at a comparatively low cost. Of course, the problem only is postponed four or five years, or whatever it takes for a nongraduate, jobless person to become a graduate one; but four or five years in the future was as far as any politician in power in Italy in the early 1970's (and is such a qualifying attribute needed?) could conceivably look. When the wave of jobless graduates hit the job market in 1976–1977, the situation became explosive, as witnessed by the tremendously disruptive student movements in 1977. Meanwhile, the universities had become parking lots for the unemployed, with the cultural consequences that such a situation inevitably entails. (It will probably take many years to fully understand the Red Brigades phenomenon, which started in the early 1970's and (hopefully) culminated at the time of writing. Yet there is no doubt that the cultural vacuum of the Red Brigades' ideology was nurtured in Italian universities, that the criminal rage and hatefulness of their behavior is related to the problem of the unemployment of young graduates, and that the nutrient broth for their growth, the "water where they swim," has been the large mass of university students who at the age of 22 suddenly realize that they have been qualified, if they have, for a professional job that doesn't exist, but have failed to qualify for whatever blue-collar job may in fact be available.)

In 1971, Congress decided that a reform of the University system was needed badly, and solemnly announced that a new law would be enacted within, at the very most, three years. Meanwhile, the old law was repealed, and Urgent Provisions were established for the three-year interval. It took three years to do what the Urgent Provisions plan required to be done in one year; nothing at all has been done since 1974. The new university law is in the program of every government, but it does not have a very high priority, and no Italian Government lasts long enough to get down to it. Italian universities, in the meantime, have come to a complete standstill.

Academic chemical engineering has not been able to extricate itself from the overall crisis. People active in the field today were already active in 1970; some have given up in desperation. No new generation is being formed today; no equivalent of a graduate program exists. Research support is becoming less and less. Single individuals may be personally as productive as they would have been without the crisis, perhaps even more so since they don't have the burden of training a new generation. But active individuals are growing old and disillusioned, and if they are doing something they are in fact divorced from the reality of Italian universities.

The historical account given above shows how parallel the evolutions of industrial and academic chemical engineering have been in Italy. If and when Italian economy will pick up again, it is to be hoped that academic life will pick up also. But by that time the generation which was responsible for the great cultural growth of academic chemical engineering in the 1950–1970 period may be too old to pass the flag to the next one, and there in fact may not be a next generation to pass the flag to.

Acknowledgments

I am indebted to L. Massimilla, Dean of Engineering at the University of Naples, for having made available the Archives of the Engineering School; to several friends in other universities and in industry for supplying useful information; to my son Tom for help in the research concerning the period up to 1860; and, last but not least, to the Chemical Engineering Department of the University of Delaware which repeatedly has offered me the possibility of being exposed to the American chemical engineering culture.

Literature Cited

1. Forbes, R. J. "Studies in Ancient Technology"; E. J. Brill: Leiden, 1964.
2. Dante, "Inferno," XXIX, 119–120.
3. Dante, "Purgatorio," V, 109–111.
4. Crombie, A. C. "Medieval and Early Modern Science"; Doubleday: Garden City, 1959.
5. Astarita, G. "Scale-up problems arising with non-Newtonian Fluids," *J. Non-Newt. Fluid Mech.* **1979**, *4*, 285.
6. Biringuccio, V., "Pirotechnia," Venice, 1540.
7. Deperais, C. "Trattato Teorico—Pratico della Fabbricazione dell' Allume. Introduzione: Storia della Fabbricazione dell'Allume," *Att. R. Ist. Incorag. Napoli* **1888**, IV-1, n. 9.
8. Merz, J. T. "A History of European Thought in the Nineteenth Century"; W. Blackwood: London, 1903.
9. Manogue, K. R. "The Concept of a University"; Univ. of California Press: Berkeley, 1973.
10. Ferrarelli, G. "Il Collegio Militare di Napoli"; Rivista Militare Italiana: 1887.
11. Masoni, U. "Sullo Sviluppo dell'Insegnamento Tecnico Superiore," *Atti R. Ist. Incorag. di Napoli* **1901**, II-5, n. 3.
12. Decreto Reale, March 4, 1811. Bollettino delle Leggi del Regno di Napoli, 1811, n. 105.
13. "Decreto che prescrive di stabilire in questa Capitale una Scuola di Applicazione per gli ingegneri di ponti e strade", November 10, 1818. Archivio di Stato di Napoli, Ponti e Strade, N.S., n.1374.

14. Russo, L. "La Scuola di Ingegneria in Napoli, 1811–1967"; Facoltà di Ingegneria: University of Naples, 1967.
15. Annali Civili del Regno delle Due Sicilie, 1837–1860.
16. Reuleaux, F. "Le Grandi Scoperte, Trattamento Chimico della Materia Prima"; U.T.E.T.: Torino, 1889, Vol. 5.
17. Foraboschi, F. P. "L'Ingegneria Chimica"; Facolta di Ingegneria dell'Universita di Bologna: Bologna, 1971.
18. "L'Industria Italiana alla metà del Secolo XX", Confederazion Generale dell' Industria Italiana, 1953.
19. Compendio Statistico Italiano, 1936.
20. Compendio Statistico Italiano, 1940.
21. "Il Centenario del Politecnico di Milano"; Politecnico di Milano: Milan, 1963.
22. Russo, L. "La Scuola di Ingegneria in Napoli, 1811–1967"; Facolta di Ingegneria: Naples, 1967.
23. Lori, F. "Storia del R. Politecnico di Milano"; A. Cordani: Milano, 1941.
24. Hougen, O. A. "From Plumbers to Professionals: Development of the Chemical Engineering Profession from Obscurity to Prominence," Bicentennial lecture on Chemical Engineering History, Ntl. Mtg. A.I.Ch.E., 82nd, Atlantic City, 1976.
25. Truesdell, C.; Noll, W. "The Non-Linear Field Theories of Mechanics," in "Enc. of Physics"; Springer-Verlag: Berlin, 1965; Vol. III/3.
26. Malquori, G. "Il Politecnico," unpublished report, Library of the Institute of Industrial Chemistry, University of Naples, 1944.
27. "Organization Europeenne de Cooperation Economique: L'Industrie Chimique en Europe"; Paris, 1956.
28. "Mediobanca, Ricerca e Sviluppo"; L'Industria Chimica: Milano, 1970.
29. Pasquon, I. "Orientamenti Moderni della Ingegneria Chimica," Chim. Ind. 1968, 50, 648.
30. Astarita, G. Chim. Ind. 1968, 50, 1040.
31. Pasquon, I. Chim. Ind. 1968, 50, 1040.
32. Astarita, G. "L'Evoluzione dei Fondamenti Teorici dell'Ingegneria Chimica," Q. Ing. Chim. Ital. 1972, 8, 112.
33. Massimilla, L. "Il Dottorato di Ricerca Nelle Discipline Chimiche di Orientamento Ingegneristico Anche Nella Prospettiva della Preparazione e Selezione dei Futuri Docenti Universitari," Q. Ing. Chim. Ital. 1972, 8, 120.
34. Pasquon, I. "Stato della Ricerca nelle Università Italiane nel campo della Tecnologia e dell' Ingegueria Chimica", Q. Ing. Chim. Ital, 1972, 8, 74.
35. Bird, R. B.; Stewart, W. E.; Lightfoot; E. N. "Transport Phenomena"; JohnWiley and Sons: New York, 1960.
36. "Alta Tecnologia Chimica: Processi di Scambio, V. Corso Estivo di Chimica, Varese 26 Settembre–8 Ottobre 1960"; Accademia Nazionale dei Lincei, Fondazione Donegani: Rome, 1961.
37. Sayno, A. "46 anni di vita del R. Istituto Politecnico di Milano"; Milan, 1910.
38. Astarita, G. "Mass Transfer with Chemical Reaction"; Elsevier: Amsterdam, 1967.
39. Carrà, S. "Introduzione alla Termodinamica Chimica"; Zanichelli: Bologna, 1972.
40. Foraboschi, F. P. "Principi di Ingegneria Chimica"; U.T.E.T.: Torino, 1973.
41. Sebastiani, E. "Termodinamica dell'Ingegneria Chimica"; Sidereo: Rome, 1969.
42. Trevissoi, C. "Elementi di teoria dei Reattori Chimici"; Litopress: Bologna, 1970.
43. Astarita, G. Chim. Ind. 1974, 56, 585.

RECEIVED May 7, 1979.

A History of Chemical Technology and Chemical Engineering in India

DEE H. BARKER[1] and C. R. MITRA

Birla Institute of Technology and Science, Pilani, Rajasthan, India

The development of a chemical industry in India, traced over a period of 4000 years, shows a rapid rate of growth since independence in 1947. Examples in selected industries are given which illustrate this growth and which also show the coexistence of both ancient and modern technology. The government and social system play a vital part in the development. Chemical engineering has been utilized only since independence with the establishment of universities and a professional society. Research has had only a small effect on the development but currently is receiving much support. Most technology has been imported with the research and development effort being directed toward adaption using indigenous materials and accounting for local conditions.

Т he history of the use of chemicals for the betterment of mankind covers a period of over 4,000 years in India. Many books and papers have been written concerning the history of science and of chemical technology *(1, 2, 3, 4).* Much of the knowledge about the early chemical technology must be inferred from archeological finds. These include the existence of water and sewer systems in ancient towns, the finding of cosmetic cases and bottles and pottery. Pottery shows the progression of knowledge in chemical processing relating to the firing of the vessels. The use of chemicals progressed from black magic and art to science as the years went by. Most of the early treatises and available information relate to the preparation of medicines and of metals used in pursuing war. Since independence (1947), the chemical industry has grown at a faster rate in India than elsewhere in the world.

[1] Current address: Chemical Engineering Department, Brigham Young University, Provo, Utah 84602.

The long period of time and the vast range of chemicals manufactured make it impossible to present a comprehensive history in a short
chapter.   Selected areas which illustrate the growth and development
have been chosen and discussed.   This neglects many important areas.
Basic research has not played an important role in the past but currently
is being given greater emphasis.   The major part of the technology has
been imported.   Early failures of some processes were caused by failure
to recognize local conditions and materials.   A large portion of the development effort has been directed towards material substitution and making processes more labor intensive.   This has resulted in few new processes being developed.

The government and the social system play a critical role in the
overall development.   The progress, or lack of it in some areas, is a
direct result of government control.   Examples of this effect are shown
below.

Most historians divide the Indian historical scene into several different ages or periods.   These periods include the Pre-Vedic Age—all of
the time prior to 1500 B.C., the Vedic Age from 1500–600 B.C., the
Classic Period from 600–1200 A.D., the Medieval Age from 1200 A.D. to
the end of the 18th century, the British Period through most of the 19th
century, the Pre-Independence Period from 1900 to about 1947 and the
Post-Independence Period from that time on.

India from ancient times was known as a storehouse of treasures,
spices, gold, metals, and many medicinal plants.   It was looked at as a
source of raw materials, not of finished goods.   For this reason, the
emergence of chemical technology was slow.   The general pattern was
for invasion, shipping of raw material from the country, and finally as-
similation of the invading population.   This was repeated many times
during India's history.   When India gained its independence, its leaders
set forth to build a democratic socialistic nation in which progress was to
be the key word.   Planning was undertaken to develop self-sufficiency in
the field of chemical technology as well as in many other fields.

The chemical technologies considered cover a broad field including
pharmaceuticals, metallurgical industries, heavy chemicals, petroleum,
etc.   While not all fields are discussed, all have had spectacular increases
since the coming of independence.

A Land of Contrast

India remains a land of contrast.   The modern methods of chemical
technology exist almost side by side with the ancient ones.   Descriptions, drawings, and pictures of ancient operations resemble many of the
ones currently still being used.   Three examples of these are given
below.

Figure 1. Charcoal burning: (A) wood-stack; (B) preparing the heap of wood; (C) covering it; (D) a freshly lit heap; (E) a nearly burnt-out heap; (F) uncovering a carbonized heap (5).

Singer (5) describes the chemical industries' use of wood for charcoal around the middle of the 15th century.   Figure 1 (5) shows the process of making charcoal for use in iron works.   Wood is cut and stacked in conical piles, covered with clay, and then burned under controlled-combustion conditions.   Figure 2 shows the method for making bricks near Pilani, Rajasthan, India.   Although the process is not identically the same, the means and methods are much the same as in the early times and easily could be mistaken for the drawing shown in Figure 1.   The bricks are sun-dried, piled in alternate layers with wood in a conical pile, covered with dirt, and then fired.   Subsequent to the firing, the bricks are removed for use in construction in homes, etc.   Not all of the bricks are fired in such kilns, but most are hand-formed and fired in trench kilns.

Another process which has not changed much over the many years is the making of salt.   Agricola (6) describes the process of making salt and

Figure 2. Brick kilns, central India, 1974

presented a wood block print shown as Figure 3.   Figure 4 shows salt being made in southern India in 1974.   Agricola's description and his line drawing closely match that current practice in India.   More modern methods are used also.

Drying is carried out mostly with the use of the sun.   This is particularly true of agricultural projects.   In driving through southern India, it is not uncommon to see the sides of the road lined with fiber mats covered with rice drying in the sun—the level and raised-road bed making it an ideal place for drying.   Figure 5 shows a field of bright red chiles drying in the sun in the western part of India.   Thus, the practice of drying has not changed, although some efforts are being make to build more efficient solar dryers.

Location of Industries in India

In general, the industries, particularly the chemical industries, tend to become congregated in a few areas.   These are areas in which the raw materials and the water and labor supplies are plentiful.   Large concentrations of industry can occur near Bombay, Calcutta, Ranchi, Delhi, Madras and Cochin.   The original plan of the Indian government, at the inception of independence, was to spread the industrial complex throughout India to meet the social needs of providing jobs (7).   However, local political influences have prevailed and resulted in a concentration of industries.   The primary exception of this is the production of cement and sugar which are located more or less uniformly across India. Current planning calls for a wider distribution of chemical complexes.

Government and Planning

The development of the chemical industry in India always has depended on the ruling government, whether this was in ancient or in modern times (8–13).   Until the 17th century there was little, if any, organized chemical technology within India.   The technology primarily concerned itself with gathering plant materials and using these materials locally.   However, some metallurgical practices, such as the production of iron, were well known before that time and the iron products were highly prized throughout the world.   India was considered to be primarily a source of raw materials, such as agriculturally produced jute, raw cotton, raw silk, indigo, and raw drugs such as opium or in some cases minerals.   Textiles and cotton were one of the most widespread industries in India (13).

The chemical industry and industry in general both increased and decreased during the British presence in the 18th and 19th centuries (14).   This was particularly true with respect to sugar, steel, and tex-

Figure 3. Making salt in ancient times: (A) sea; (B) pool; (C) gate; (D) trenches; (E) salt basins; (F) rake.

Figure 4. *Making salt in southern India, 1974*

Figure 5. *Drying chilies in western India, 1968*

tiles. The trade laws and the encouragement of industry depended on the needs in Great Britain. For example, in 1880, approximately 60% of the people were engaged in agriculture with the rest being in industry while in 1921, 73% of the people were engaged in agriculture, indicating a decline in fraction of people employed in the industries.

With the coming of independence in 1947, the government has played a major role in the development of the chemical industry. In 1948, the government announced a policy of planned development and regulation of industry. The objective was to arrive at a "mixed economy," which overall would be beneficial to the nation. Since India is dedicated to a form of social control through democratic procedures, mixed economy refers to ownership of industrial units partly under public funds and partly under private funds. Modifications have been made from time to time in the policy and in the actual administrative structure, but the control remains with the government and its policies.

Under the industrial policy, the Indian industry is classified in three groups (8). The first category, to be under the control and management of the central government, included arms, ammunition, and atomic energy. The second category considers heavy industries such as coal, iron, steel, ship building, etc.; these are also the responsibilities of the central government. The final group of industries are those which are to be developed by private enterprise. In addition, the industries are classified as to large and small scales, depending on the amount of money being invested in the industry. An extensive list of the various industries available to each sector of the economy is published periodically (8, 15, 16).

The overall planning concerning the amounts of moneys to be invested in each industry is under the direction of a central planning board. The country operates under five-year plans which are prepared by this board and presented to the parliamentary body for ratification. There have been five plan periods plus a sixth plan which is being prepared currently (17, 18).

The industrial policy is under the control of the central government and is the responsibility of cabinet-level ministers who work through a director general and various other agencies. A central advisory council advises the government on all matters concerning the development and regulations of industries. In addition to the central advisory council, there are development councils set up for each individual industry. These include not only the ones for the chemical industries such as inorganic chemicals, sugar, drugs, and pharmaceuticals, etc., but also for all other industrial segments. The control of research and development, which will be discussed later, is also under the central goverment.

Control is carried out by using a licensing procedure in that all manufacturing units have to be licensed either by the state or the central government. This licensing arrangement controls not only the permission to manufacture, but also the ability to obtain funds. The major problem in the licensing process has been that firms would apply for a license and then not utilize that license for production, but only for obtaining funds and materials which were diverted then to other markets.

Thus, since licensing, obtaining funds, materials, and, in some respects, the availability of market are controlled by the government, the development is a function of the political philosophy of the government officials.

Development of Selected Industries

As pointed out earlier in this chapter it is not possible to cover all areas of chemical technology here. A few areas will be (including production rates which indicate the growth of that particular industry). Other areas such as the newest areas of petrochemical, oil, and gas production are not covered. They are very small at the present but are finding increasing importance in the expansion of India.

The development of chemical engineering covers only a period of 30–40 years. Its development follows that of research, education, and the formation and growth of the professional society—the Indian Institute of Chemical Engineers. These also will be discussed in this chapter.

Drugs and Pharmaceuticals. Man's earliest strivings were the preservation and lengthening of life. This was also true in India and the earliest literature indicates an interest in this field. It is in the field of pharmaceuticals, cosmetics, and drugs that the science of chemistry had its beginning (1). The earliest mention of medicines and healings are in the literature of some 2000 years before Christ. Archeological evidence would indicate that the science of medicine was known long before this. In the early times, the art of medicine and drugs was surrounded with mystery and passed on from father to son, very little being written down.

Medicines were made from natural, raw materials which were compounded in many, many different ways. A large portion of the medical treatment available in India today is the so-called Ayurvedic system (science of life) which is based on these ancient cures. Specific plants or combinations were used for specific ailments. Mention of these are given in the great epic poems of India, the Mahabharata, and the Rayman. Some of the medical research being carried out in India today is based on finding the active ingredients in the plants being used by the Ayurvedic physician.

The establishment of the laboratory or work place of the physician or chemist followed elaborate ritual. An artist's representation of a chem-

Figure 6. The Rasasala—an artist's impression (1)

ical laboratory, the Rasasala, is shown in Figure 6. The laboratory is described as follows:

> *"The laboratory is to be erected in a place rich in medicinal herbs. It should be spacious, furnished with four doors and decorated with the portraits of divine beings. It should have several types of apparatus or contrivances. The phallus of mercury in the east, furnaces in the southeast, instruments in the southwest, washing operations in the west and drying operations in the northwest—these and other ingredients necessary for alchemical operations should be installed with chantings. There should be the Kosthi apparatus (or the extraction of essences), pair of bellows, pestle and mortar, sieves of varying degrees of fineness, earthen materials for crucibles, dried cow dung cakes for heating purposes, retorts of glass, iron pans, conch shells, etc."(1).*

Several of these apparati can be seen in Figure 6. The central instrument is a still, while another is being used for the washing of solids. The vertical member to the right is the lingam which is the symbol of the god watching over the laboratory. From evidence such as this we know that there were extensive equipment for evaporation, sublimation, prolonged heating, steeping, distillation, and other chemical processes. All of these were used in small scale and not the large scale which we know today.

Drugs and pharmaceuticals were produced mainly at the village level until well after independence. The chief production of drugs was in the agricultural field, in which they were given primary processing and then shipped to other countries for further processing. Quinine, opium, and belladonna were among the raw materials exported. Oils were produced also in great quantity in India and involved crushing and expressing, all carried out at the village level. The growth of the pharmaceutical industry has been very great since independence in 1947 (*19, 20, 21*). Over 30% of the chemical industry in India is pharmaceutical. For the world as a whole, this averages out to be only 12%.

The production of pharmaceuticals can be measured best in terms of retail pharmaceutical sales rather than actual amounts of the drugs, since in some cases the measure is not a matter of weight but of units of production. The domestic sale of drugs in 1948 was $12,000,000, in 1976 it was $771,000,000. This represents an increase of 540% in a little less than 30 years. The potential for increase in the pharmaceutical industry is great when the per capita consumption of drugs is considered to be in the order of $1.00 (based on retail sales in India) as compared with almost $40.00 in the United States. The planning for the next plan period, the 6th, calls for an increase of $750,000,000 at current prices, which is nearly double (*18, 20*).

At the present, there are more than 2,900 units making drugs and pharmaceuticals. Of these, 116 are large-scale units which produce more than 80% of the total production. Of these, 82 produce basic pharmaceuticals and formulations, while the others depend on the 82 for most of their raw materials. Some of the important drugs currently in production include antibiotics such as penicillin, tetracycline, sulfa drugs, antimalarials, antihistamines, and sex hormones. India produces almost the entire range of modern pharmaceuticals. Part of the technology has been imported from the United States and a larger part from Russia.

In order to control the price of drugs, the government instituted the drug (price control) order of 1970. This law tries to keep the cost to the consumer at a reasonable level. The position of the pharmaceutical industry is difficult since it has to work between fixed prices and growing costs. This requires the strict control on spending and also on instituting steps to see that productivity increases. Another basic need of the industry is a larger expenditure on research and development. At the present time, this accounts for only about 2% of the total turnover. Plans are being made to increase this to approximately 5%.

The pharmaceutical industry furnishes a strong base for employment in India where it is estimated that approximately 5,000,000 people obtain their livelihood directly or indirectly from this source. In addition, great strides are being made in exporting drugs to foreign countries, particularly the developing nations. In view of the low consumption per capita, this may or may not be a desirable situation.

Chemicals and Dyes. Practically every chemical being used is produced within India, some in small quantities, others in large. India also has the capacity to export chemical plants, on a turnkey basis, for such materials as sulfuric acid, caustic soda, soda ash, and fertilizer. In the following discussion, the history of two chemicals, once in great demand from India and which are no longer produced, will be set forth as well as three whose production rate is continuously growing.

INDIGO. Accounts of indigo production occur as early as 1607 (5). A drawing showing the production of indigo is presented in Figure 7 (5). After cutting and gathering the plants, the process consisted of steeping the plants in water, putting the liquid in great pots where it was beaten to allow oxidation to take place, and then settling the resultant material. This was repeated over a period of days, after which the remaining solid was taken out, formed into balls, and laid on the sand to dry. Indigo was grown widely over most of northern India, usually with two crops a year. Water supply was considered in the factory location. The water was carried to the processing plant by gravity. Later, in an improvement of the process, the materials were refined by mixing with more water boiling, filtering through heavy canvas sheets, and forming into thick cakes in a press. After the product was dry it then was packed for shipment. In 1894, over 2,400,000 lb were exported from India, primarily to England. The advent of modern dye-making techniques, developed in Germany, destroyed the indigo industry.

SODIUM NITRATE. In early times, India was one of the chief producers of saltpeter or potassium nitrate in the world. It was exported and used in local industry for making gun powder and pyrotechnics. Most of this saltpeter was produced in northern India in the state of Bihar. This highly populated area of India has climatic conditions which favor the growth of nitrifying bacteria in manure. In addition, the population used wood and cow dung for fuel and ashes of these were scattered over the fields. Following a monsoon (heavy perennial rains), the soil would effloresce and the top ½ in. would be scraped up and placed in long shallow pits where it was leached with water. The resulting crude potassium nitrate then was transferred either to an iron evaporating pan about 5 ft in diameter, or to a solar pond. The rough material contained

Figure 7. Tropical indigo manufactory, 1694: (A) the white overseer; (B) cutting the indigo plant; (C) infusing it in water; (D) carrying away the dye (5).

about 60% potassium nitrate, with common salt as a major impurity.   The purification process, which was conducted in about 400 places in 1905, consisted of redissolving, boiling, removal of scum, and first-formed crystallized material, then recrystallization in a vat containing a bamboo lattice work.   The resulting white crystals were placed in bags, water was trickled through to wash out the final impurities, and the resulting material was spread out and dried.   In 1905, about 2000 tons of saltpeter were produced in India.

SULFURIC ACID.   Until the 1940's the manufacture of heavy chemicals in India was insignificant.   The production of sulfuric acid was only 18,000 tons as contrasted with 7,000,000 tons in the United States at the same time.   Until the establishment of the first contact acid plant in 1948, all production used the chamber process.   Installed annual capacity in 1948 was 175,000 tons, distributed among 49 different units. Because India does not have a source of free sulfur, the production of sulfuric acid is one of the critical chemical industries.   India, however, has large deposits of iron pyrite in the northern part of the country and development of these is being undertaken to alleviate the problem of the lack of sulfur.

By 1951, the production of sulfuric acid had increased to 107,000 metric tons.   There has been a steady increase in the production of sulfuric acid by the addition of many plants.   Some of these are produced within India using completely Indian materials.   In 1977, the estimated production was 1,900,000 metric tons. The production since 1951 through 1977 is shown in Table I.   Also shown in this table is a figure of percent utilization.

One of the problems in Indian industry, particularly in the public sector, is the poor utilization of facilities. The average for most industries is close to 50–55% utilization. The utilization shown here is somewhat higher. The higher figures are primarily due to those plants within the private sector. The poor utilization is caused by two primary factors, both of which are the responsibility of the central government. The first relates to the availability of materials and power. Inadequate attention has been given to transportation (moving by rail), so that basic materials such as coal pile up at the mine head or point of origin. The lack of power results in long periods of shortages. The second relates to the use of labor. The central government tries to hire as many people as possible without due regard to the efficiency of use. This results in low pay, poor working conditions, and poor morale. There are many strikes with the consequent loss of production.

By 1968–69, all of the sulfuric acid plants were of the contact type.

Table I. Sulfuric Acid Production (1000 Metric Tons) (21,22)

Year	Capacity	Production	% Utilization
1951	201	107	53.1
1952	192	96	50.1
1953	189	109	57.7
1954	193	151	76.5
1955	208	166	74.9
1956	245	165	67.4
1957	273	196	71.8
1958	290	227	78.1
1959	374	292	78.1
1960	476	354	74.4
1961	564	423	74.9
1962	702	470	66.9
1963	821	568	69.2
1964	1011	580	67.2
1965	1082	685	63.3
1966	1328	690	51.9
1967	1529	805	50.8
1968	1955	1008	51.5
1969	1921	1121	58.4
1970	1930	1051	54.4
1971	1963	1022	52.1
1972	1963	1028	52.4
1973	2112	1434	67.7
1977	2640	1900	72.0

The sixth five-year plan *(18)* shows a planned increase of sulfuric acid production to 3,790,000 metric tons per year. This is an increase of 44%.

CHLOR-ALKALI INDUSTRY. Soda ash and caustic soda are among the fastest-growing, heavy-chemical industries within India. The production of soda ash was about 48,000 metric tons in 1951 and increased to an estimated 530,000 metric tons by 1977. Soda ash in India is produced primarily through the Solvay process and the locations of the plants are heavily dependent upon the location of raw materials of limestone and salt. Since 1966, licenses have been issued for soda ash plants which are in noncoastal regions, which means that the availability of salt is not as good as in other locations. The soda ash industry has nearly achieved self-sufficiency and has potential for export. Sufficient know-how has been developed to design and fabricate plants within the country and to establish similar plants in other developing countries.

Caustic soda also has shown a very good annual increase of about 17% per year. The production in 1951 was 14,000 metric tons and had increased to 530,000 metric tons by 1977. Small units using the electrolytic process were established first in 1941 in Calcutta and Metur. Later several plants were established using caustification of soda ash. Mercury cells went into production in 1952. The country achieved self-sufficiency in 1967. Nearly all of the states in India now have caustic soda–chlorine plants. Complete caustic soda plants using mercury cells now are designed and produced entirely within the country. All of the parts are manufactured indigenously within India and some plants are being designed and constructed for other countries.

Metals. IRON. The major metals produced within India are iron, aluminum, and copper. Many others are produced but not covered here. Iron made its first appearance in India sometime prior to 1500 B.C. It probably was brought to India by the Aryans who entered from the north and eventually settled over most of the northern part of India. The use of iron is evidenced by slag piles and some furnace remains which are scattered over much of India. Iron objects found encompass the entire range of tools, weapons, and utensils. Following 800 B.C., Indian iron and steel objects were recognized and praised for their quality throughout the western world. When Alexander reached India, one of the gifts to him was 100 talents of steel, a prized commodity. The Indians were adept also in using steel in cutlery and armor. The early technicians of iron and steel in India were able to make some of the largest castings in the world. The outstanding example of this is an iron pillar located near New Delhi which has a height of over 24 feet and weighs over 6 tons. By chance or by design this material is made of nonrusting iron. Even after all of these years and after standing in the open for centuries, there is no sign of rust. It is covered with a uniformly blackish-colored coating except where near the base it has become bright and shiny from hands of countless visitors. Other evidences of the use of iron in construction also are found in many of the ruins within India.

The modern steel industry is about 75 years old. The first fully successful iron and steel mill was established in 1907 by the Tatas. There were many earlier attempts to establish iron works but these used western technology unsuited for the materials locally available and were all failures. In about 1918, two more iron mills were established within India, founded by British interests. These are still in production today. During World War II, steel was produced in greater and greater quantities. India became self-sufficient in steel about 1954. The steel plants in India are modern and have been built primarily with the help of Russian collaboration. Part of these mills are still in the construction stage. Production of steel in 1951 was about 1,000,000 tons and it had increased to 13,000,000 tons by 1977.

At the present time, 1978, there are six integrated steel plants and two specialty plants with a total installed capacity of 10.6 million tons of steel and pig iron and 137,000 tons of speciality steel. India now stands 13th in steel production in the world.

Construction is expected to start in 1979 on a 6,000,000-ton plant in the southern part of the country. This plant will be entirely of Indian design. Two other plants are also under consideration and should be started during the sixth planned period (18).

ALUMINUM. Aluminum is used extensively in modern India particularly in house wiring and the transmission of electrical power, about 50% of the production going for this use. All of the technology has been imported primarily from the United States. Production has increased from about 3000 metric tons in 1951 to 180,000 metric tons in 1977. It is estimated that the demand for aluminum in 1984 will be 400,000 metric tons. The demand is likely to increase dramatically since the per capita use in India is 0.4 vs. 22 kg in the United States and of 2.9 kg in the rest of the world. Aluminum today is produced in five producing plants, four of them in the private sector and one in the public sector. The most serious problem with the aluminum industry is the pricing policy. By law, 50% of production goes to the government at a fixed price, so-called levy metal. The current levy price is $903 a metric ton vs. a production cost of $1,084 per metric ton.

COPPER. Copper is of ancient origin in India and some of the earliest uses of the metal in the world seem to originate from the India subcontinent. The method of production was crude and consisted of very small furnaces. An example of a furnace is shown in Figure 8. The ore was charged to the small cylindrical furnace, mixed with charcoal or with cow dung, and then heated by fire from the bottom, the fire being driven by goatskin bellows. Goatskin bellows still are used today to make a small, hot fire for use in blacksmithing.

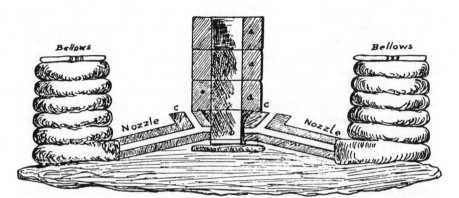

Figure 8. Schematic of a native copper-smelting furnace at Singhara near Khetri in Rajasthan (1831). (a) Kothi of three separate annular parts made of fire-clay and placed one upon the other firmly: exterior diameter of each part, 15"; height, 9"–10"; thickness, 3". Quantity of the charge: 2½ maunds (200 lb) of the ore balls (pindi) and 3 maunds (240 lb) of charcoal along with some iron-bearing material to act as flux. (b) Chamber for burning some quantity of charcoal to drive out the moisture from the newly molded furnace. (c) Openings for poking the fire from time to time, being closed with moist clay after the poking operation (1).

At Khetri, the same location that used the smelting equipment shown in Figure 8, a modern copper complex is being built. This complex consists of the latest in foreign technology for the smelting and refining of copper. It includes all of the phases of copper production from mining, benefication, smelting, refining, recovery of sulfuric acid, and the making of fertilizers from the resulting sulfuric acid. The remains of the ancient mines and slag dumps can be seen adjacent to the most modern copper smelters.

Fertilizer. Fertilizer continues to be a critical need for the progress of India *(21, 22, 23, 24)*. All fertilizer production is of modern origin with very little production occurring before 1961. Most of the manure is used as fuel so that there has been little use of fertilizers within India. In India, the application rate is only 13.2 kg/ha as contrasted to the 47.4 kg/ha in the world as a whole. This means that the production levels of fertilizer are going to have to increase dramatically in the future. A large portion of investment in the next five-year plan will be in fertilizers. The production of total fertilizer content (N, P, K) in 1961 was 347,000 tons, by 1977 it was estimated to be 2,430,000 metric tons. The largest problem with fertilizers is a very low utilization factor. In 1977, this was less than 55%.

India currently has the ability to manufacture fertilizer plants entirely within the country and is exporting some plants to other developing countries.

Research. In ancient times, research was carried out in the laboratory by private individuals and was concerned mainly with alchemy. Under the many phases of occupation of India, no research was planned. With the coming of independence, research was included in the planning. The overall research effort within India is very small. There never has been a great incentive for the Indian industrialists to expend funds for the research and development necessary to maintain a lead or develop the chemical industry. The licensing policies make it more advantageous to hire foreign technology and buy foreign equipment rather than to develop it internally. These policies are currently under review and hopefully more research will be done in the future.

The bulk of the industrial research within India is controlled through government funds and a separate organization has been developed for the pursuit of research and development (25). The primary research organization in India is the Council of Scientific and Industrial Research (CSIR) which is an autonomous body. This and other bodies are organized in various areas and have the primary responsibility for research. The actual research is being carried out in the national laboratories supported by the government. The results from these laboratories are made available to industry through licensing agreements. The research organization is carried on at two levels—both at the federal level and at the state level.

Most research is carried out either in educational research institutions or in the national research laboratories (mentioned earlier), both of which are structured and operated much like a graduate research facility in an American university. The use of processes developed in these institutions is small compared with the total production in the chemical industry. This is caused, primarily, by the charges for use of the process and by the burdensome licensing procedures involved. Steps recently have been taken to overcome these deficiencies through the formation of advisory boards.

The National Committee on Science and Technology (NCST) has made an extended study to determine the types of research which are now necessary. Many questionnaires were circulated throughout the industry and the results were analyzed. From this study the NCST technological plans (26, 27, 28, 29, 30) have been published which outline the research needs of the industry. There are two reports which pertain particularly to the chemical industry (29, 30). These documents are being used for national planning and are influencing the allocation of research funds.

The National Chemical Laboratory (NCL) at Poona in the western part of India is charged with the primary responsibility for the development of expertise in the manufacture of a large number of chemicals and materials. There are also a number of other related national laboratories.

Education.   In ancient times, education consisted of small groups of scholars studying under a single teacher.   The teacher or guru accepted a small number of students who lived, worked, and studied in a secluded place called an ashram.   Most of the study was in the fields of philosophy and religion and did not relate to medicine or chemicals.   The practices of chemistry were limited to making drugs and cosmetics and were closely held and guarded family secrets, the skill being passed from father to son.   During the British era, education consisted mostly of training Indians to function in governmental service.   For this reason, science and other related fields were more or less neglected although some very famous scientists emerged during this period.   There was little training in engineering until after independence.   The education problem in India is complicated by the sheer size of the country.   At the present time there are over 625,000,000 people in India with an annual increase of 12,000,000–13,000,000 per year.   The age distribution is dominated by the younger ages, about 42% being less than 15 years old *(15).*   The college-age population represents about 17% or over 106,000,000 persons.   There are 118 universities having a total enrollment of approximately 3,200,000 *(31).*   This can be compared with the United States where there are over 9,000,000 in the colleges and universities.   Education on the scale practiced in the United States would pose a problem not only in expenditure of funds but also in the problem of employment of the students after graduation.   The economy is not such that it could assimilate this many people.   There presently are many who cannot find suitable employment.

The pattern of education is varied and changing.   At present the *(15)* plan being implemented by the state and central governments is designated as 10 + 2 + 3.   This consists of ten years of school followed by a state-administered examination covering a prescribed number of subjects (these may or may not include science).   Successful students then complete two more years of schooling and take another state (or central) examination.   Graduation is contingent on passing these examinations and not on the work done in the course of study.

The students then compete, on the basis of the examination score, for a place in one of the universities.   A three-year program leads to a Bachelor of Science (BS) or Bachelor of Arts (BA) degree.   A five-year program leads to a Bachelor of Engineering (BE) degree.   The number of students admitted to each program of study is limited and the competition is very great, the greatest competition being for a seat (admission) in engineering.

The course of studies available is very rigid.   These are set up on a year-by-year basis.   A student passes or fails by taking a final examination at the end of the year's study.   He must pass all of the courses or repeat the entire year.   There are a number of schools now adopting the U.S. system of semesters and continuous assessment.

There are 32 chemical engineering departments with an annual capacity of approximately 1400 students. In 1948 there were only 8 chemical engineering departments with a capacity of 200 students. The fraction of students in college studying chemical engineering is on the order of 7000 students—a very small fraction. For comparison, the estimated BS degrees granted in chemical engineering in the United States in 1978 was 4,600 with a total undergraduate enrollment of over 30,000 *(32)*.

At the post-graduate level, 25 institutions offer a Master of Engineering (ME) degree, with 20 of these offering PhDs. Due to the competition from jobs and of going to a bigger university, the post-graduate programs are limited and the available students are not always the best. In addition, the wage structure does not favor a person with an ME degree.

There are five large, all-Indian universities which offer chemical engineering degrees. These are the Indian Institutes of Technology located at New Delhi, Bombay, Madras, Kharagpur, and Kanpur. Each of these was started with the help and collaboration of different foreign governments. For example, the IIT at Kanpur was formed through the help of a consortium of ten American universities. Current plans are to restrict the number of chemical engineering students undergoing study to match the current job potential in India.

Numerous educational experiments in chemical engineering are being carried out within India. An example of this is the one carried out at the Birla Institute of Technology and Science (BITS) located at Pilani, Rajasthan, India. This is a small engineering institution of about 2200 students. It includes faculties in engineering, science, and the humanities. They have been developing a program designated as practice school. This is similar to the practice school which is in operation at M.I.T. and is modeled, in part, after it. All students spend three different sessions in practice school locations, one of about two months, another of about six months, and a final one of about four months in a design firm. In this the students and faculty are posted at a factory location and work on real problems. These problems are of interest to the local industries and form a good educational background. The unique part of this practice school program is that it encompasses not only chemical engineers but all engineering, science, and humanities students within the institution. It is well received by the students and by the faculty and industry. Other Indian schools currently are watching the development of this program.

Indian Institute of Chemical Engineers. Just as chemical engineering is a newcomer to the Indian scene, so is the Indian Institute of Chemical Engineers (IIChE), the counterpart of the American Institute of Chemical Engineers (AIChE). The first chemical engineering department was near Calcutta at Jadavpur University. There, one of the leading chemical engineers was Professor Hira Lal Roy. He felt the need for having a chemical engineering society to represent the profession in India Professor Roy held the first formal meeting of the IIChE on May 18, 1947 at the Indian Science Congress held in Patna. Dr. Roy gave the Institute distinguished leadership over a period of 18 years during which it became a national platform for the profession.

In 1948, there were 101 members and in 1958, 384; by 1978 the membership was over 2800. This enrollment represents about 10% of the 18,000 chemical engineers in India. At present, the country is divided into 18 regional centers (sections) *(3)*. The Institute moved into its own facilities on the Jadavpur University campus in 1973. The Institute was recognized by the AIChE as well as other professional societies in 1958. Publication of the society journal, *Indian Chemical Engineers*, was started in 1959.

The IIChE is a dynamic and growing organization. In 1959, an associate membership was started. To obtain this membership, a rigorous examination is given by the Institute to chemical engineers who have not had the benefit of a formal education from a recognized university. The associate membership is recognized as equivalent to a degree by the government of India. This institute is a member of the Federation of Engineering Institutions, representing the chemical engineers. It promotes excellence among students by sponsoring an essay contest and aiding in the formation of student chapters. It is also active in setting standards for education, setting up continuing education programs and programs designed to promote the development of appropriate technology, and in advising the government.

Literature Cited

1. Bose, D. M.; Sen. S. N.; Subbarayappa, D. B. "A Concise History of Science in India"; Indian National Science Academy: New Delhi, 1971.
2. De Sonsa, J. P. "History of the Chemical Industry in India"; Technical Press Publications: Bombay, 1961.
3. "Twenty-five Years of Chemical Engineering in India"; Editorial, *Indian Chem. Eng.* 1973, 15(1).
4. Watt, Sir George "The Commercial Products of India"; Today and Tomorrow's Printers and Publishers: New Delhi, 1966; 1908 reprinted edition.
5. Singer, Charles; Holmyard, E. J.; Hall, A. R.; Williams, Trevor I. "A History of Technology"; Oxford University Press: London, 1964.
6. Agricola, Georgius "De Re Metallica," translated by Herbert Clark Hoover and Lou Henry Hoover; Dover Publications, Inc.: New York, 1950.
7. Chaudhuri, M. R. "Indian Industries Development and Location," 4th ed., in "Indian Economic Geographic Studies"; 1970.
8. Malik, K. B.; Dhawan, C. L. "Main Industries for Indians"; The Youngmen's Own Institute, 1933.
9. Mukherjee, Radajkmamal; Dey, H. L. "Economic Problems of Modern India"; MacMillion Co., Ltd.: London, 1941; Vol. II.

10. Mukherjee, S. K. "Our Developing Economy: Areas of Concern to Chemical Engineering," *Indian Chem. Eng.* **1978**, *20*(1).
11. "The Imperial Gazetteer of India," in "The Indian Empire"; Economic, Oxford Clarendon Press: 1908; Vol. 3.
12. "Twenty Years of Indian Chemical Industry, 1949–1969," *Chemical Age of India* **1969**, *20*(11).
13. Chaudhuri, Rohinimohan "The Evaluation of Indian Industries"; University of Calcutta: Calcutta, 1939.
14. Basu, D., Mager, D. Chatterjee, R. "Ruin of Indian Trade and Industries, Calcultta"; 1939.
15. "India 1977–78," Publications Division, Government of India: New Delhi, 1978.
16. "Sanctioned Capacities in Engineering Industries," National Council of Applied Economic Research: New Delhi, 1971.
17. "Fourth Five-Year Plan 1969–1974," Planning Commission, Government of India: New Delhi,
18. "Draft, Five Year Plan 1973–1983," Planning Commission Government of India: New Delhi, 1978.
19. "Indian Pharmaceutical Guide, 1978"; Pamposh Publications: New Delhi, 1979.
20. "Report of the Committee on Drugs and Pharmaceutical Industry," Ministry of Petroleum and Chemical Government of India, New Delhi, April 1975.
21. "Kothari's Economic and Industrial Guide of India," 31st ed.; Kathari and Sons: Madras, India, 1976.
22. Sharma, Lal "Directory and Year Book, 1978," *The Times of India*, Times of India Press: Bombay, 1978.
23. Sahar, Ankulkarui C. "Fertilizer Statistics," Fertilizer Association of India: New Delhi, 1972.
24. "Fertilizer Statistics, 1973–74," The Fertilizer Associates of India: New Delhi, 1979.
25. Rahman, Bhargava, R. N., Qureshi, M. A., Pruthi, Sandarshan "Science and Technology in India"; Indraprasthan Press News: Delhi, 1973.
26. "Science and Technology Plan," National Committee on Science and Technology, Government of India, New Delhi, **1974**, Vol. 1.
27. "Science and Technology Plan," National Committee on Science and Technology, Government of India, New Delhi, **1974**, Vol. 2.
28. "Science and Technology Plan," National Committee on Science and Technology, Government of India, New Delhi, 1974.
29. Tilak, B. D. "Report on Science and Technology Plan for the Chemical Industry, *A General Overview*," Government of India, New Delhi, **1973**, Vol. 1.
30. Tilak, D. B. "Report on Science and Technology Plan for the Chemical Industry, Status Report on Chemical Industry," Government of India, New Delhi, **1973**, Vol. 2.
31. "Report of the Education Commission," Department of Education, Government of India, New Delhi, 1966.
32. Matley, Jay; Ricci, L. "New Chemical Engineers, Too Many, Too Soon," *Chem. Eng.* **1979**, *86*(2).

RECEIVED May 7, 1979.

14

The Separate Development of Chemical Engineering in Germany

KARL SCHOENEMANN[1]

Technische Universität Darmstadt, D 6100 Darmstadt, Federal Republic of Germany

Whereas in the United States chemical engineering has been developed as an autonomous discipline towards the end of the last century, it generally has been rejected in Germany until approximately 1960. Process development and plant design were done by teams of conventional chemists and mechanical engineers. During the last two decades international exchange of experience increased and finally brought about an approximation of standpoints. In Germany today the field of American chemical engineering is covered by several professions with different educational backgrounds.

During the three and a half decades after World War II chemical engineering, as developed in the United States, has been under much discussion in Germany *(1–11)*. Until about the first part of the 1960's the American concept generally has been rejected.

The United States, which was at the beginning of chemical industrialization in the 1880's, already realized that conventional chemistry and mechanical engineering would not suffice for the novel tasks of developing chemical production processes and that the gap had to be closed by means of a new autonomous discipline. In Germany these tasks were solved through the teamwork of conventional chemists and mechanical engineers, both thinking in their own philosophy and working according to their own methods.

As late as 1952, E. L. Piret *(12)* of the University of Minnesota, with his large amount of experience in education and industry, complained about the European lack of understanding with regard to the importance of chemical engineering for chemical industry:

---

[1] Current address: D 8730 Bad Kissingen, Heinrich von Kleist–Str. 2, West Germany.

0-8412-0512-4/80/33-190-249$05.75/1

"On the continent one can count on the fingers of one hand the educational centers giving what approaches our concept of chemical engineering education. At not one of these the student receives the integrated training . . . which we consider important. The professional status . . . existing in the field of chemistry and in the other branches of engineering on the continent and in America does not present nearly as sharp a contrast as in chemical engineering" (12).

Piret considers:

". . . the retarding influence of strongly entrenched interests and rigid tradition working within the universities and technical schools and even in the industries themselves to be worse . . . than the effects of World War II" (12).

The federation of the German chemical industry still rejected the introduction of autonomous chemical engineering in 1954 (13). The contrast between the American and the German concept became obvious at the First European Conference of Chemical Engineering Education in London in 1955. At this conference the author was the only European representative who advocated the American concept (2, 5). In 1956, a group of American Fulbright lecturers gave their impression in a report in Chemical and Engineering News (14) with the headlines: "Chemical Engineering New to the United Kingdom, Unknown in Germany—Germany Does Not Agree—Darmstadt Differs in Aims."

After this culmination the contrast has turned more and more into a distinct approximation of the differing standpoints as a consequence of the growing international scientific exchange and of the increasing similarity of industrial tasks leading to similar technological solutions in the various countries (15). However, the necessary linguistic and institutional measures have not been taken yet. As before, two great professional institutions exist in Germany that represent chemical engineering scientifically, economically, and socially: the Deutsche Gesellschaft für Chemisches Apparatewesen (DECHEMA) and the Gesellschaft für Verfahrenstechnik und Chemisches Ingenieurwesen (GVC). DECHEMA is concerned more with the chemical aspects of chemical engineering while GVC with those of mechanical engineering (16). Both institutions have established very active working parties for the manifold branches of chemical engineering. Some of them have developed into joint units, each with a common chairman.

At the universities both traditional lines of education, namely that of engineering with its mechanical, civil, and electrical divisions, and that of chemistry with its organic, inorganic, and physical divisions, have been continued, of course. During the last few decades the departments of both faculties have, though maintaining their leading philosophy, contributed to bridging the gap by establishing specialized branches that are designated by the chemists as "technical chemistry" and by the engineers as "verfahrenstechnik." At the various universities the right understanding of chemical engineering differs considerably.

The German word "verfahrenstechnik," literally to be translated as process technics, can be misunderstood easily; Americans should not try to translate it. When this expression was coined prior to World War I, it was practically identical with unit operations. Now it is used by engineers to express that their methods can be applied to any type of industry such as mechanical, chemical, food, oil, and other industries. The German technical chemist by using this word means the development of a process whose core is the reactor which is decisive for the chemical conversion and thus for the whole arrangement of the plant for production cost and profitability.

Even in industry, research and development still are organized mostly in separate departments for technical chemistry and Verfahrenstechnik whose members cooperate, but may also be rivals.

Only about ten years ago three German universities—Erlangen, Karlsruhe, and Dortmund—founded independent departments of chemical engineering according to the American concept. During the past several years the novel problems arising from the shortage of raw materials etc. have enhanced the demand of chemical engineers considerably. Unemployed organic chemists are placed in temporary courses into chemical engineering—circumstantial evidence that the rejecting attitude described above has changed completely (17).

The long-range development in the German Democratic Republic does not greatly differ (16).

An Illustration of the Historical Development of German Chemical Engineering

We may now ask: what are the reasons for this separate development in Germany and what are its consequences? The emergence of the facts is so manifold and often accidental that a survey of theory and practice (taken as a whole) can be given only intuitively and subjectively. Such a picture then can be compared with the more systematic development in the United States as for instance described by O. A. Hougen in his report *Seven Decades of Chemical Engineering* (18). According to the author's experience gained during nearly 60 years of activity, half in industry and half at the university, the development of American and German chemical engineering presents itself in Figure 1.

About 1890, which was prior to the synthesis of alizarin and indigo, the United States enjoyed a position of organizational superiority because they recognized the importance of chemical engineering as an independent discipline.

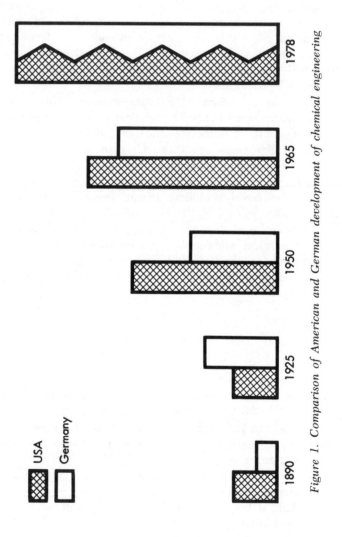

Figure 1. Comparison of American and German development of chemical engineering

At that time education in pure chemistry had reached a high standard in Germany where since the beginning of last century the technical schools teaching trade methods in order to support the textile industry and their auxiliary productions (lead-chamber sulfuric acid, hydrochloric acid, the Deacon process, etc.) had affiliated chemical departments that took up Liebig's system of chemistry, and later on conventional physical chemistry and mathematics.   In time these schools developed to such a degree that towards the end of last century they were technical schools of university level.   The same branches had been introduced in all of the traditional universities.

The decisive impetus for chemical engineering came from the dyestuff industry.   As shown by the formulae of alizarin synthesis in Figure 2, fuming sulfuric acid was required which at that time was produced by roasting ferrosulfate.   In view of its exorbitant price, Rudolf Knietsch of BASF, the leading chemist of that time, was directed to the production of SO_3 by catalytic oxidation of SO_2 with air *(19)*.   The son of a blacksmith he first became a locksmith, then an engine driver; later on he found the time to graduate from high school and finally to study chemistry.

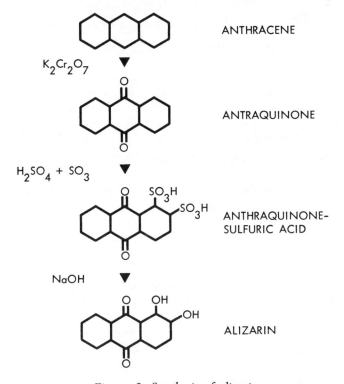

Figure 2. Synthesis of alizarin

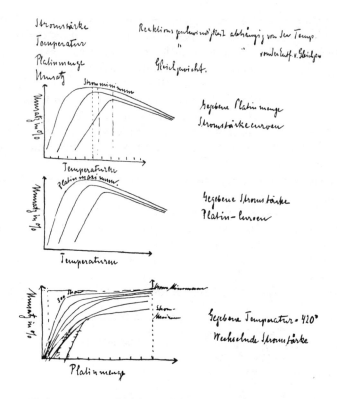

Figure 3. Knietsch's original diagrams of SO₃ catalysis

Although Knietsch's original diagrams of his laboratory experiments in
SO₃ catalysis are nearly 100 years old, they look rather modern. The
sequence (*see* Figure 3) representing the degree of conversion in its
dependence on the various reaction factors proves that he had a clear
insight into the complex interaction between reaction velocity and equilibrium. For this reaction he invented the ingenious tube-bundle
reactor whose feed for the fixed bed inside the tubes was preheated in
countercurrent by the hot reaction mixture outside of the tubes.

Supported by the profits made with alizarin, Heinrich von Brunck of
BASF (20, 21) could tackle the synthesis of the "king of dyestuffs" of that
time—indigo. Several processes were tried out, and success came only
after 17 years when 18,000,000 marks had been spent on this development, an amount of money that equaled the capital stock of the company.
The indigo synthesis demonstrates how in an admirable effort a company
summons all of its energy for years in order to achieve one single objective. The intermediates of the abandoned indigo processes became the
origin of independent productions thus prompting industrial diversification (*see* Figure 4).

The success achieved with indigo encouraged BASF to cope with larger tasks and helped to raise funds for developing the catalytic high-pressure synthesis of ammonia, then the big world problem. This synthesis, as realized by Fritz Haber and Carl Bosch *(20–25)* is known by everyone. Its fundamental achievements mark the turning point on the way leading from the traditional batch processes to modern continuous catalytic mass production. Today reactor capacities have reached 1700 tons/day at 2.4 m in diameter and 34 m in length and at a pressure of 330 atm and a temperature of 500°C. Historically it is of interest that in 1905 Carl Engler, head of the Chemical Department of Karlsruhe Technical School, in his position as member of the board of BASF, invited the attention of that company to the ammonia experiments of his assistant Fritz Haber. It is important for the further development of chemical

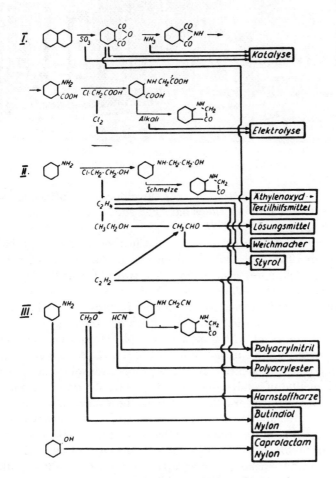

Figure 4. Consecutive products of the indigo syntheses

Figure 5. Coal hydrogenation plant in 1927

engineering that Alwin Mittasch (22) developed the principle of the mixed catalysts which in our times have had such a far-reaching influence on chemical syntheses.

The columns of Figure 1 indicate that further progress was made in Germany from 1925 until 1950. During this period the development of high-pressure techniques proceeded along two different lines.

One was the hydrogenation of coal and crude oil bottoms into motor fuel as fundamentally solved by Friedrich Bergius (26), who together with Carl Bosch was awarded the Nobel Prize in 1932. In 1927 the first industrial gasoline plant with a capacity of 100,000 tons/year (*see* Figure 5) was erected by BASF, at that time part of the IG Farben group (27).

It is impossible to mention all of the problems involved such as those of the steel quality, of catalysts resistant to sulfur, of pumping ground coal in the form of slurry, of exactly controlling the enormous reaction heat, and of control technique, etc. However, coal hydrogenation was no economic success as a consequence of the competition of the cheap excess production of motor fuel caused by the improvement of oil field exploration and crude oil cracking. In comparison with the indigo synthesis, development cost reached an even higher level, amounting to 40% of the share capital of the whole IG Farben group, which was comprised of six big works *(28)*. In the following decades, however, when high-pressure hydrogenation was applied to the oil industry, great successes were achieved, for instance the increase of gasoline yield by hydrogenation of residues, the aromatization of gasoline, and the production of pure aromatic and unsaturated hydrocarbons.

In 1921, Fritz Winkler discovered, in connection with the high-pressure syntheses, the principle of fluidization, a further valuable contribution to chemical engineering. Starting from Winkler's first rough sketches *(see* Figure 6) the realization of a generator gas capacity of 50 000 m³/hr *(see* Figure 7) required only ten years *(20, 21)*. The application of oxygen from the rational mass production by the Linde–Fränkl air liquification allowed the production of water–gas and hydrogen. The whole field of synthesis gas production from coal—the only raw material being available in Germany—is a typical German development. The fluidized-bed principle was extended to the catalytic cracking of petroleum, the roasting of pyrite, the oxidation of naphthalene into phthalic acid, and the production of olefins, etc.

The other line of development in high-pressure techniques were the syntheses of methanol and isopropyl alcohol. Later on, in combination with olefin and acetylene chemistry these syntheses opened up the great field of highly selective catalytic processes for pure organic compounds.

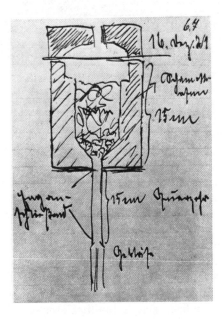

Figure 6. Fritz Winkler's sketch of his discovery of the fluidized bed

Figure 7. Fluidized-bed gas generator of 50,000 m³/hr capacity

First there was the chemistry of olefins which owing to Germany's oil deficiency could not be obtained by cracking, but had to be produced by hydrogenation of acetylene originating from coal via calcium carbide. At the beginning of the 1920's, acetylene was used first for addition reactions at normal pressure; the most important product was acetaldehyde which via aldol, 1,3-butanediol, and butadiene led to synthetic rubber. Other important products were ethylene oxide, acrylic esters, styrene, etc.

At the end of the 1920's, Walter Reppe (*29, 30, 31, 32*) started his experiments on catalytic reactions with acetylene under pressure. On the basis of his studies, which soon were known all over the world as Reppe chemistry, it was possible to construct complicated organic compounds of high value from simple building stones. From the standpoint of the chemical engineer the greatness of his achievement was that the dangerous manipulation of acetylene under pressure from which every-one had recoiled could be realized and steadily improved by means of minute experiments, often carried out contrary to the safety instructions of that time.   The conditions on which the various kinds of acetylene decomposition, as e.g. slow reaction, explosion, and detonation, originated were investigated for each synthesis reaction.   Thus safe operating conditions were created, for instance by always maintaining definite partial pressures and avoiding empty volumina of vessels and tubes.   In 1940, a large plant for 30,000 tons/year of synthetic rubber was erected on the basis of acetylene and formaldehyde (*see* Figure 8).   It was the first industrial application of the trickle-bed reactor.   After the production of synthetic rubber had been abandoned, butynediol became the

Figure 8. Butynediol plant

Table I. Oxo and Reppe Syntheses

1. $CH_3-CH=CH_2 + CO + H_2$ $\begin{cases} \longrightarrow CH_3-CH_2-CH_2-CHO \\ \longrightarrow CH_3-CH_2-CH_2-CH_2OH \\ \text{BUTANOL} \end{cases}$

2. $CH_3OH + CO \xrightarrow{\text{Co/J}} CH_3COOH$
     ACETIC ACID

3. $CH \equiv CH + CO + HOH$ $\xrightarrow[\text{Cu J}]{\text{Ni Br}_2}$ $CH_2 = CH - COOH$
4.   respect. $+ HOR$ $- COOR$
     ACRYLIC ACID

5. $CH_2 = CH_2 + CO + HOH$ $\xrightarrow[\text{200°C \quad 300 at}]{\text{Ni-propionate}}$ CH_3-CH_2-COOH
     PROPIONIC ACID

6. $CH_3-CH=CH_2 + 3CO + 2H_2O$ $\xrightarrow[\text{<100°C \quad 12 at}]{\text{Fe(CO)}_5 + \text{org base}}$ $CH_3(CH_2)_2 CH_2OH + 2CO_2$
     BUTANOL

starting material for a large number of syntheses. The most important ones were derived from butanediol, such as γ-butyrolacton, α-pyrrolidon, and tetrahydrofuran (THF).

With the same aim in mind of obtaining valuable products from low molecular compounds, Otto Roelen (33) of Ruhrchemie discovered the oxo synthesis in 1935 (see Table I).

According to the various reaction conditions, especially those of temperature and pressure, aldehydes or alcohols were produced with cobalt catalysis by reaction of olefins with carbon monoxide and hydrogen, as shown by Equation 1 of Table 1. Acetic acid is formed by a specific reaction according to Equation 2.

In continuation of Reppe chemistry, catalytic reactions of acetylene with carbon monoxide and compounds containing hydrogen atoms, e.g. water or alcohol, were developed in the process of carbonylation (31), as seen in Equations 3 and 4.

Under more severe conditions, at 200°C and 300 atm, these reactions could be executed not only with acetylene, but also with olefins, as shown by Equation 5.

By industrially manipulating acetylene or olefins under pressure in this way, chemical engineering helped to bring about very important productions, for instance that of acrylic acid from acetylene with a capacity of 130,000 tons/year and of butanol from propylene with 30,000 tons/year. For each of these reactions very specific and laborious developments were additionally required. The decomposition of iron carbonyl by nascent CO_2 forming $FeCO_3$ had to be prevented by an increase of CO partial pressure.

The whole development described above had coal as a basis, Germany's only raw material. There is no doubt that the standard reached in chemical engineering at about 1950 shows that much progress has been made in comparison with 1925 (see Figure 1). This important progress, however, does not so much concern an increase in chemical engineering theory and methods, but more so in practical experience gained in definite processes. The design of industrial plants was carried out mostly by empirical scaling up. The descriptions of chemical processes in *Ullmann's Encyklopädie der Technischen Chemie* edited in 1951 *(34)* and in the FIAT reports on German chemical processes written after World War II on command of the Allied occupation army don't contain appreciable chemical reaction engineering knowledge. No textbook existed comprising the entire field and the intrinsic viewpoints of modern chemical engineering.

Though collected editions of chemical engineering theory existed since the early 1930's, as for instance by Eucken and Jakob *(35)* and by Berl *(36)*, their generalizing kind of thinking did not meet widespread response. The importance of retention time distribution *(37)* and of Damköhler's fundamental work on the influence of diffusion, fluid flow, and heat transfer on the degree of conversion *(38)*, which still today represents the basis of chemical reaction engineering, was not realized properly until 1950. Ever since then the fields of heat transfer *(39)* and distillation generally have been dominated by chemists.

In 1950 it was recognized in Germany that one had a lot to learn, and it wasn't until 1965 that the American standard had nearly been reached with regard to unit operations and chemical reaction engineering *(see* Figure 1).

All this was made easier when, with the beginning of the computer age, the traditional graphical methods of plant design were replaced by numerical methods which in Germany were adopted as quickly as in the United States. During the last two decades *(see* the columns for 1978 in Figure 1) the keystone of chemical engineering progress was laid in Germany or in Europe, respectively, so that both the American and the German sides were on a par, as has been expressed in British periodicals *(11, 12, 13, 14, 15)*. At present one may say that there is a complete equilibrium with an intensive interchange of theoretical and experimental findings. There no longer is a difference in the level of manufacturing companies and chemical constructors *(40)*. Recently expensive developments such as the resumption of large-scale coal hydrogenation experiments for training purposes have been achieved by joint ventures. [The adaptation of the chemical industry to oil as a raw material that occurred during the period under review could be done without any difficulty.]

Interaction Between Theory and Practice in Germany

In considering all of the contributions made in Germany by theory (from education) and by practice (from industry) one arrives at the present standard of chemical engineering *(see* Figure 9).

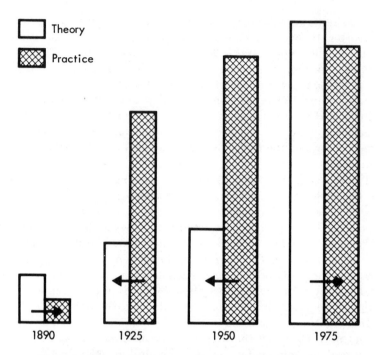

Figure 9. Interaction between theory and practice in Germany

Until about 1890—that means prior to alizarin synthesis and SO_3 catalysis—the column of industry is lower than that of education, so that at that time industry could not contribute so much to the progress of theory. This appears to be justified with regard to the high level of education in chemistry, especially in physical chemistry and physics. Until 1925, the emphasis shifted from theory to practice, which reflects the enormous technical progress during this period resulting in the development of dyestuffs, SO_3 and ammonia. O. A. Hougen wrote in his report on *Seven Decades of Chemical Engineering (18)* that "Germany was 50 years ahead of the rest of the world in organic chemical processing."

The height of the 1950 column for industrial practice represents the extension of German high-pressure engineering into such new fields as coal hydrogenation and Reppe chemistry, where, however, industry did not contribute to the theory of chemical engineering; the column of education still lags behind. As shown by the arrows on the diagram in Figure 9 in 1925 and in 1950 theory could profit considerably from the industrial know-how. The theoretical conception of catalysis can in the last analysis be traced back to the work done by Knietsch and Mittasch. From 1950 chemical engineering theory gains more and more in importance; chemical processes mainly are executed catalytically and not so much with the aid of auxiliary chemicals.

For 1975 we can notice the whole breadth of theory and practice with a certain advantage of theory of which industry is now profiting. New developments, such as the introduction of the computer into chemical industry, got decisive impulses from the universities at their start; methods for revealing reaction mechanisms have been developed in university institutes and were adopted by industry *(41, 42)*. Modern separation processes such as extraction by means of supercritical gases *(43)* were realized for the first time at the universities and were later applied by industry. The basic knowledge on hydrodynamics of the fluidized bed has been developed at the universities as well, and the same is valid for modern analytical methods, e.g. high-pressure liquid chromatography (HPLC). Computer programs for reactor design are sold by the universities to industrial companies. The distribution of concentration and temperature in fixed-bed reactors, which formerly has been determined graphically, can be calculated now numerically even if the systems concerned are much more complicated.

Types of Professions in the German Chemical Industry

Considering the above described great successes of the German chemical industry it is obvious that the German way of solving industrial problems through teamwork by the two independent professions of chemists and engineers did not have too much of an adverse influence. Germany always has had a considerable share in chemical exports, a fact that would be unintelligible if the wrong way had been taken.

Figure 10 shows in approximate figures, based on statistics of the German chemical industry *(44)*, that at present the German chemical industry is run by about 24,000 university-educated employees of whom about one half *(above)* has a chemical and the other half *(below)* a mechanical engineering background.

The curves reflect the strong rise after World War II. At the turn of the century the German chemical industry exclusively employed chemists. Only with the rise of the ammonia synthesis did a larger number of engineers take up activities. They increased with the further extension of continuous catalytic techniques to other products whereby the importance of equipment also increased. Nowadays 50% of the personnel with university educations are composed of engineers; in 1950 their share amounted to only one third.

This diversification also has its influence on chemists whose numbers have doubled from 1950 to 1975. To this group belongs a growing number of so-called technical chemists who now make up one-fifth of the chemists. Formerly, after having been trained in pure chemistry, they had acquired a certain knowledge of chemical engineering from either industrial practice or training at technical universities. Since 1971, by agreement of the scientific societies, for all chemists at both conventional

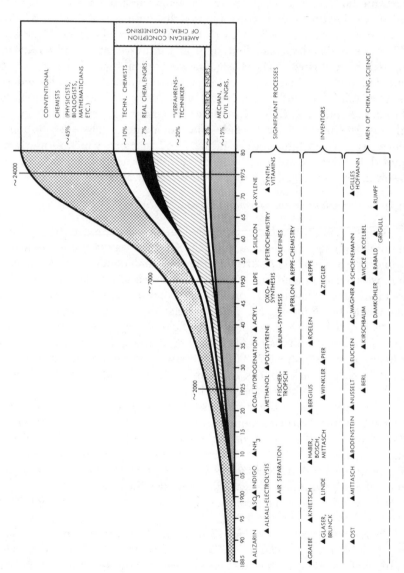

Figure 10. The development of professions in Germany's chemical industry

and technical universities, an introductory course of chemical engineering is compulsory and technical chemistry is admitted as the fourth optional branch of advanced training besides organic, inorganic, and physical chemistry.   At present the heads of the production plants are still such specialized chemists.   The majority of conventional chemists are engaged in research, while a smaller number are in applications, etc. with a number of biologists, physicists, and mathematicians being included, but they are strongly increasing as a consequence of the growing demand for the novel problems arising from the scarcity of raw materials etc.   Since the past year unemployed organic and inorganic chemists have been placed into chemical engineering via short-term courses initiated by the union of the chemical industry.

Chemical engineers in the proper sense did not appear any sooner than in the middle of the 1960's and actually constitute only 7% of the university-trained personnel.

In conclusion, the various tasks of industry such as research, development, design, operation, and management, which in the United States are the various fields of chemical engineers, are attended to in Germany by different categories of personnel.   These people are trained in special disciplines and with regard to the type and scope of their activities can, however, be compared with American chemical engineers as indicated by the vertical stripe on the right side of the graph.

Experience with the Training of Technical Chemists at the
Technical University of Darmstadt, Germany

The author was the only Professor of Chemical Engineering at Darmstadt Technical University.   Prior to his retirement in 1968 training in chemical technology was compulsory for the diploma examinations of all chemists.   His training was distributed over three terms; however, it required only one term of the entire ten terms necessary for the final diploma.   Graduates who had chosen chemical engineering as their main line of studies and as the subject for their doctor's thesis additionally enhanced their knowledge in special lectures.

In contrast to conventional teaching educational emphasis was put on the proper aim of industry, the development of new processes, and the design of novel plants.   This line of thinking starts from the industrial conception of the process and the proposed methods of manufacture, leading to reactor design, auxiliary equipment, energy consumption, etc., and finishing with the calculation of production cost and the estimation of profitability.   In this way the manifold viewpoints of the chemical industry are represented in a comprehensive and logical scheme into which new knowledge can be incorporated easily *(2–4, 44, 46–48)*.

Although the syllabus was not as broad as that of chemical engineering training in America, it was sufficiently profound for the solution of difficult problems. The close connection with basic chemical training and with the survey of the development of industry and their combination into an entirety proved to be very useful for the real understanding of the various branches of chemical engineering.

The subjects for research were taken from very different fields, with the purpose of combining the investigation of a scientific problem with the design of an actual industrial plant which later on permitted the checking of the precalculated results. Research by contract was necessary to increase the insufficient budget of the bomb-damaged institute. The variety of projects is shown by the following examples:

- selective hydrolysis of wood *(50, 51, 52)* using the easy separation of the $HCl-H_2O$-azeotrope by distillation under two different pressures *(53)*;
- increase of furfural yield by liquid–liquid extraction of the reaction mixture *(48, 54)*;
- synthesis of the explosive trimethylene trinitramine *(48)*;
- catalytic synthesis of phenol from benzene *(46, 47)*
- economic heavy water production by the H_2S-process *(55)*—later on confirmed by Du Pont Comp. *(56)*; crystallization of urea in cubic shape *(57)*—proved by Grace Comp., urea feeding of ruminants *(58)* applied in USA: 1957 125,000 tons/y, 1978 1 million tons/y, high pressure polymerization of ethylene *(59, 60, 61)*;
- the chemical reactors as control system and its dynamics *(62)*;
- optimization of a stirred-vessel cascade *(63)*.

Out of these, Perry's *Chemical Engineer's Handbook (64)* deems two examples to be interesting because industrial plants were constructed, based on very exact laboratory experiments only without the need for a semiscale stage.

With the very complex synthesis of the explosive trimethylene trinitramine (Hexogen, RDX) by nitration of hexamethylene tetramine *(see* Figure 11), part of the yield was obtained by nitration of the triazine ring (K_1), part by hydrolysis of hexamethylene tetramine into fractions forming to some extent methylene nitramine (K_2) which trimerized into additional explosive. A side reaction was the very exothermic formation of N_2O gas (K_3). Kinetically the desired synthesis could be carried out only at the low temperature of 75° Celsius *(see* Figure 12) whereby the dangerous formation of N_2O was suppressed sufficiently. From the mother liquor of the hexogen the nitric acid was separated by crystallization in the form of ammonium trinitrate which was recycled. The polymerized formaldehyde (K_4) was reconverted into hexamethylene tetramine and recyled too. Batch experiments yielding 30 g were scaled up in one step to a continuous industrial plant with a capacity of 200 tons/month.

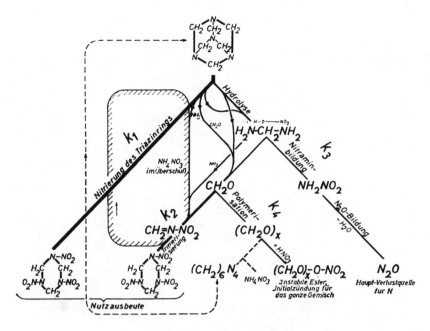

Figure 11. Reaction mechanism of RDX

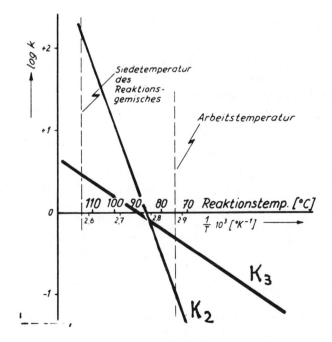

Figure 12. Kinetics of RDX formation

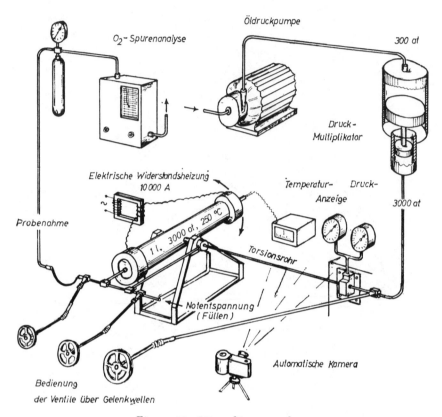

Figure 13. PE rocking autoclave

A typical example of the mathematical penetration of problems supported by computers was the development and design of a high-pressure polyethylene (PE) plant with a capacity of 24,000 tons/year in 1958 based on laboratory batch experiments only without a pilot plant by three doctoral candidates. (*58, 59, 60*).

Compressed ethylene gas containing traces of oxygen was preheated moderately in a rocking autoclave (*see* Figure 13). Starting from the measured temperature (*T*) and pressure (*P*) sequence, the reaction path was plotted in a P–T vs. time diagram (*see* Figure 14, *top*). The difference between the decreasing pressure and the isochore (Figure 14, *bottom left*) represents the conversion of ethylene and permits determination of the conversion–time diagram (Figure 14, *bottom right*).

From this diagram the constants of the reaction velocity equation were calculated assuming the most simple mechanism of radical polymerization. The improved kinetic model resulted in differential equations that were integrated numerically along the length of the reactor to predict the polymerization behavior in the large-scale plant and, after expanding the model, even the structure and the quality of the polymer. The plant functioned at first go and was repeated several times.

From the author's 17 years as a professor (1948–1965), 10% of his doctoral students became full professors. This was a further indication of the change in attitude toward chemical engineering.

Outlook

The concept of chemical engineering as described in this article will continue to imprint itself on future industrial development. More than ever before chemical engineering will be confronted with a rapidly changing world. During the last few decades mass productions have increasingly been shifted to countries with cheap raw materials and cheap energy. The exhaustion of resources, the shortage of energy, the need to use self-reproducing raw materials, and the protection of environment represent new impacts.

In the industrialized countries with high wages the production of fine chemicals will have to be pushed through to a much larger extent. At present, for instance, the production of vitamin A for animal nutrition and

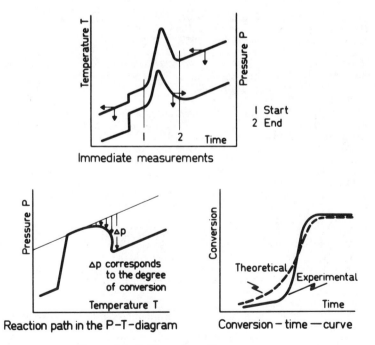

Figure 14. P–T vs. time diagram

vitamin E for human purposes is a new climax of industrial processes. Here, too, complex partial processes that have been developed over long periods of time were combined into economic syntheses. Chemical thinking will gain in importance in chemical engineering.

In the same way as in Reppe chemistry the reaction conditions could be mitigated by including organic bases in the catalysts, more and more organic molecules will be added as ligands to central metal atoms. Thus catalysts will be built up according to the pattern of such biocatalysts as chlorophyll and porphyrine which are effective even at mild conditions.

All this means is that new branches such as biochemistry, photo-chemistry, and electrochemistry have to be incorporated into chemical engineering training programs so that the present fields of knowledge will have to be condensed by about one-third. As a consequence there again will be a change in the aims and therefore also in the entire concept of chemical engineering.

Acknowledgment

I am indebted to Drs. H. Hofmann of Erlangen and N. Kutepow of BASF in Ludwigshafen for fruitful discussions.

Literature Cited

1. Reppe, W. Angew. Chem. 1949 61, 432.
2. Hartley, Sir Harold Chem. Ztg. 1955 79, 462.
3. Schoenemann, K. Chem. Ind. 1953, 7, 529.
4. Schoenemann, K. Chem. Ind. 1955, 7.
5. Bartholomae, E. Chem. Ing. Tech. 1976, 48, 917.
6. Miessner, H. "Verfahrenstech. im In- und Ausland, Verfahrenstech. Ges.," In Ver. Dtsch. Ing. 1961, 160.
7. Bartholomae, E.; Sinn R. Conf. Chem. Eng. Educ., Cambridge: U.K., 1968.
8. Blenke, K.; Dialer, K. Conf. Chem. Eng. Educ., Cambridge: U.K., 1968.
9. Franck, H. G. Chem Ing. Tech. 1978, 50, A110, A112.
10. Günther, R. Chem. Exp. Technol. 1977, 3, 209.
11. R. D. Chem. Ztg. Chem. Ind. 1960, 84, 752.
12. Piret, E. L. Chem. Eng. Prog. 1952, 48, 29.
13. Fond der Chemischen Industrie, Frankfurt Rundschreiben vom 9.4.52, 14.7.52, 8.11.54.
14. Piret, E. L. Chem. Eng. News 1956, 34, 212.
15. Piret, E. L. Chem. Process Eng. 1960, 41, 257.
16. Gruhn, G. Chem. Tech. 1976, 28, 428.
17. Kirkpatrick, L. D. Verband der Chemischen Industrie, "Fahrenberichs 1979/80", McGraw–Hill: New York, Frankfort, 1980.
18. Hougen, O. A. Chem. Eng. Prog. 1977, 73, 89.
19. Knietsch, R. Ber. Dtsch. Chem. Ges. 1901, 34, 4069.
20. Oberdorffer, K. "Ludwigshafener Chemiker"; Econ Verlag: Düsseldorf, 1958; Vol. 1.
21. Ibid., 1960; Vol. 2.
22. Mittasch, Alwin "Geschichte der Ammoniaksynthese"; Verlag Chemie: Weinheim, 1951.
23. Mittasch, Alwin "Kurze Geschichte der Katalyse in Praxis und Theorie"; Springer-Verlag: Berlin, 1939.

24. Holdermann, Karl "Im Banne der Chemie"; Econ Verlag: Düsseldorf, 1954.
25. Haber, L. F. "The Chemical Industry During the Nineteenth Century"; Clarendon: Oxford, 1969.
26. Schoenemann, K. *Brennst. Chem.* **1949**, *30*, 177–181.
27. Krönig, W. "Die katalytische Druckhydrierung von Kohlen, Teeren und Mineralölen"; Springer-Verlag: Heidelberg, 1950.
28. Jähne, F. "Der Ingenieur im Chemiebetrieb"; Verlag Chemie: Weinheim, 1951, p. 17.
29. Reppe, W. "Neue Entwicklungen auf dem Gebiete der Chemie des Acetylens und Kohlenoxyds"; Springer-Verlag: Berlin, 1949.
30. Reepe, W. "Chemie und Technik der Acetylen-Druck-reaktionen"; Verlag Chemie: Weinheim, 1952.
31. Reppe, W. "Carbonylierung," in *Ann. Chem.* **1953**, *582*, 1–162.
32. Reppe W. "Polyvinylpyrrolidon"; Verlag Chemie: Weinheim, 1954.
33. Roelen, O. *Nat. Forsch. Med. Dtschld.* **1948**, *36*, 166.
34. "Ullmanns Encyklopädie der technischen Chemie," 3rd ed.; Urban und Schwarzenberg: München/Berlin, 1951.
35. Eucken, A.; Jakob, M. "Der Chemie-Ingenieur"; Akadem. Verlagsgesellschaft: Leipzig, 1933–1940.
36. Berl, E. "Chemische Ingenieur-Technik"; Springer-Verlag: Berlin, 1935.
37. Schoenemann, K. *DECHEMA-Monogr.* **1952**, *21*, 203.
38. Damköhler, G. In "Der Chemie-Ingenieur"; Akadem. Verlagsgesellschaft: Leipzig, 1937; Vol. 3, pp. 359–485.
39. "VDI-Wärmeatlas;" VDI-Verlag: Düsseldorf, 1954–1957.
40. Herbert, W. In "Ullmanns Encyklopädie der Technischen Chemie," 4th ed.; Urban und Schwarzenberg: München/Berlin, 1967; Vol. 4, pp. 70–158.
41. Hoffmann, U. *Habil. Schr. Fortschr. Ber. VDI-Zeitschr.* **1977**, *3*(49).
42. Hofmann, H. *Chem. Ing. Tech.* **1979**, *51*, 257.
43. Schneider, G. M. *Angew, Chem.* **1978**, *90*, 762–774.
44. Fond der Chemischen Industrie, "Statistische Übersichten zum Bestand und Bedarf an Chemikern in der chemischen Industrie der Bundesrepublik Deutschland," Frankfurt, 1976.
45. "Wer Ist's?", *Nachr. Chem. Techn.* **1975**, *23*, 5.
46. Schoenemann, K.; Hofmann, H. *Chem. Ing. Tech.* **1957**, *29*, 665.
47. Schoenemann, K. *Chem. Ztg.* **1962**, *86*, 247.
48. Schoenemann, K. *Chem. Eng. Sci.* **1963**, *18*, 565.
49. Bartholomae, E. *Chem. Ing. Tech.* **1976**, *48*, 917.
50. Schoenemann, K. *DECHEMA Monogr.* **1959**, *34*, 92.
51. Schoenemann, K. *Chim. Ind.* **1958**, *80*, 140.
52. Schoenemann, K. *Chem. Eng.* **1954**, *61*, 138,
53. Riehm, T. Dissertation, Darmstadt, 1950.
54. Schoenemann, K. *Accad. Naz. Lincei Corso Estivo Chim.*, 5th **1960**, 251.
55. Connemann, J. Dissertation, Darmstadt, 1960.
56. Proctor, J. F.; Thayer V. R. *Chem. Eng. Prog.* **1962**, *58*, 53.
57. Raskob, W. Dissertation, Darmstadt, 1957.
58. Schoenemann, K.; Kilian, E. F. *Arch. Tierenähr.* **1950**, *10*, 37.
59. Thies, J.; Schoenemann, K. In "Chemical Reaction Engineering," *Adv. Chem. Ser.* **1972**, *109*, 86.
60. Thies, J. Habil. Eidgen. Techn. Hoschschule Zürich, Switzerland, 1977.
61. Thies, J. *AIChE Symp. Modeling High Pressure Polyethylene Reactors*, Houston, TX, 1979.
62. Gilles, E. D. *Regelungstech.* **1965**, *13*, 361, 493.
63. Kreth, W. Diplomarbeit, Darmstadt, 1962.
64. Perry, R. H.; Chilton, C. H. "Chemical Engineer's Handbook," 5th ed.; 1973; pp. 4, 19.

RECEIVED May 7, 1979.

The History of Chemical Engineering in Japan

GENJI JIMBO—Department of Chemical Engineering, Nagoya University, Nagoya 464, Japan

NORIAKI WAKAO—Department of Chemical Engineering, Yokohama National University, Yokohama 240, Japan

MASAHIRO YORIZANE—Dean of Engineering, Hiroshima University, Hiroshima 730, Japan

The prehistory of chemical engineering in Japan before its formal and systematical establishment in 1920–1930 is discussed. Chemical engineering education and the growth of Japanese chemical industries are also outlined.

R ecently it was found that early in 1903 Kotaro Shimomura had tried to translate G. E. Davis' book, *A Handbook of Chemical Engineering*, into Japanese, and he named, for the first time, this new engineering "Kagakukogaku" as is used now *(1)*. It was only two years after the original publication of Davis' handbook. The letter from Davis to Shimomura was found recently and shows that he got this handbook before March of 1902. Such a quick purchase of the book was rather mysterious in Japan in those days, but even more interesting was his active response toward the new engineering. Here it must be pointed out that even in 1911 O. A. Hougen could not find Davis' handbook in the library of Washington University *(2)*.

After graduating from Doshisha, a private college based on Christianity, Shimomura studied organic chemistry at Worcester Polytechnic Institute, in the United States where he received his BS and PhD. He then moved to Johns Hopkins University where he did research in organic chemistry under Professor Ira Remsen. His stay at Johns Hopkins was discontinued due to the proposal of a millionaire, J. N. Harris, to donate $100,000 toward the founding of a school of science. It was said that Harris asked Shimomura to come back to Japan to take the responsibility for organizing and managing the new school, which was named the Harris School of Science (Harisu Rikagakko). From its founding in 1890 until his unexpected and unhappy resignation in 1896,

0-8412-0512-4/80/33-190-273$05.00/1

Shimomura was a Director and Professor of Chemistry at the school, which was the first private educational institute in Japan to specialize in science and engineering. Unfortunately the school closed shortly after his resignation.

Shimomura then focused his attention to a career in industry and selected the development of the coke-manufacturing process with its recovery of gas, tar, and ammonia by-products as his new area of research. To investigate this process he was sent to the United States and Europe in 1896–1897 where he selected Semet–Solvay coke ovens to be the most adequate.

Therefore there was a possibility that he received some information on chemical engineering or Davis' activities during his trip. (Incidentally Shimomura was a member of the Society of Chemical Industry, London.)

He perhaps was struggling at that time with the construction of the coke-manufacturing process (Japan Patent No. 2907 (1901)), which later led to the development of a new coke-manufacturing process which utilized Japanese low-grade coal (Japan Patent No. 13583 (1908)). This newly invented process was highly engineering-oriented, especially with the development of an agitator for agitating and conveying coking coal in the furnace. It is therefore quite probable that he became aware of the importance of mechanical and process design of the chemical process. It is reasonable to say that his interest toward chemical engineering may have been the result of such experiences in industry. Actually the coke-manufacturing process was one of the key technologies of Japan, because it had a close connection with the construction of the Government Steel Works of Yawata, then the biggest national project of technology, to which Shimomura contributed by constructing the Semet–Solvay coke ovens.

Unfortunately Shimomura did not complete the translation of Davis' book and it was neither published nor shown to other people.

Another very early trial of the education of chemical engineering by T. Nishikawa was very influential to those who were interested in chemical engineering. After returning from industry to Tokyo Imperial University as an Assistant Professor, Nishikawa started a course in Chemical Plant Design (or Chemical Machinery) ca. 1910. One year later he moved to Kyushu University, but the influence of his course was believed to be very great.

While he gave lectures at the Department of Applied Chemistry of Tokyo Imperial University, J. Inoue, Y. Tanaka, and G. Kita were staff members in the department. Later on Inoue moved to Tohoku Imperial University, where he established the educational system of chemical engineering (3). Y. Tanaka became a leading educational figure in this new engineering program at Tokyo Imperial University, although his speciality was applied organic chemistry (3). G. Kita later moved to

Kyoto Imperial University, where he was active in establishing a program in chemical engineering *(4, 5)*.

Nishikawa's influence on Kyushu Imperial University was naturally very strong because his lectures on Chemical Plant Design and/or Chemical Machinery continued from 1911 to 1924, at which time he was succeeded by T. Kuroda *(6)*.

Since the Imperial Universities of Tokyo, Kyoto, Tohoku, and Kyushu were all very important sources of the pioneers of chemical engineering, it can be said that almost all of the founders were more or less influenced by Nishikawa.

From 1896 to 1908, Nishikawa worked in an alkali-manufacturing company (Nihon Seimi Co.) as an Engineering Director, and he strongly realized the very severe state of Japanese chemical technology. It was the turning point of the soda-manufacturing process from Le Blanc to Solvay, and the Japanese chemical industry was struggling to establish this new industry on a technologically independent basis. Unfortunately the level of sophistication of the Solvay process was too high for the Japanese chemical industry at that time, and the conditions of raw materials were also very poor. Another severe factor was the monopoly of the Solvay process; no patent and no knowledge of this process were sold without ruling capital investment. Therefore, the development of the soda process had to be another big national project, and in fact soon after the outbreak of the First World War, the Japanese government started an Organization for the Investigation of the Chemical Industry (Kagaku-kogyo–Chosakai), in which the development of the soda-manufacturing process was the main subject. This means that the development of the soda process became a sort of national project. Nishikawa was always a leader in that organization and he was engaged throughout his entire life in the development of a Japanese soda process. It is undeniable that his pioneering contribution to the founding of chemical engineering in Japan was based on his experiences in the soda-process development.

Chemical Engineering Education

Japan is a densely populated country with 111,000,000 people in an area of approximately 377,000 sq km. All children from 6 to 14 years of age are enrolled in compulsory schools (six years in elementary school followed by three years in junior high school), with an enrollment rate close to 100%.

After junior high school, most of the youths enroll in a senior high school. This school is not compulsory, but the enrollment rate is about 90%. There are two types of senior high schools: general course (63%) and vocational course (37%). Students in both courses are qualified equally to advance to the institutions of higher education—junior colleges and universities.

Junior colleges offer two-year programs in which practical and professional education is emphasized. About 91% of the students are females (as of 1977).

All of the universities offer four-year programs (six years for medical school) leading to a bachelor's degree, and many universities carry two-year graduate programs leading to a master's degree. In some universities more advanced professional studies are pursued in three-year doctorate programs that are available for graduate students who already have master's degrees. An engineer in industry who has only a BS degree also can obtain a doctorate of engineering by writing a special paper.

In quite a good number of universities, the School of Engineering to which a Chemical Engineering Department belongs has some other Chemistry (such as applied and/or industrial chemistry) Departments. In addition, the school of science has a Chemistry Department. Note that female students (undergraduate and graduate) in schools of engineering and science account for less than 5% of the enrollment.

In most universities, the Applied Chemistry and Industrial Chemistry Departments produce roughly four times as many students as the Chemical Engineering Department. This ratio is considered to be a result of the demand from the chemical industry.

Table I shows the increase in the number of chemical engineering graduates and of institutions having a Chemical Engineering Department or Subdepartment since 1932. In 1940, chemical engineering was taught at only three institutions: Kyoto University, Tohoku University, and Tokyo Institute of Technology. The number of graduates from these institutions in 1940 was only nine. In contrast, in 1960, 18 universities (17 national and 1 municipal) had Chemical Engineering Departments and eight universities (4 national, 2 municipal, and 2 private) had Chemical Engineering Subdepartments. Altogether 263 chemical engineers were graduated in 1960. In 1969, 25 Chemical Engineering Departments (22 national, 1 municipal, and 2 private) produced 810 graduates, and ten Chemical Engineering Subdepartments (4 national, 3 municipal, and 3 private) had 394 graduates; altogether this meant 1,204 graduates in 1969.

Since 1969, the figures seem to have remained almost unchanged. Besides this, about 300 graduate students are enrolled in the master's program and about 30 in the doctorate program each year. The increase in the number of Chemical Engineering Departments in the 1960's was in accord with the sharp growth in the chemical industry, particularly in petrochemicals, which resulted in a large demand for chemical engineers. However, the increase has ceased since 1970.

Before 1950, unit operations was the major course taught in the Chemical Engineering Department. However, since 1960, in most universities, process design, chemical reaction/reactor engineering, transport phenomena, and process control, etc. also have been included in the

Table I. Number of Chemical Engineering Departments and Subdepartments and Graduates[a]

	School						Year			
		−1931	1932	1940	1950	1960	1969	1970"
Chemical Engineering Department	National	0(0)	2(1)	9(3)	109(9)	204(17)	706(22)	almost the
	Municipal	0(0)	0(0)	0(0)	25(1)	23(1)	27(1)	same as
	Private	0(0)	0(0)	0(0)	0(0)	0(0)	77(2)	1969
Chemical Engineering Subdepartment	National	0(0)	0(0)	0(0)	0(0)	10(4)	231(4)	
	Municipal	0(0)	0(0)	0(0)	0(0)	9(2)	57(3)	
	Private	0(0)	0(0)	0(0)	1(1)	17(2)	106(3)	
Total		0	2	9	135	263	1,204	

[a]The first figure indicates the number of graduates; the second figure in parentheses indicates the number of Chemical Engineering Departments and Subdepartments.

curriculum. There are a few Chemical Engineering Departments that have the same curriculum as those of the Applied and Industrial Chemistry Departments, while some departments emphasize mechanical courses.

However, the average curriculum is considerably different from that of the Applied and Industrial Chemistry Departments. Polymer and metallurgy courses of the type often taught in Chemical Engineering Departments in North America are covered only in the Applied and Industrial Chemistry Departments in Japan.

In the Chemical Engineering Departments of Japan, on the average 225 hr is spent teaching fundamental engineering courses, 186 hr in unit operation, 73 hr in reaction and reactor engineering, 75 hr in process design, and 287 hr in chemical engineering experiments and drawing. (Note that a course is defined as one 2-hr lecture per week for 15 weeks.)

Undergraduate students are required to complete general education courses for the first 1–1.5 years, followed by the specialized courses. The students attend these specialized courses for about two years.

The entrance examination for a university, in general, is very strict in Japan. We probably should shift the narrow gate to each year—or semester—end examination as well as to the final examination. The university door should be kept open as widely as possible (as much as the facilities accept) so that more people can become at least first-year students. The competition then should be continuous throughout the four-year period, not just at the time of entry to a university.

As far as the Engineering Departments are concerned, the graduate school, particularly at the master's course level, should have more people who have worked for several years in industry. This is popular in North America. We believe that the exchange of people and knowledge between universities and industries is very important for both sides.

Growth of Japanese Chemical Industries

The Beginning of the Chemical Industry. The origin of the Japanese chemical industry was at the beginning of the Meiji Era. It was actually an outgrowth of knowledge from the Tokugawa Era. Scholars who learned the Dutch language or Dutch scholars thoroughly systematized the chemical industry in an occidental way.

The Proclamation of the Commercial Law was made in 1868 by the Meiji government in order to modernize Japan. The aim was to import modern industries from advanced countries and to stimulate the domestic commercial and industrial activities of Japan. Prior to this, domestic industries had the traditional and historical background of guild socialism which maintained order and control. Once freed from feudalistic control, a very large number of small industries started up almost everywhere. Sulfuric acid started in 1872, caustic soda in 1877, and the cement industry in 1873.   These were all run by government hands. Soap (1873) and matches (1875) were run by private hands.   The competitive position in world markets was obtained through the lower wages of child and female labor.   The Kaiseisho (1863), founded by Shogunate during the Tokugawa Era, was one of the predecessors of Tokyo University which was established in 1877 and included a Chemistry Department.   The Japan Chemical Society was organized in 1878.

In 1877, the Le Blanc process was tried but it failed: it was tried again in 1895.   When Solvay production of soda exceeded more than 70%, it is curious to note that they should try to establish the Le Blanc process with such a small demand as existed in 1895.   However, both the Le Blanc process and the electrolysis method were rather manual industries and were easy to industrialize on a small scale.

On the contrary, the ammonia soda process is a combination of unit operations and is a fairly large continuous process.   More purification of raw material is required to manufacture the highly purified product, and reactions have to be performed with more precision.   The construction and operation of the process requires a much more sophisticated type of technology.   Therefore, the advantage of scale is that it is larger, but it also requires more investment.   Moreover, the process requires highly purified salt as raw material, and this must be imported.   Because of these circumstances, it is important to consider the amount of risk involved in the industrialization of the ammonia–soda process.

Early Growth.   The following charts the beginning of each synthetic ammonia process in the world:

Haber Bosch Germany (1910), England (1924), France (1927),
             TNRI–Japan (1929)
Claude   France (1919), Italy (1924), United States (1924),
             Japan (1924), Germany (1928)
Cassale   Italy (1922), Japan (1924), France (1925), Germany
             (1928)
Fauser   Italy (1923), Germany (1928), Japan (1929)
NEC   Germany, France, Japan (1930)

TNRI stands for the Temporary Nitrogen Research Institute.   (This is now the Government Chemical Industrial Research Institute.)

Original Japanese research started with only one report of Haber et al. in *Zeitschrift für Electrochemie*.   In order to establish this process, the following technology was indispensable:

1. synthesis of ammonia; industry based on chemical theory;
2. analysis of high-pressure reactions; chemical equilibrium theory in physical chemistry (thermodynamics);

3. High-pressure process, starting from synthetic dye stuff;
4. Continuous process (based on ammonia–soda process);
5. Catalytic chemistry—contact sulfuric acid process.

The researchers of TNRI had developed a modification of the Haber process—the Tokoshi Process—which was composed of all domestic machines and apparatus except for the Linde air-separating machine. Industrialization of this process was tried by Showa–Denko in 1929; production was recorded at 20 tons per day.

The Hikoshima plant started by importing technology for ammonia synthesis. A pilot plant having a daily production run of 5 tons was constructed utilizing the Claude process from France at a cost of 5,000,000 yen. The commercial process at Hikoshima started in 1924, but this system was new and needed many improvements. There were several accidents, causing explosions and casualties. Steady production was not reached until 1928.

The Reconstruction of Japanese Industry

In 1945, the United States' economic policy for Japan and West Germany included a decision to reconstruct Japan and West Germany as the active factories for Asia and Europe. Special and strong methods were required to resume production and get out from under a paralyzed economy.

The first phase of recovery was started with what was called an "inclined production theme." "It is to put all economic resources into the production of the basic material, coal, which was only able to be controlled in our hand . . . It is most important to support industrial production by increasing production of this primary material urgently" (7).

Then heavy oil (which was allowed to be imported by GHQ), and coal (which was domestically produced), were supplied to the iron and steel industry. Steel products, in turn, were supplied to the coal industry for its production.

There was a weak policy controlling price structures at that time, but there was no strong organization to support the new production theme. However, it was one way to make heavy chemical industries predominant over the industrial structure.

Reconstruction of the Japanese chemical industry began with the chemical fertilizer (ammonia sulfate) industry, with the government supporting the basic chemical industries such as ammonia, carbide, and sulfuric acid. During the establishment of these basic industries, chemical engineering unit operations provided the technical basis for this industrialization.

As the inclined production theme progressed, rationalizations were required for industry, and the chemical industry changed in quality as gas sources changed.   Those consuming hydraulic power, coal, and coke changed energy sources to heavy oil, natural gas, waste gas from the iron industry, and gasification of crude oil.

The expanding vinyl chloride market helped industry to escape from inclined production.

Higher Growth

Petrochemicals played a major role in generating higher growth for the chemical industry.   From the view point of industrial policy, innovation of new materials started in 1961.   This innovation was accompanied by processing technology and manufacturing processes.   Machine tools, industrial machines, the apparatus and equipment industry, computer control, and the changing ammonia–soda process contributed to this growth.

The consumption revolution happened with technical innovation and equipment investment.   Production of color televisions, automobiles, and air conditioners (prices around 200,000–500,000 yen) are one side.   Aluminum, paper, plastic, and transistors are the other side.   In 1970, energy distribution was 8% coal and 71% oil compared with 47% coal and 18% oil in 1953.

Literature Cited

1. Shimao, N. Doshisha-jihyo University, Jihyo, 1977, Nov., 62, No. 53, p. 4.
2. Hougen, O.A. *Chem. Eng. Prog.* **1977**, Jan., 73, 89.
3. Hatta, S. *Kagakukogaku* **1968**, *32*, 390.
4. Kamei, S. *Kagakukogaku* **1962**, *26*, 5.
5. Kamei, S. Tokyo Institute of Technology, Kagaku Kogakuka Dosokaishi, 25 Shunen Kinengo **1963**, p. 3.
6. Nishikawa Torakichi Tsuisoroku, Kyushu University, 1951; p. 411.
7. Arisawa, H. Hyoron, *Hyoron-sha* **1947**, Jan.

RECEIVED May 7, 1979.

16

Du Pont and Chemical Engineering in the Twentieth Century

VANCE E. SENECAL

Manager, Engineering Research, Engineering Department, E. I. Du Pont de Nemours & Co., Inc, Wilmington, DE 19898

The Du Pont Company's business decision in the early 1900's to diversify into chemical manufacturing came at a time when U.S. technology was young and the U.S. chemical industry was in its infancy. The process of growing into the largest U.S. chemical company encompasses a microcosm of chemical engineering history in the United States. We can follow, through the years, an increasing number of new challenges and new roles for the chemical engineer, some evolutionary, others of necessity more radical in nature. Looking towards the end of the century, we can see the developing environmental, energy, and feedstocks situations adding to the existing challenges and opportunities for chemical engineers.

Over 4,300 of Du Pont's 132,000 employees are chemical engineers. The reason for this relatively high percentage is the broad spectrum of career opportunities available to these chemical engineers, reflecting the inherent flexibility in their educational background and the myriad of opportunities offered by a chemical company having Du Pont's size and diversity.

Historically, as in a number of other companies, the need for, and indeed the practice of, chemical engineering actually predated the formal emergence of the profession. Recognition of industry's need in this respect led to the evolution of chemical engineering curricula in a number of the leading universities both in England and the United States around the turn of the century. This interplay between industrial needs and university response has been and continues to be a key factor in providing proper training in a dynamic discipline that encompasses an ever-growing number of technical challenges and career opportunities.

0-8412-0512-4/80/33-190-283$05.00/1

Although the emphasis of this chapter is on the 20th century, events of the prior century shaped the course of the more formal chemical engineering that was to come in to Du Pont and so will be covered briefly.

The 19th Century

The Du Pont family fled a turmoil-torn France in 1800, seeking a new start in a young and weak United States. Irenee Du Pont, only 28 years old, had been a pupil of Lavoisier—"The Father of Modern Chemistry" —from whom be learned the latest gunpowder-making techniques. He established E. I. du Pont de Nemours and Company in 1802 on the banks of the Brandywine River near Wilmington, Delaware with the intent of supplying America with a much-needed local source of quality gunpowder. Good powder, which had to be imported (mostly from England), was a necessity on the frontier for land clearing, mining, engineering, and hunting. This young Frenchman had the instincts of a chemical engineer with his unusual concern for product quality control and safety. Mills were divided into well-spaced small units to limit the effects of accidental explosions (*see* Figure 1). Demands were imposed on quality and uniformity of raw materials and all processing steps. Achieving these goals helped his mills survive the hazards of business and intense foreign competition—a unique accomplishment in those days. He inculcated his company with the importance of a continuous effort to improve the old and invent the new.

Practicing this philosophy, the young company constantly improved manufacturing procedures and instituted better methods of management and marketing, which led to a healthy growth. A patent issued to Irenee on a "Machine For Granulating Gunpowder" attested to his interest in improved methods of manufacture. Lammot Du Pont invented soda powder in 1857 with "a formulation of 72 parts refined Chilean sodium nitrate, 12 parts sulfur, and 16 parts charcoal" (1). Not only did a better strictly industrial explosive result, but the process also broke the centuries-old dependence on potassium nitrate, obtained principally from India.

In 1880, Du Pont began the manufacture of nitroglycerin and dynamite. These much more powerful commercial blasting agents were needed in ever-increasing quantities for building new roadbeds and tunnels for railroads, mining of coal and minerals, and quarrying rock. They gave new impetus to the industrial growth of the nation. Concern with safety in the manufacture of dynamite and the need to improve the practice of mixing ingredients by hand rakes and shovels led Lammot and William Du Pont to develop a power-driven wheel mixer, which was used for many years throughout the industry.

Figure 1. Since earliest days Du Pont has built its own plants and mills. Design of early powder mill, above, a counterpart of French powder mills, was widely followed later.

Despite improvements, the inherent danger in powder manufacturing was re-emphasized when Lammot was killed in 1884 while attempting to prevent an explosion in the nitrating house.   To cope with such problems, a new approach was introduced: teaming engineers and chemists to replace rule-of-thumb methods in explosives manufacture with applied chemistry.   This approach was a forerunner of chemical engineering in Du Pont.

In the 1890's Pierre S. Du Pont undertook to perfect another new explosive, smokeless powder, based on nitrocellulose.   This development presaged a turning point in the company's history because interest was generated quickly in the many chemical possibilities presented by the raw material cellulose.

1901–1925

The turn of the century brought major changes in the Du Pont Company.   In 1902 it was reorganized under three Du Pont cousins: Coleman, an engineer, as president; Alfred, a black-powder expert, as vice-president; and Pierre, a chemist, as treasurer.   Laflin and Rand Powder Co. was purchased as the first step in overhauling and consolidating the American explosives industry.   The Engineering Department was organized in 1902 to develop and build machinery and plants for the company.   Du Pont also traces its formal research roots back to 1902 when it established a research facility called the Eastern Laboratory in Repauno, New Jersey.   That lab, headed by noted industrial chemist Dr. Charles L. Reese, specialized in explosives research and developments, including low-freezing dynamites and a class of dynamites that could be

Figure 2. The Du Pont Experimental Station, established in 1903, occupies a site in the rolling country near Wilmington, Delaware. By 1912, when this picture was taken, it included several buildings. Some of them were refurbished powder mills on the banks of the Brandywine. At this time, the entire personnel of the station was less than 70.

used safely in gaseous and dusty mines.   One of the first industrial research laboratories in this country, it is believed to represent the earliest organized research effort in the American chemical industry.

Management was centralized under an executive committee, established in 1903, and composed of the corporation's ablest executives. This committee recognized that, while the U.S. chemical industry was in its infancy, unprecedented opportunities could be realized through the expansion of American chemical research. Consequently, in 1903, the Experimental Station was established on the banks of the Brandywine River (*see* Figure 2) near where the original powder mills had been built a century before. Its primary objective was to further research in cellulose chemistry so that Du Pont could branch out from explosives into new fields. Although Du Pont was already 100 years old, it had only about 6,000 employees and a single product line—explosives. It so happened that 1903 was also the year that William H. Walker (often referred to as the "Father of Chemical Engineering") left his position as consultant with Arthur D. Little, Inc. to join M.I.T. In 1904 Du Pont began producing a special type of nitrocellulose for lacquers, leather finishes, and other industrial uses. Nitrocellulose-coated fabrics, then known as "artificial leather," were added in 1910 with the purchase of Fabrikoid Co.   A central Chemical Department was established in 1911 to better administer the growing research (mostly applied) effort directed at developing new products.   Note, however, that industrialists of the time agreed that fundamental research was a thing best left to professors in the universities.

World War I clearly pointed to the need of an independently strong American chemical industry, which added impetus to Du Pont's acquiring companies in various product areas to diversify and broaden its product base *(1)*.   Growth by this route was guided by four policies: (1) the fields entered were either new with good growth potential, or, if old, amenable to further growth through innovative improvements; (2) Du Pont-trained personnel were added to the staff of the acquired business; (3) research programs were launched immediately to improve the processes/products and to develop new and better products; and (4) a venture, once launched, was subject to continual improvement through additional modernization investments.

The Arlington Co., manufacturer of pyroxylin plastics, lacquers, and enamels, was acquired in 1915.   The following year Fairfield Rubber Co. (rubber-coated fabrics) was purchased.   The purchase of Harrison Brothers and Co. (1917) added acids, heavy chemicals, pigments, dry colors and paints to the growing list of products.

A national dye shortage led to initiation of an intensive research program on dye intermediates in 1916.   This program was a natural extension of Du Pont's experience with coal-tar crudes in compounding high explosives.   Moving rapidly, the company built its first dye plant in Deepwater, New Jersey. The first fast vat dyes were produced in 1919, and by 1923 the plant was a major producer in America's newly established dyestuffs industry.

This, then, was the Du Pont Company when its first chemical engineering graduate was hired in 1920—the same year, incidentally, that the first independent university department of chemical engineering was established at M.I.T. *(2)*.   Until this time the roles later handled best by chemical engineers had been filled with industrial chemists, chemists adding engineering skills "on the job," or chemists and engineers (of other disciplines) working together.   It was in the middle of Hougen's *(3)* second decade (1916–1925) of American chemical engineeering, when the concept of unit operations as put forth by Walker, Lewis, and McAdams *(4)* *(Principles of Chemical Engineering*, published in 1923) directed the educational system to the type of training best suited at that time for the emerging U.S. chemical industry.   The industry needed chemical engineers trained in chemistry, physics, and mathematics in a manner that prepared them to solve processing problems.   They had to be able to make things happen.   The unit operations approach was a giant step towards fulfilling this need.

With U.S. rights acquired from French interests, Du Pont started to manufacture viscose rayon yarn in 1920 and cellophane in 1923, both at Buffalo, New York. Intensive internal research programs led to marked improvements in the process, permitting a long sequence of price reductions and construction of a number of plants across the country to meet the growing sales. Duco lacquer (1923) came directly out of Du Pont's research efforts and revolutionized the finishes business. Not only did it provide a better, more durable automobile finish, but application time was reduced from days to hours. Du Pont introduced photographic films in 1924, industrial alcohol in 1925, and also acquired Viscoloid Co. (plastics) the same year to close out the first quarter of the century.

1926–1950

In spite of this marked expansion, a quote from the Manchester Guardian in this period best depicts the international situation:

> "The chemistry of today remains almost wholly the product of British and German research while chemical trade of the world is dominated by two firms—one British and the other German. These two—Imperial Chemical Industries and the I.G. Farbenindustrie—might almost be called nationalized chemical industries of their respective countries; beside them a representative American firm such as Du Pont de Nemours & Co., despite its growth last year, is a mere pigmy" (5).

The momentum to expand the Company's chemicals base continued into the second quarter of this century. Three expansion routes were followed. Other businesses were acquired through outright purchase: National Ammonia Co. (1926); Grasselli (1928), which provided a position in the midwest in acids and heavy chemicals; Krebs Pigment and Chemical (1929), maker of lithopone; Roessler and Hasslacher (1930), specialists in electrochemicals, ceramic colors, peroxide, sodium, and insecticides; Commercial Pigments Corp. (1931), makers of titanium pigments; Newport Co. (1931), producer of dyes and synthetic organic chemicals; and majority interest in Remington Arms Co. (1933).

A second expansion route was through the purchase of technology: an ammonia plant was built at Belle, West Virginia based on high-pressure synthesis technology purchased from French and Italian interests (1926); and cellulose acetate rayon production (1928) was based on the purchase of U.S. rights to foreign patents. Thus acquisitions, started in 1917, permitted rapid expansion of the product base. The basic philosophy of purchasing with intent to improve markedly both the processing steps and the product itself provided a continuous flow of new challenges to the technical staff and new opportunities for chemical engineers in

Figure 3. Thomas H. Chilton headed first Chemical Engineering Group established in Du Pont in 1929 at the Experimental Station. This group evolved and expanded into the current Engineering Research and Development Division, which has been the corporate leader in conducting basic chemical engineering programs covering first unit operations, then a systems approach to processing problems involving interdisciplinary teams, and then supplemented with the use of computer and mathematical modeling.

research and development, process engineering, production, design, technical service, management, etc.

The third route to expansion of the product base, namely by creating new proprietary products internally through research and development, gradually took hold during this quarter century and then accelerated. The huge success of this brilliant business strategy created new technical challenges, forcing accelerated maturing of the chemical engineering profession.   Research organizations had been formed in each of the company's industrial departments in 1921 (a plan still followed today).

A formal program of fundamental research in chemical engineering, physical and organic chemistry, and physics began in 1927 in the Central Chemical Department at the Experimental Station, and in 1929 a Chemical Engineering Group was set up under the direction of Thomas H. Chilton (*see* Figure 3).   The chemical engineering portion of the program encompassed investigation to provide authoritative information for the design of chemical process equipment and selection of materials of construction. Fields under study included fluid friction, heat transfer, absorption, drying, distillation, solvent recovery, crystallization, mist and dust removal from gases, and properties of fluids under high pressures. In 1931 a second research group was established in the Engineering Department, designated as the Technical Division, and located at the Experimental Station. Studies more of a mechanical nature (vs. physico-chemical) were undertaken by this group such as crushing and grinding, mixing and agitation, and filtration, as well as metallurgical studies on stainless steels, heat–resisting alloys, properties of lead, and materials resistant to acids.

The two groups were consolidated in 1935 by transfer of personnel from the Chemical Department to the Technical Division of the Engineering Department first under the guidance of H. B. Du Pont and later under T. H. Chilton. Specific studies were aimed at corporate needs which were determined by numerous visits to company plants and laboratories to consult with those who were in the best positions to help direct activities in the right channels. Some expressed concern that the planned programs sounded "somewhat academic and like pure research." Actually the basic research studies were done to increase the fund of accurate scientific information upon which chemists and chemical engineers could draw to solve specific practical problems and to design new processes. They soon learned that problems uncovered in applied research studies provided excellent guidance for the fundamental studies. The Technical Division's objective was to serve all departments of the company by developing and making available fundamental data of chemical engineering operations. The unit operations information developed from these studies was documented well in internal company reports, used to prepare Engineering Department Design Standards, and frequently submitted to national and international technical journals for publication. Close ties were established with a number of universities, particularly by retaining as consultants professors recognized as leaders in their fields—a practice still followed.

The now familiar *Chemical Engineer's Handbook* first was issued in 1934. John H. Perry, Editor-in-Chief, and W. S. Calcott, Assistant Editor, were Du Pont chemical engineers. The preface states that:

> "This handbook is intended to supply both the practicing engineer and the student with an authoritative reference work that covers comprehensively the field of chemical engineering as well as important related fields. To insure the highest degree of reliability the cooperation of a large number of specialists has been necessary; this handbook represents the efforts of 60 contributing specialists" (6).

Most of the 30 sections covered unit operations as practiced at that time. Approximately one-third of the contributors were members of various university staffs, one-third were Du Ponters, and the remainder represented other industrial organizations. Almost half of the contributors were chemical engineers. The wide variety of other disciplines represented reflected Du Pont's recognition that the growing complexity and refinement of chemical processes required the input of specialists from many fields working cooperatively.

Results of the basic research programs launched in the late 1920's were perhaps more spectacular then even the proponents of the program had expected. Synthesis and polymerization of 2-chloro-3 butadiene led to the 1932 commercialization of neoprene, the first general-purpose synthetic rubber. Large-scale production of synthetic camphor started in 1933. Basic research on viscose spinning resulted in commercialization (1934) of the first high-tenacity rayon tire cord. Crystalline urea was marketed in 1935 and methyl methacrylate was introduced in cast sheets in 1939. Research on the synthesis and polymerization of tetrafluoroethylene (TFE) during the early 1940's produced Teflon TFE–fluorocarbon resin. New and more effective herbicides were developed. Polyester resins were introduced in 1942, polyethylene went into production in 1943, and a process was developed for making metallic titanium in 1948.

In 1928 Wallace H. Carothers (a former professor at Harvard) and Julian Hill decided to concentrate their groups' fundamental studies on polymers known to be the building blocks of many natural substances. Learning how and why certain molecules hook together end-to-end to form chains (linear polymers) guided the team to the development of 66 nylon. Following repeatable demonstrations on a laboratory scale, the challenges to chemical engineers and other technical personnel to scale up to commerical facilities (Seaford, Delaware) were new and formidable. The high melting point of nylon was a temperature at which Carothers said "the world of organic chemistry begins to cease to exist" (7). Some of the problems included developing processes to produce, in quantity, high-quality raw materials (later known as polymer-grade materials) needed to make nylon, hexamethylenediamine (at that time, a laboratory curiosity), and adipic acid (available in commercial quantities only in Germany); developing special equipment to transport the raw materials in a hot form; developing reactors for making the highly viscous polymer, forming it into chips, remelting, holding at a constant temperature (285°C) in an oxygen-free environment, pumping, filtering to remove impurities, spinning into a fiber, drawing precisely, and winding up reproducible packages of yarn. A team of some 230 chemists, chemical engineers, physicists, and other technical people were assembled to solve this multitude of seemingly insurmountable problems. Over 200 patents dealt with novel technical devices required to get the fiber produced. Introduced at the San Francisco and New York World's Fairs in 1939, nylon hosiery created a national sensation. Few major inventions have been as successful from the outset. Chemical engineers, with colleagues from other disciplines, have since introduced innumerable other uses for nylon and have improved markedly all of the processing steps involved in its production. The success and continuing impact of these developments are reflected in the fact that some 57,000 people in the United States are employed in the production of nylon fibers, resins, and filaments. Worldwide employment is estimated to be about 250,000 and downstream employment at over 3,100,000.

The changing role of the chemical engineer is reflected in a statement made by Chilton in a (November, 1939) Chemical Directors' Meeting.

> *"Twenty-five or 30 years ago it was more or less customary in the chemical industry for the research chemist to work with the master mechanic and/or with an engineer or two to carry out between them all of the necessary steps to establish an operating plant unit. During recent years large numbers of highly trained chemists and engineers have become available and specialization has been introduced. We have today not only rather clear-cut lines between chemist, engineer and chemical engineer, but have introduced further subdivisions of activity under the name of engineering specialists who limit themselves to particular unit processes or similar fields which need to be covered in detail. This process of subdivision has been necessitated not only by the tremendous increase in activity in the chemical industry but also by a corresponding increase in the knowledge a man must have to make sure that nothing is overlooked as his work progresses"* (8).

Chilton recommended that chemical engineeers be involved in four steps of a new (research) process development: (1) initial selection of alternative processing steps to be considered; (2) laboratory studies of the reactions involved and selection or design of equipment; (3) pilot-scale operation; and (4) design of the initial plant unit.

Du Pont's major contribution to the country's war effort in World War I was the manufacture of explosives. By the time World War II erupted, explosives was just one of the many Du Pont products needed in the war effort. The insatiable appetite of the war machine imposed numerous demands on technical personnel to build new process lines (over 50 plants were built) in record time and to develop specialty products "by yesterday." Chemical engineers played a key role in meeting these demands. Since natural rubber was no longer available, plants to manufacture synthetic rubber such as neoprene had to be built. Rayon and nylon were used in tire cord. Industries producing combat equipment needed heavy chemicals. The emergence of air power required special chemical materials such as plastic enclosures, finishes, ingredients for high-octane gasolines, high-tenacity rayon for self-sealing gas tanks, and nylon to replace Asian silk previously used for parachutes.

The number of Du Pont employees involved with the manufacture of military explosives increased about one hundredfold over the peacetime level. The company's total production of 4,500,000,000 lb of explosives was three times the World War I output. Yet this large amount represented only 25% of the company's total output during World War II vs. 85% during World War I.

In the fall of 1942, Du Pont, at the request of Major General Leslie R. Groves, undertook an important phase of the development of atomic energy for military explosives. Du Pont was asked to manufacture plutonium and agreed to do so under the conditions that the company receive no profit for the work and that all patents resulting from the work become the property of the government. Chemical engineers were involved heavily in a well-coordinated team effort which took on the unique task of developing a process for producing large quantities of plutonium. Up to this point, this element, discovered in 1941, had been made only in microscopic laboratory quantities. Although this work represented a marked departure from chemical manufacturing, the team developed the process and designed and built first a semiworks then the main plant at Hanford, Washington on schedule.

During the war years earnings per share had declined from the prewar level. Once the war was over, technical efforts were directed toward reversing this trend. Two important fibers were introduced, Orlon acrylic (1948) and Dacron polyester (1950). In 1949 Du Pont became the first chemical company with sales greater than $1,000,000,000.

Some of the toughest processing problems ever undertaken by chemical engineers were those involved in developing the chloride process for making TiO_2 pigments. This (then new) route promised to be more efficient, minimize waste disposal problems, and produce a whiter pigment. All of these advantages were realized with plant start-up in late 1949, but not before Du Pont scientists and engineers underwent the trials of Job to make the process work (9). Two basic chemical reactions are involved—chlorination of impure titanium oxides to form titanium tetrachloride, which is oxidized then to form titanium dioxide. But these reactions are just the bare bones of the complete process. Materials must be preheated before each step. Separation and purification steps are complex. The reactants are volatile, the materials corrosive and sticky, and temperatures reach 1550°C. After four years of work in the laboratory and miniworks stage, a full-scale plant was built. Not 1 oz of usable product was produced, however, during the first year of start-up. Sometimes plant problems appear in a full-size plant that were nonexistent in small-scale experiments. This was one of those times. Over 50 separate sources of trouble were pinpointed, few of which could have been anticipated by further intermediate development. Thus chemical engineers used the plant as its own testing ground. The plant was made to operate not through legerdemain, but by the painstaking process of licking the problems one by one—in other words, by solid engineering skill.

During the second quarter of the 20th century a number of chemical engineers under Chilton's technical leadership became known internationally through their publications covering specific unit operations. A. P. Colburn's 1933 paper, *A Method of Correlating Forced Convection Heat Transfer Data and a Comparison With Fluid Friction*, is considered to be a classic *(10)*. Colburn developed the analogy between heat transfer and fluid friction inside tubes using a "j-factor" to account for differing values of the fluid properties associated with fluid motion and heat flow. Chilton and Colburn jointly issued "*Mass Transfer (Absorption) Co-efficients—Prediction from Data on Heat Transfer and Fluid Friction*" in 1934. Mass transfer in an absorption system was correlated via the dimensionless Schmidt number. These are only two of a continuous flow of papers that had a major impact on chemical engineering practice and university teachings. Subject matter included: flow in pipes, across tube banks, through packed columns, and in two-phase systems; mixing of gases and liquids; heat transfer including boiling and condensation; mass transfer including distillation, absorption, solvent recovery, extraction, and drying; filtration; agitation and mixing; mist and dust removal; crystallization; crushing, grinding, and classification; and fluid properties.

It is worth noting that during this period there was a movement of chemical engineering talent and knowledge from Du Pont to academe. Ideas acquired while engineers were in industry served to stimulate and modify university thinking, courses, and graduate research studies. These initial moves included Allen P. Colburn (1939), University of Delaware; Thomas B. Drew (1941), Columbia University; James O. Maloney (1946), University of Kansas; Shelby A. Miller (1947), University of Kansas; W. Robert Marshall (1947), University of Wisconsin; Robert L. Pigford (1947), University of Delaware; and Charles E. Lapple (1950), Ohio State University.

1951–1975

The third quarter of the century started with a renewed corporate dedication to research and development. A comprehensive building program initiated in 1948 at the Experimental Station culminated in the 1951 dedication of new research facilities for Central Research (the former central Chemical Department), Engineering, Industrial and Biochemicals, Pigments, Polychemicals, and Textile Fibers Departments. During the 1955–1960 period, additional laboratories were added to the

Figure 4. This complex of buildings, just north of the Brandywine Creek near Wilmington, Del., is Du Pont's Experimental Station, one of the world's largest and most diverse research laboratories. With its establishment in 1903, Du Pont embarked on a wide-ranging research program which, with the discovery of neoprene and nylon, introduced a new era in polymer chemistry. Today about 1,400 scientists and engineers work at the station on projects ranging from theoretical investigation and the research for new structures to applied studies and new-product support.

station complex for Elastomer Chemicals, Electrochemicals, Engineering, Explosives, Fabrics and Finishes, Film, and Organic Chemicals Departments (*see* Figure 4).

In 1950, the Du Pont Company accepted an assignment from the government to design, build, and operate new production facilities for nuclear materials as part of the Atomic Energy Commission's effort in support of national defense and security. The facility, known as the Savannah River Plant, is located on 300 sq mi near Aiken, South Carolina, and would cost upwards to $10,000,000,000 if built today. As in the Hanford project, the company, at its request, received a token fee of $1. Basic data had to be developed in record time to form a sound technical basis for building five nuclear production reactors, two chemical separation areas, a raw materials fabrication plant, waste tanks, a heavy-water plant, etc. Interdisciplinary teams working on a systems approach were formed to solve the many processing problems encountered because of operating in processing regimes never encountered before. Chemical engineers, after receiving "Q" security clearance from the government, worked a six-day-week schedule behind closed doors at the Experimental Station's Engineering Research Laboratory to meet the rigorous schedule. The resulting plant, run by Du Pont under contract with the Department of Energy, produces plutonium-239 and is the sole source of tritium. A large on-site laboratory continues to work on ways of increasing production, defining and developing more versatile chemical processes, and developing know-how for making and separating a number of radioisotopes for peacetime uses. Materials produced include californium-252, americum-241, uranium-233, curium-244, and cobalt-60.

By the mid 1950's, emphasis of Du Pont's chemical engineering research had started to shift away from studies of unit operations to a more fundamental approach, requiring a better understanding of interaction among the basic mechanisms of heat, momentum, and mass transfer. In a number of technical areas, Du Pont was generally well ahead of competition and only a minor research effort was necessary to maintain the edge. These research studies produced a number of experts, many of whom became internal corporate consultants. Individuals had the know-how to deal with most of the problems occurring in the normal areas of distillation, spray and pneumatic drying, dust and mist separation, and ion exchange. Research was refocused on other areas. Fluid mechanics studies were aimed at developing an understanding of "fine-grain happenings in all parts of process operations, and this requires intimate knowledge of flow patterns and their interactions with other transport processes" (12). This work played a key role in successful design of the Savannah River Plant and in film and fiber processing, viscous processing systems, and reactor technology. Fluid models proved to be an important tool in such work and great strides were made in the use of experimental facilities related to prototype systems through analogy and dimensionless groupings of process variables. Since an increasing proportion of Du Pont products required processing of highly viscous melts, solutions, or suspensions, intensive chemical engineering studies focused on developing more efficient ways to carry out viscous mixing and transfer operations.

Automatic computers, first commercially available about 1950, found use in repetitive calculations of ballistics and astronomy. The possibility that they also might have application in the chemical process industry led the Engineering Research Laboratory to acquire the first Du Pont computers for technical appraisal—a Philbrick analog in 1950 and an IBM CPC digital in 1953. The potential for broad applicability in the field of chemical processing was demonstrated by devising special computational techniques for nonrepetitive problems and identifying specific areas suitable for quantitative calculation. Early applications included optimization of heat exchanger design for Savannah River (1951); polyethylene (PE) process control simulation (1952); and design of monovinyl acetylene (MVA) columns for neoprene (1954). To accelerate recognition of the role of computers, the Laboratory conducted numerous seminars, participated in computer application to trial problems to demonstrate the utility of mathematical methods, and helped to guide the formation of new computing groups. Technical computers soon were recognized as an essential ingredient in both the research and engineering function.

The systems approach was applied to the design of reactors, considered the heart of the chemical process. Recognition was given to the various mechanisms and interactions existing simultaneously in a reaction system—mixing, dispersion, heat addition or removal, diffusion of reacting species, and chemical reaction rates and equilibria. Extensive use of comprehensive math models, depending in turn on high-speed computers, was a prime factor in allowing chemical engineers to select and size some reaction equipment, tailored to promote desired reactions while suppressing undesirable ones. The systems approach proved to be a valuable one. Chemical engineers could assess the effects of changes in reactor operation on output, including yield and conversion, and on downstream operations, including the separation and purification steps, as well as process economics. The development of a process math model, when feasible, during the process development stage greatly facilitated the systems approach, often helping to pinpoint needed key experimental data. The 1960 publication of *Transport Phenomena* by R. Byron Bird, Warren E. Stewart, and Edwin N. Lightfoot *(13)* provided a text with the scientific approach to chemical engineering needed in the systems approach to chemical processing. The impact on chemical engineering educational curricula was timely.

The corporate effort behind the development of polyester films such as Cronar photographic film base (1951) and Mylar film (1952) typifies procedures applied to hundreds of Du Pont developments. Sixteen years of development work were required before the fundamental chemistry was translated into commercial success. Along the way numerous chemical engineers and other technical experts were needed to complete studies on such tangential subjects as the crystallization and orientation of polymers, chemical kinetics, and reaction mechanisms. The process of manufacturing the polymer continuously, rather than by separate batches, required new methods of control. New processes had to be developed to manufacture the chemical intermediates.

In the 1950's and 1960's Du Pont's fibers business was a major source of sales and earnings, and significant investment was made in fibers both to increase capacity and to develop new and/or improved products. With a combination of a chemical engineer's systems approach, better instrumentation, refined analytical techniques, and mathematical modeling, key experimental programs were formulated to obtain an understanding of the composition–time–temperature–deformation history from the raw materials stage through the finished products. Relationships of processing variables to differences in key product properties provided the basic data needed to redesign segments of the processes. This work illustrates the value of the flow back and forth between industrial needs and laboratory research—the need to improve process economics, stimulating more basic research.

Research and development in fluid jet technology led to its widespread use in film and web processing, mixers, reactors, and fiber processing. Applications to fibers alone includes Taslan bulked yarns, jet piddlers, draw-jets, interlaced yarns, bulk-crimped carpet fibers (BCF), and spunbonded products such as Reemay polyester, Typar polypropylene (PP), and Tyvek olefins. New fibers also were introduced: Nomex aramid, especially suited for high-temperature applications; Lycra spandex, an elastic textile yarn; Qiana nylon, with a silklike appearance; and Kevlar aramid, used in products ranging from tires to bullet-resistant vests.

By 1960 the outstanding performance of Du Pont's fibers business led to a dichotomy: the financial success was welcomed, but a concern developed about the potential excessive dependence on the fibers business. Therefore technological capabilities and other resources were directed into new fields and new markets. Extensive R&D efforts led to a prolific expansion and diversification period. Since, on the average, between three and four new ventures were introduced every year, the list is too long to be all inclusive. A few of the developments that required particularly important inputs from chemical engineers included Hypalon synthetic rubber based on chlorosulfonated PE, Teflon polytetrafluoroethylene fiber based on research concerned with methods of handling intractable polymers; Delrin acetal resin, a stable linear polymer formed from very pure monomeric formaldehyde; Corfam poromeric material (a technical success later abandoned for business considerations); Dycril printing plates, one of a series of products resulting from fundamental research on photopolymerization (these nonsilver, light-sensitive materials can be processed with nonpolluting chemicals and little energy consumption); families of biological innovations such as highly selective herbicides, fungicides that work systematically and have low toxicity to man, and quick-acting insecticides that rapidly break down in the field to innocuous byproducts; the "aca," automatic clinical analyzer which performs both kinetic enzyme and conventional chemical analyses on body fluids in minutes; Nafion perfluorosulfonic acid products for high-efficiency electrochemical cells; Viton fluoroelastomer, resistant to oils and fuels at high temperatures; and Surlyn ionomer resin, a thermoplastic polymer with ionic cross-links.

During the last ten years of this period, the company accelerated its growth in its major new lines—electronics, pharmaceuticals, and instruments—through acquisition of Endo Laboratories (ethical drugs), Berg Electronics (precision electrical and electronics connectors), and Ivan Sorval (biomedical laboratory equipment). The combination of these acquisitions with R&D efforts and marketing capabilities produced a synergistic effect, quickly establishing Du Pont in new markets.

The 1970's brought great turbulence in the business environment. In addition to worldwide economic problems, there were a host of other external developments that impacted on the chemical industry.   Government regulations, based on good intentions (e.g. pollution control and safety), became increasingly burdensome, with some arbitrary policies giving little or no consideration of cost-benefit–risk analyses.   Then came the energy crunch and subsequent escalation of petrochemical prices, that rapidly changed the world's price structure for raw materials.   All of these events affected the economics of chemical processes.   An increasing portion of the chemical engineers' efforts was directed towards keeping established lines healthy through process modifications, reacting to these ever-changing external forces.

1976–2000

As we enter the last quarter of the 20th century, Du Pont's continuing commitment to research and development is reflected in a $400,000,000 1979 R&D budget.   As demonstrated in the first three quarters of this century, R&D forms the cutting edge of progress, leading to new and expanding business.   Career opportunities are opened for chemical engineers in a cascading fashion.   R&D itself offers a variety of challenges ranging from basic research to process development, with a growing interplay between the two.   Chemical engineers are suited particularly well to the design function and to serve as project engineers, coordinating the efforts of a variety of technical disciplines required to translate basic technology into successfully operating plants.   Many become production engineers, who take over the next step, making the plant run as required on a day-to-day basis.   Those in plant technical groups solve plant problems and introduce new technology or modified facilities to improve performance.   In the Engineering Department many chemical engineers become specialists and corporate consultants to the industrial departments.   Supervisory and managerial assignments appeal to many; others like the challenge of marketing, or the field of patents, etc.   Each of these general areas has many subdivisions; hence the chemical engineer's career opportunities are as broad or broader than any other discipline.

Over the past 30 years the computer explosion has revolutionized information processing systems.   These are powerful tools in the hands of innovative chemical engineers teamed with those of other disciplines, and the variety of applications is an impressive, growing list.   Radical advances in solid-state microelectronics and their rapid exploitation by the computer industry are making very low-cost microcomputers available.   This continuing trend will open an even wider variety of industrial applications ranging from control of individual instruments and machines to systems for total processes.

The excellent fit of the chemical engineer into industry is not by coincidence. Indeed one would be hard pressed to find a better example of academic response to an industrial need than the history of the chemical engineering profession. One practice well established in the chemical industry, but seldom practiced in universities, is the use of chemical engineers as coordinators of interdisciplinary teams to tackle problems on a systems basis. A few universities have broken down the barriers between departments and others are encouraged to do the same. When any subdivision of an institution attempts to operate in an autonomous manner, it is saddled with inefficiencies and invites total failure. Using best resources is a necessary approach for conducting meaningful long-range research.

Like Mark Twain, chemical engineers are concerned about the future because that is where they will spend the rest of their lives. Continued radical changes in the social and economic climate are predicted. The political and complex international aspects may be frustrating, but the resultant technological challenges are real and represent opportunities for the chemical company that copes with them best. Thus today's climate in Du Pont, as well as tomorrow's, is tailor-made for creative chemical engineering. Conditions demand new ways of doing things and the years ahead will be unusually hospitable to technical innovation.

Literature Cited

1. Haynes, W. "American Chemical Industry"; D. Van Nostrand Co.: New York, 1949; Vol. 6, p. 125.
2. Landau, R. "The Chemical Engineer—Today and Tomorrow", *Chem. Eng. Prog.*, 1972, 68(6), 9.
3. Hougen, O. A. "Seven Decades of Chemical Engineering," *Chem. Eng. Prog.* 1977, 73(1), 89.
4. Walker, W. H.; Lewis, W. K.; McAdams, W. H. "Principles of Chemical Engineering"; McGraw-Hill: New York, 1923.
5. "Chemical & Engineering News 50th Anniversary," *C&EN* 1973, 51, No. 3, 21 (Jan. 15).
6. Perry, J. H. "Chemical Engineers' Handbook"; McGraw-Hill: New York and London, 1934.
7. *Nylon—The First 25 Years;* E. I. du Pont de Nemours & Co.: Wilmington, 1963; p. 9.
8. Chilton, T. H. "The Technical Division—Engineering Department" (Internal Booklet), E. I. du Pont de Nemours & Co., Dec. 29, 1939.
9. *The D of Research and Development*, No. 30 in "This is Du Pont" Series; E. I. du Pont de Nemours & Co.: Wilmington, 1966.
10. Colburn, A. P. "A Method of Correlating Forced Convection Heat Transfer Data and a Comparison with Fluid Friction," *Trans. Am. Inst. Chem. Eng.* 1933, 29, 174.
11. Chilton, T. H.; Colburn, A. P. "Mass Transfer (Absorption) Coefficients—Prediction from Data on Heat Transfer and Fluid Friction," *Ind. Eng. Chem.* 1934, 26, 1183.

12. "Advances in Technology Through Engineering Research" (Internal Report), E. I. du Pont de Nemours & Co., Engineering Dept., 1958.
13. Bird, R. B.; Steward, W. A.; Lightfoot, E. N. *"Transport Phenomena"*; Wiley: New York, 1960.

RECEIVED May 7, 1979.

The History of Chemical Engineering at Exxon

EDWARD J. GORNOWSKI

Exxon Research and Engineering Co., P.O. Box 101, Florsham Park, NJ 07932

From W. K. Lewis, one of our most valued consultants, we learned early on to go for the fundamentals, to understand what we're doing, to build on a solid foundation—but not to wait until we understand everything before being willing to make a decision.

Major chemical engineering contributions from Exxon include continuous thermal cracking processes; putting fractional distillation on a sound basis; processes for tetraethyllead, 100-octane avgas, and raw materials for synthetic rubbers and chemicals; fluid cat cracking, fluid coking, and the Flexicoking process. Fluid hydroforming was a disappointment. Reactor engineering and separations technology have seen major advances. Coal conversion is again a focus of activity. This chapter deals with refining technology although chemical engineering has been essential to many other Exxon activities.

Chemical Engineering at Exxon goes back to the founding of its corporate predecessors and underlies much of the technology that provides the basis of Exxon's multi-faceted operations. Today Exxon is involved in chemicals, minerals, nuclear fuel assemblies, solar photovoltaic cells, and in many other fields—and chemical engineering has played an important role in the history of each of them. I wish that I could delve into the historical contributions of chemical engineering and chemical engineers to all of these fields but space limitations have persuaded me to limit myself to just one—the refining of fossil fuels.

Relevant commercially oriented technology existed even before Drake drilled the first oil well in 1859. In 1850 James Young, a scientist from Glasgow, Scotland, patented a distillation process to produce naphtha, kerosene, lubricating oil, and paraffin wax from coal tar and oil shale. We shouldn't forget that what we now call the "potential synthetic fuel industry" actually predates the petroleum industry.

The infant petroleum industry soon saw the benefit of technological expertise and the Standard Oil Alliance, formed in 1875, engaged William G. Warden, inventor of an oil-distilling process and an improved railroad tank car, Henry Rogers, inventor of a widely used petroleum distillation process and a naphtha separator, and Eli Hendrick, a noted lubricating oil specialist.

By the time the Standard Oil Company was organized in 1882, it held over 20 patents for general refining processes and equipment as well as some 30 lubricating oil patents. Although several extended R&D programs added to this base of technical knowledge after 1882, the work was directed towards the solution of very specific problems. F. W. Arvine set up the company's first engine lab, known as a "power and machinery room," at 128 Pearl St., New York City, in 1882 to help improve lubricating oil and grease quality. George M. Saybolt set up his standard inspection lab at the company's headquarters at 26 Broadway in New York City in 1883 to develop product quality tests and uphold the meaning of the word "standard" in the company's name. Herman Frasch set up a lab at the Whiting, IN Refinery in 1886 to develop a process for converting the sour, sulfur-laden crude oil from the Lima, OH field into marketable petroleum products.

When the old Standard Oil Company was split into 34 companies by the Supreme Court antitrust decision of 1911, the Standard Oil Company (New Jersey)—which later changed its named to Exxon—lost the services of the research laboratory at the Whiting Refinery, then headed by William M. Burton. It also lost the benefit of a new cracking process developed at that lab in 1913.

During World War I, Jersey Standard's management was too busy supplying petroleum products to the allied armies to press far into new fields. But by 1919 the company's top management had become concerned about the technological base for its operation.

Jersey Standard President Walter C. Teagle wrote an associate in June, 1919, "I have felt more than ever before the need for a thoroughly organized and competent research department under an able executive, such a department not to be confined to chemical research, but to general research in connection not only with production and refining of our products, but with the sales end of our business as well." A formal notice announcing the formation of a development department was issued on September 27, 1919. The department's 26 employees worked in labs at the Bayway Refinery in New Jersey.

When we undertook to set up an organization to apply chemical engineering to the oil industry as rapidly and efficiently as possible, it was absolutely necessary to turn to others for help. The original group which advised and assisted in creating the technical organization was made up of Ira Remsen, President Emeritus of John Hopkins University and perhaps the leading organic chemist of his day, Robert A. Millikan of the California Institute of Technology, one of our greatest physicists and later a Nobel Prize winner, and Warren K. Lewis, professor of chemical engineering at M.I.T. We also relied heavily on Charles A. Kraus of Brown University and Carelton Ellis of Montclair, NJ.

The first major chemical engineering refining problem we faced was the development of a satisfactory continuous cracking process for converting heavier oils into gasoline to replace the Burton process. Two processes were developed. Both were engineering applications of the principle that the cracking of petroleum is a function of time and temperature. The first process was the Double Coil Process, in which the oil was heated rapidly to a high cracking temperature in pipes exposed to the radiation in the furnace, and then "soaked" or held at this high temperature for a further period in pipe coils that were exposed only to the mild heating effect of low-temperature combustion gases. The heat flux was so small that there was little danger of burning the tubes when they became coated with carbon inside.

The second was the Tube-and-Tank Process, in which a relatively large pressure tank on the exit end of the pipe coil provided the time necessary for the desired cracking reaction to complete itself.

These two operations almost immediately became the standard operations for the cracking of heavy oils for Exxon, and also were used widely by licensees, replacing the older batch-type cracking stills and effecting enormous economies, both in the first cost and in the operating cost, as well as greatly extending the range of practical feedstocks that could be used. The attention of the world's entire oil industry was focused on this development, and there were endless patent controversies.

Our next important problem was the modernization of the basic operation of petroleum refining—fractional distillation. The oil refinery of the early 1920's was made up of batteries of horizontal cylindrical batch stills with a capacity of 200 to 1,000 bbl each, sometimes arranged for operation singly, and sometimes connected in cascade by overflow pipes. The stills were mounted in brick settings that exposed the bottom third of the still to the direct radiant heat of the furnace. To assist the separation carried out by the distillation itself, the stills usually were equipped with a series of partial condensing towers in which heavier components of the vapors condensed seriatim before the final condensation in the water-cooled worm or pipe coil. The separations effected by this combination of fractional distillation and fractional condensation were very poor, and it was necessary for the refiner to do an enormous amount of redistillation. A refinery such as Bayway would have several batteries of stills engaged in redistillation for every battery engaged in crude distillation. The solution to this problem was found during the 1920's by combining the pipe coil type of furnace first used on a large scale in the continuous cracking processes with fractionating columns designed in accordance with principles of fractional distillation, which were understood well in the scientific world, but which the oil industry never had applied before on a large scale.

The oil industry in general owes more to Warren K. Lewis than to any other individual for the quick and successful application of the scientific principles of fractionating column design to the oil industry. We have continued to develop the chemical engineering technology for fractional distillation and Exxon now has continuous distillation units capable of handling up to 275,000 bbl/day (40,000 tons/day) of crude oil.

One of the lessons that Doc Lewis taught us, along with teaching it to whole generations of chemical engineers, was to go for the fundamentals, to understand what you're doing, to build on a solid foundation— but not to wait until you understand everything before you're willing to make a decision. This has been as important a lesson for us as any other. We still are shooting for that optimum balance of theoretical understanding, engineering knowhow, common sense, intuition, and guts that characterize a good chemical engineer.

During this same period, i.e. the 1920's, we were engaged in another extremely important chemical engineering development. Messrs. Kettering and Midgley of General Motors had investigated the phenomenon known as knocking in internal combustion engines and identified it as detonation of the fuel charge in the engine cylinder. Finding a way to upgrade gasoline to prevent engine knock became an important research goal during this period as the horseless carriage gained in popularity and as automobile engines became more powerful. Midgley discovered that a mixture of tetraethyllead (TEL) and an alkyl halide, in the astonishingly small proportions of .1 of 1%, would prevent knocking of ordinary gasoline. But there was no known method of making TEL economically in the required large amounts. With the help of Dr. Kraus, our chemical engineers solved this problem by developing a simple and practical method of producing TEL from an alloy of sodium and lead treated with ethyl chloride. The ethyl chloride required for the production of TEL was made also from refinery ethylene by a process that our chemical engineers worked out.

During the 1920's fears of a future petroleum shortage caused Jersey Standard to become interested in synthetic fuels. This interest led to an agreement in 1927 between Jersey Standard and the German firm, I. G. Farben, for a cooperative research program to develop a coal hydrogenation process in the United States. A group of 18 American oil companies was organized. This group exchanged technical information but left the principal burden of technical effort to us.   The technology was to become a cornerstone of the company's chemical engineering expertise.

A month after signing the Farben agreement, Jersey Standard took another step which signaled the start of an era of expansion in chemical engineering.   On October 27, 1927, Jersey established the Standard Oil Development Company and transferred the development and general engineering departments, formed in 1913, to the new company.   The December, 1927 issue of *The Lamp* reported that the new company "will have personnel and facilities suited for handling all varieties of engineering and chemical work of general interest, for carrying on major projects of laboratory research, and for the technical and financial direction of major projects."   Even today Exxon Research and Engineering Co., the successor of the Standard Oil Development Company, retains much of the basic structure established in 1927.

The magnitude of the technical problems involved in high-pressure coal hydrogenation seemed staggering.   Preparing to meet this situation in 1927 we drafted R. T. Haslam, an associate of Doc Lewis at M.I.T., to organize an entirely new technical group at Baton Rouge.   Robert P. Russell, assistant professor of chemical engineering at M.I.T., became manager of the new labs.   Russell recruited a staff composed largely of young M.I.T. faculty members and graduate students.   This crew of chemical engineers without any previous refinery experience was greeted with some consternation by the established old-timers on site, but the dedication, willingness to learn, competence, and undoubted successes of the newcomers soon established their credentials as a valuable chemical engineering asset.   This role has been maintained for over 50 years by Exxon's Research & Development Labs at Baton Rouge.

Discovery of additional prolific oil fields postponed any need for hydrogenating coal, but hydrogenation made 100-octane aviation gasoline possible.   At first, 100-octane fuel was made by the dimerization of refinery isobutylene.   This hydrogenated diisobutylene which was, in effect, commercial isooctane, was mixed with selected natural gasoline of the highest quality and with TEL to produce the 100-octane gasoline on which the U.S. Army Air Corps did its original high-compression engine development work.   Soon the supply of natural gasoline of this quality became inadequate and it was necessary to resort to hydrogenation to produce a synthetic gasoline of high enough octane number for blending with the synthetic isooctane.   At the outbreak of the war in 1939, the Baton Rouge hydrogenation plant was producing both the synthetic blending agent and the synthetic base, and was the largest single source of 100-octane fuels in the world.   Availability of 100-octane avgas, and the power boost it provided, allowed the Royal Air Force's Spitfire fighters to outperform the Luftwaffe's planes that were performance limited by Germany's 87-octane fuel.

In the production of aviation gasoline the original hydrogenation methods soon were supplemented by two other processes—alkylation and catalytic cracking. We engineered the first commercial alkylation plant and also the first plants for the isomerization of normal paraffins to produce isoparaffins for alkylation purposes. The catalytic cracking process was pioneered commercially by the Houdry Company, but an entirely new development, fluid catalytic cracking, which represented the contribution of our chemical engineers to a cooperative effort participated in by several American and foreign companies, enormously advanced the economic frontiers of catalytic cracking.

The development of fluid catalytic cracking was a real chemical engineering challenge. There were two basic problems: the catalyst was deactivated rapidly by coke deposits and the cracking reaction was very endothermic—huge quantities of heat had to be supplied to the catalyst. Houdry's fixed-bed process had overcome these two problems by switching the flow through the catalyst bed every few minutes; first hydrocarbons to crack and deposit coke, then air to burn the coke, regenerate and heat up the catalyst, with steam purges in-between to prevent disasters. We believed that this cumbersome process, with its constant need to shift large flow streams, could be improved by continuously moving the catalyst between a reaction zone and a regeneration zone.

But how to move the catalyst? Bucket elevators? Screw conveyors? Early on we thought of the pneumatic transport used to move grain and we considered a powdered catalyst to make the transport easier. Fortunately, or unfortunately, we did some lab work on the coke-burning reaction rate and, assuming pneumatic transport would maintain a solids density of about 1 lb/cu ft (16 g/L), we calculated that we would need a regenerator some 7 ft (2 m) in diameter and about 7 miles (11 km) long. We set out, therefore, to raise the reaction rate by finding a catalyst that could stand a higher regeneration temperature. And we put a long regenerator on our 100 bbl/day (10 L/min) pilot plant in Baton Rouge. In order to fit as much regenerator pipe as we could into a reasonable space, we built what we called a "snake"; it went up and down and up and down and up and down. . . To our surprise, we found that the solids density was higher in the upflow legs than in the downflow ones. Doc Lewis quickly put some of his students to work measuring solids density as a function of flow direction, gas velocity, and solids flow rate in glass apparatus . . . and shortly the dense fluid bed regenerator became more than a gleam in the eye.

Dense fluid beds were known—the Winkler gas generator had been around for years—but it had always been thought that you had to operate at gas velocities below the free-fall velocity of your particles; that at higher velocities the particles simply would be blown away. Of course, practical regenerator gas velocities eventually will blow away (or entrain) the fine cat-cracking catalyst particles, but it came as a great surprise that before they blew away they acted just like a Winkler bed.

We capitalized on this surprise and quickly converted the Baton Rouge pilot plant to dense fluid-bed operation. There were still many things we didn't understand, but we went ahead. We developed a whole host of ancillary technology, from catalyst standpipes that built up hydrostatic pressure to improved cyclones. And while we were theorizing and experimenting in our pilot unit, we were designing and building commercial plants. Commitments were made to build 30 fluid cat-cracking plants before the first commercial unit started up.

Many names are associated with the development of fluid cat cracking. Along with those of our consultants, Doc Lewis and Ed Gilliland of M.I.T., the names that stand out particularly were those of four of our chemical engineers: Homer Z. Martin, C. Wesley Tyson, Donald L. Campbell, and Eger V. Murphree. It is hard to convey the magnitude of the challenge and the sense of accomplishment that the teams of chemical engineers felt in creating a whole new technology and in seeing it go on-stream practically simultaneously.

The cat plants provided more than components for high-octane gasoline. They provided raw materials that were essential to the U.S. Government's synthetic rubber program during World War II—butylenes directly and butadienes indirectly via dehydrogenation. Chemical engineers at Exxon were involved deeply in the synthetic rubber program (for Buna S as well as butyl rubbers) and in such other wartime programs as the synthesis of nitration-grade toluene and the development of the steam-cracking process to produce chemical raw materials. Again, space limitations do not permit detailing the many chemical engineering contributions made on the chemicals side of the fence.

After WWII, chemical engineering continued to provide opportunities for technological advances in refining for Exxon—and for an occasional retreat. One of the less successful developments was fluid hydroforming, a process to raise the high-octane aromatics content of naphthas. The desired high conversion and yield levels, coupled with the relatively slow catalyst deactivation rate and modest heat requirements, really did not make catalytic reforming a good candidate for fluid-bed operation. Fixed-bed reforming with noble metal catalysts has supplanted fluid hydroforming.

More successful was the extension of fluidized-bed technology to fluid coking, a process to upgrade heavy petroleum residues. Fluid coking presented many chemical engineering problems. Coke is laid down in layers on existing coke particles which thus tend to grow. Therefore, it is necessary to provide small seed particles. There is no natural seed formation as in crystallizers, so seed particles must be generated by grinding. Relations between grinding, growth, and agglomeration rates must be worked out. Coking of tar deposits in the cyclones must be prevented, but places for tar deposition must be provided. And so on. Our chemical engineers provided adequate solutions to all of these problems and fluid coking became an established process.

Fluid coking technology has been extended further recently to the Flexicoking process, where the coke is gasified with steam and air to form a low-BTU fuel gas and there is only very little net coke production.

Two chemical engineering applications based on fluidized-bed technology are currently under development at Exxon that we believe will make the history books of the future: pressurized fluid-bed combustion (PFBC) and magnetically stabilized beds (MSB). PFBC shows particular promise for the clean, compact combustion of coal in power plants or process heaters. Capital investments should be less and environmental protection easier than with regular coal-fired furnaces. MSB uses externally imposed magnetic fields acting on beds of magnetizable particles to collapse gas bubbles as soon as they start to form in a fluidized bed. An MSB is thus a calm, quiescent bed without the gas bypassing and the back mixing that gas bubbles ordinarily engender. Our chemical engineers are working to use these characteristics of MSB to help solve some long-standing process problems.

I have stressed fluidization technology in my recounting of chemical engineering at Exxon over the last 40 years. Fluidization has been an important area of chemical engineering activity but it is by no means the only one. We have been very active in all aspects of what has become known as reactor engineering—the design and operation of reactor systems to optimize a combination of heat, mass and momentum transfer, chemical kinetics, and control strategy to permit ready, safe, and profitable operation with a variety of feed and product constraints. The power of modern computers, along with our innovative software, allows our chemical engineers to handle far more variables than would have been thought possible just twenty years ago.

We have continued to pioneer in the development and application of many separation techniques from heatless drying (the removal of moisture by intermittent absorption and desorption at different pressure levels), through the use of molecular sieve absorbers, advanced lube oil extraction media, and CO_2 absorption promoters, to the use of liquid membranes (containing an encapsulated absorbing phase) and lasers to activate only specific molecules.

Some 50 years after the flurry of interest in coal hydrogenation and 120 years after James Young's patent, we again are immersed in the chemical engineering problems of turning coal into more attractive liquid and gaseous fuels.

And, of course, chemical engineering at Exxon has, along with the Corporation's expanding interests, branched out into areas far removed from the refining of fossil fuels.   There are difficult chemical engineering problems associated with the manufacture of petrochemicals, minerals recovery, advanced batteries and other forms of energy storage, the economic utilization of solar energy to replace other energy forms, and even with information processing technology.   Exxon's chemical engineers are active in all of these areas.   But, as I said, I am restricting myself to refining technology.

The Exxon Research and Engineering Company has long had a prominent role to play in the development and application of Exxon's refining technology with the help of chemical engineering.   We continuously aim to hone our technical skills through challenging assignments, thorough peer review of our work, continuing education, and ample contact with our academic colleagues.   I trust that our future accomplishments will warrent Exxon again receiving an invitation 50 years from now to prepare a paper discussing "Chemical Engineering at Exxon from 1979 to 2029."

RECEIVED May 7, 1979.

Three Decades of Canadian Nuclear Chemical Engineering

H. K. RAE

Chalk River Nuclear Laboratories, Atomic Energy of Canada Limited, Chalk River, Ontario, KOJ 1JO Canada

Major contributions to uranium refining, heavy-water production, reactor coolant technology for heavy water, boiling-water and organic-cooled power reactors, waste immobilization, and irradiated fuel processing are described. Included is a brief outline of the evolution of the highly successful Canadian nuclear power program based on natural-uranium-fueled, heavy-water-moderated reactors. The most important chemical engineering achievement in the nuclear field in Canada was the establishment of an industrial capability to produce heavy water. Two other key contributions were (a) the control of the build-up of radiation fields caused by the activation of corrosion products and their transport by the coolant, and (b) pioneering work on the immobilization of fission products in glass which included a field test now in its 20th year.

The Canadian nuclear power program has brought the natural-uranium-fueled, heavy-water-moderated reactor design to the stage of a demonstrated, commercially competitive power source. This design has the acronym CANDU—CANada Deuterium Uranium. A brief outline of the evolution of this reactor concept provides useful background to the discussion of the history of nuclear chemical engineering in Canada.

The choice of heavy-water moderation for Canadian nuclear power reactors was influenced strongly by the wartime decision (1) that an Anglo–Canadian team would build a heavy-water reactor in Canada to make plutonium. This became the NRX experimental reactor at the Chalk River Nuclear Laboratories (CRNL) which is now in its 32nd year of operation. This project paralleled the plutonium-production route in graphite reactors used at Hanford by the United States Manhattan District Project. Thus Canada acquired early operating experience with heavy-water reactors and a sound appreciation of the inherent advantages of this unique moderator (2). Its extremely low neutron capture cross section permits a sufficiently high thermal power density to be achieved

0-8412-0512-4/80/33-190-313$05.25/1

with natural uranium fuel for an economically viable system. In contrast, although ordinary water has even better moderating properties permitting high power density, its relatively high neutron capture cross section means that the fissile uranium-235 content of the fuel must be enriched to about four times the natural value. Thus the light-water reactor design commercially developed in the United States and widely adopted elsewhere requires uranium isotope enrichment while the CANDU design requires hydrogen isotope separation.

Another early decision which shaped future events was taken in 1950—to build the NRU reactor at CRNL. This large experimental reactor (3) developed confidence in our ability to minimize leakage in complex heavy-water coolant circuits and pioneered the technology of changing fuel at full reactor power. NRU provided large irradiation test facilities which remain unique in the world. Full-scale fuel channels and fuel bundles are studied at power reactor conditions in separate, individually controlled coolant circuits (loops). Similar smaller-scale experiments are done in the NRX reactor. This work provided the main focus for the reactor research and development program. Such experiments were the basis of the evolutionary design of two prototype reactors (4)— NPD at 22 MW_e which started up in 1962 and Douglas Point at 208 MW_e which started up in 1967. This experience provided a firm foundation for Canada's first commercial nuclear power project—the Pickering Generating Station consisting of four reactors with a total capacity of 2056 MW_e, designed by Atomic Energy of Canada Limited and built and operated by Ontario Hydro. These four reactors came into service between 1971 and 1973 and have achieved an outstanding record of high-capacity factor ever since (5). Ontario Hydro has embarked upon a major nuclear power program and now has 5 GW_e in service with a further 9 GW_e under construction (5). Quebec and New Brunswick have initiated CANDU nuclear power programs, and outside Canada units are in operation or under construction in four countries for a total committed capacity of 3 GW_e(6) in addition to Ontario's 14 GW_e.

The strategy adopted to develop nuclear power in Canada was set by 1955. Simplicity and early viability were emphasized together with the ability to manufacture the majority of the equipment and components in Canada. The simplest fueling arrangement was selected—irradiation of natural uranium followed by interim storage of the irradiated fuel. Recovery and recycle of plutonium or the use of thorium was postponed for future development. A low fueling rate and an attractive fueling cost were estimated to be possible without recycle (7), and this has proved to be the case (5). Our uranium supply was large. It was clear that Canada would take several decades to reach a nuclear industry large enough to support the minimum economic size of a plutonium recovery plant (8). And finally, interim storage of the irradiated fuel appeared to be quite feasible.

Thus, for nearly two decades from 1955, the nuclear power program in Canada could concentrate virtually completely on the supply of nuclear materials and on reactor development.

Chemical Engineering in the CANDU Program

Key contributions to the Canadian nuclear power program in the field of chemical engineering have occurred in three major areas, as indicated in Figure 1. These are the production of nuclear materials, the improvement of reactor operation, and the nuclear fuel cycle. I have selected nine topics within these areas to highlight in this historical review. Figure 1 shows the periods during the past three decades of major chemical engineering activity on each topic. I have distinguished between periods of research and development activity and those of commercial operation, and as one would expect, there is considerable overlap between these two types of activity. Each period of commercial operation of course must be preceded by design and construction; although not explicitly included in Figure 1, these activities also have important chemical engineering components.

There are three essential nuclear materials for the CANDU reactor: uranium, heavy water, and zirconium. The latter material has involved little chemical engineering activity in Canada and will not be considered further in this review.

Uranium milling and refining has become an important industry in Canada, mainly dependent on the export market. At present, domestic requirements for uranium are about 20% of Canada's total refining capability. This industry began with small-scale, wartime operations to provide fuel for the world's first reactors. Then it expanded rapidly to meet the needs of large military programs in Great Britain and the United States (9). This was followed by a period of much lower demand until commercial nuclear power became established; now we are at the beginning of a new period of expansion (10). The three major products are UO_3, UO_2, and UF_6. Large-scale production of high purity UO_3 began in the mid-1950's (11), mainly for export. Then capability for ceramic-grade UO_2 was added in the early 1960's, mainly for the CANDU program (12). Finally, facilities for conversion of UO_3 to UF_6 were added by 1970 (13) to better serve the export market by providing feed for the toll enrichment of customer's uranium.

Heavy-water production in Canada began on a small scale in a plant operated by Cominco for the United States Atomic Energy Commission (USAEC) in 1944 (14) and continued until 1956. This was also a period of initial research into heavy-water processes at CRNL (15). By the mid-1950's the USAEC and the E. I. Dupont de Nemours and Company had put into operation two large, heavy-water plants using the GS process. The

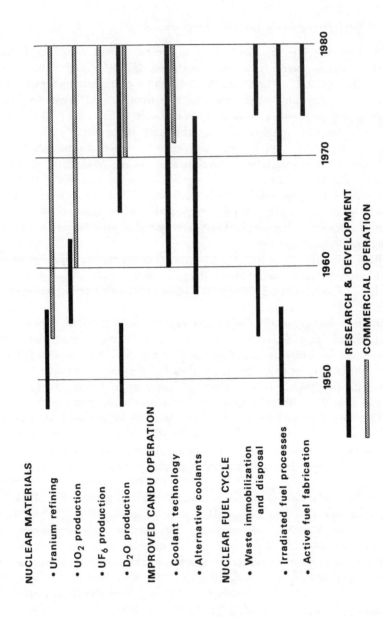

Figure 1. Major areas of Canadian nuclear chemical engineering

United States–Canada agreement on the exchange of atomic energy information provided for Canadian access to this technology when it was needed, and meanwhile for the purchase by Canada of the heavy water for the first CANDU reactors. Therefore, development work was suspended until Canada began to put its own heavy-water production capacity in place in the late 1960's (16). There are now four large plants in operation in Canada and two others being constructed, representing a total investment of $1,500,000,000 over the past 15 years. A wide range of chemical engineering contributions has been involved. Research and development has been divided about equally between improving plant performance and advancing alternative processes.

An important area where improved reactor-system design and operation have been achieved is in the control of all aspects of coolant chemistry. The major development here has been the identification of the factors controlling movement of corrosion products by the coolant into the reactor core where they are activated, and the subsequent deposition of these radioactive species on out-reactor components causing radiation fields that may interfere with maintenance work during shutdowns. In commercial CANDU reactors the fields from such long-lived radioactivity have been controlled successfully to low values (17).

The basic concept of the CANDU reactor with separate moderator and heat-transport (coolant) systems offers the opportunity to substitute another coolant for pressurized heavy water without extensive design changes (18). Two alternative coolants have been investigated through to the stage of operating prototype reactors. These are boiling light water (BLW) and an organic fluid. Both offer higher overall thermal efficiency, and the organic fluid also offers very low radioactivity in the coolant circuit. Control of coolant chemistry and corrosion-product transport in both cases has involved extensive chemical engineering research and development.

The various parts of the nuclear fuel cycle were studied extensively in the early days of the nuclear program to explore the technical feasibility of the various options. Processes to recover fissile plutonium or uranium-233 from irradiated uranium or thorium were investigated (19). This work was brought to a close about 1956 after the decision to adopt simple natural uranium fueling for CANDU reactors. In this period it was recognized that the fission product wastes eventually would need to be immobilized and isolated from man's environment. CRNL pioneered the incorporation of fission products into glass in the late 1950's, taking the process to the stage of demonstrating technical feasibility (20). Further work was postponed until the CANDU program had grown sufficiently so that long-term waste-disposal needs could be better forecast. We now can see the need to have a demonstrated technology in place early in the next century, and a large research, development, and demonstration program (21) has been initiated at the Whiteshell Nuclear

Research Establishment (WNRE). Wastes arising from reactor operation, which contain less than 1% of the total radioactivity generated but
are much larger in volume than the fission product wastes, must be
concentrated and immobilized for disposal. Development of this technology began in the mid-1970's and will be demonstrated in a waste
treatment center at CRNL (22). Coincident with these new activities in
waste disposal is a renewed interest in recycling fissile material recovered
from spent fuel, and in particular, in the thorium fuel cycle (23). Thus
new work in processing irradiated fuel and on the remote fabrication of
recyle fuel began in the early 1970's.

Uranium Refining

The Canadian uranium industry began in 1942 and rapidly grew
during the 1950's to reach a peak production of 12 Gg U/a. Production
had declined to about 3.5 Gg U/a by 1970 as military requirements were
fulfilled; many of the early milling plants had been dismantled (9). By
1978 the industry had expanded again and production was 6.4 Gg of
uranium in the form of U_3O_8. Rapid growth to over 12 Gg/a is forecast
for the mid-1980's (10).

The first Canadian plant for leaching of uranium ore commenced
operation at Port Radium, Northwest Territories in 1952 by Eldorado
Mining and Refining, Limited—later Eldorado Nuclear Limited. This
used a flowsheet pioneered by the Radioactivity Division of the Mines
Branch, Department of Energy, Mines, and Resources, Ottawa (24). Acid
leaching was followed by reduction and precipitation of the uranium. Following early work by the Oak Ridge National Laboratory, Eldorado piloted
a solvent-extraction process using a tertiary amine for the uranium recovery
and purification step. A plant using mixer–settlers was built at Port
Radium and commenced operation in 1958 (25).

Virtually all of the Canadian mills have used a variant of the acid-
leaching process developed by the Mines Branch to extract uranium from
the ore, and all but Port Radium have followed this by ion-exchange
purification of the leach solution with strong-base resins. The uranium
usually is precipitated with ammonia to produce ammonium diuranate
(yellow cake).

Physical methods of beneficiation have proved to be unattractive with
Canadian ores. Development work has indicated that acid leaching at
high oxygen pressure may be attractive (26). In situ leaching with bac-
terial oxidation was developed also by the Mines Branch (27), and a
modification of this process is being tried now on a large scale by Agnew
Lake Mines Limited (28).

A solvent-in-pulp system using a tertiary amine in a pulse-column
contactor may be more attractive than ion exchange for uranium recovery,
and is a significant improvement over the earlier mixer–settler arrangement (29).

The only Canadian uranium refinery is operated at Port Hope, Ontario by Eldorado. Tonnage quantities of oxide (U_3O_8) were produced first in 1942 from ore concentrates. A solvent-extraction pilot plant was operated in 1950 and 1951 to investigate methylisobutylketone (hexone) and then tributylphosphate as extractants for uranium to obtain a high-purity product. The present refinery was designed and built by the Catalytic Construction Company in 1955 *(11)*. Yellow cake is digested in nitric acid, the resultant slurry extracted with tributylphosphate dissolved in kerosene, and the uranium, after purification, transferred back to water. This solution is decomposed thermally to UO_3. Capacity is about 5 Gg U/a.

Initial development of the process to produce ceramic-grade uranium dioxide from uranyl nitrate solution was done by AECL *(30)* and a small plant was set up by Eldorado in 1958 *(12)*. This has been expanded now to a capacity of about 1 Gg/a. Facilities have been constructed also to prepare enriched ceramic oxide (1 to 3% U-235) from enriched UF_6.

In the late 1960's Eldorado decided to install a plant to convert UO_3 to UF_6 so that they could provide their uranium customers outside of Canada with a product ready to feed to an enrichment plant. The Kerr–McGee process was adapted to their needs and operation at a capacity of 2 Gg U/a began in 1971 *(13)*. This plant has been expanded now to 5 Gg U/a.

Heavy-Water Production

As part of the Manhattan District Project during World War II, a small plant to produce heavy water ($\sim$ 6 Mg/a) was built by Standard Oil Development Co. at Trail, B.C. and was operated by Cominco from 1944 to 1956 *(14)*. It was based on steam–hydrogen catalytic exchange plus steam–water equilibration coupled to water electrolysis. However, by-product heavy water from this process is economical only if the electrolysis cost is borne by the hydrogen product, which at Trail was used for ammonia production. In any case, the small scale of operation imposed by electrolytic capacity and the large exchange tower volume have made this production method economically unattractive.

There have been many assessments and comparisons of heavy-water processes in Canada during the past three decades *(15, 31, 32, 33)*. Despite the wide range of alternatives studied, none that can offer unlimited production are able to compete with the GS process—deuterium exchange between water and hydrogen sulfide—which was chosen by the USAEC for their large-scale production needs nearly 30 years ago *(34)*.

When the scope of this commitment became known in the mid-1950's, further heavy-water process development in Canada was halted. Initial Canadian requirements for heavy water were purchased from the USAEC, and investment in a heavy-water industry was postponed until demand was large enough to provide an economic scale of operation.

One interesting development during the early period was a parallel plate, wetted-wall packing for water distillation (35). Although not pursued beyond the pilot scale by AECL, the same principle was developed successfully into a more practical design by Sulzer Bros. in Switzerland (36). Their packing is used at many heavy-water plants for the final stage of heavy-water production which uses water distillation from about 15% D_2O to reactor grade, 99.8% D_2O.

The initial plan to establish heavy-water production in Canada was for AECL, a crown corporation, to contract with industry for a long-term supply. The first two plants were built on this basis, one for Deuterium of Canada Ltd. at Glace Bay, N.S. and the other for Canadian General Electric (CGE) at Port Hawkesbury, N.S. Subsequently these plants were purchased by AECL. The third plant was built for AECL at the Bruce Nuclear Power Development site, Tiverton, Ontario. It was sold to Ontario Hydro (OH) soon after start-up and two more plants are being built for them at the same site. A sixth plant, LaPrade, is being built for AECL near Gentilly, Quebec. Table I lists these plants and their designer–constructors, capacities, and start-up dates. Completion and start-up of Bruce D and LaPrade have been postponed because of a forecast surplus of heavy water resulting from a decline in the rate of rise in demand for electricity.

The ill-starred original Glace Bay plant was begun in 1964. Financial problems, labor and management difficulties, escalating costs, and a variety of technical problems plagued the project. Attempts to commission the plant were abandoned in 1969. Responsibility for the plant was transferred to AECL and late in 1970 a contract was awarded to Canaton MHG Heavy Water Limited (CMHG) for rehabilitation. A new flowsheet was adopted using the existing tower shells, but with new internals. This flowsheet, developed by V. R. Thayer (37), optimized production from the existing tower volume. All of the other equipment and piping were dismantled and, to the extent possible, modified for reuse. The highlights of this challenging project, successfully completed in 1975, are reported by L. Blake (38).

Meanwhile the first two Lummus-designed plants had begun operation. They more closely followed the technology established by E. I. Dupont de Nemours and Company for the USAEC at Dana and Savannah River. One major difference from the USAEC plants was to scale-up tower volume by a factor of 20, to a 8.6-m diameter and a 90-m height containing 130 sieve trays. Operating at 2 MPa pressure, these are among the world's largest high-pressure chemical process vessels. The 800 Mg/a Bruce A plant has six such towers operating in parallel to extract 29 g of heavy water/sec from a feedwater flow of 1 Mg/sec—a yield of 1 ounce of product from a ton of feed!

The very large size of these plants, and the large quantities of hydrogen sulfide at high pressure (2 MPa) involved, has made their design and operation a major challenge.

Table I. Canadian Heavy-Water Production Plants

Plant	Owner–Operator	Designer–Constructor	Nominal Capacity (Mg/a)	Start–up
Port Hawkesbury	AECL	LCCL[a]	400	1970
Bruce A	OH	LCCL	800	1973
Glace Bay	AECL	CMHG[b]	400	1976
Bruce B	OH	LCCL	800	1979
Bruce D[c]	OH	LCCL	800	
LaPrade	AECL	CMHG	800	

[a] LCCL—Lummus Company of Canada Limited.
[b] CMHG—Canatom MHG Heavy Water Limited.
[c] Commitment of the Bruce C plant was canceled.

G. D. Davidson *(39)* describes the performance of the Bruce plant with regard to safety, environmental impact, reliability, and manpower development.   Total production to the end of 1978 is 3.7 Gg for an overall capacity factor of 0.77.   The total Canadian output to this date is 5.5 Gg, worth about one billion dollars.

The construction of three plants at the Bruce site has permitted some evolutionary improvements in design, but of limited extent because of the strong desire for standardization of operations and maintenance.   These are described by R. I. Petrie *(40)*.   Perhaps the most important one is an extensive system for hydrogen sulfide recovery from the flare and vent headers, drains, tanks, and strippers which significantly will reduce releases of hydrogen sulfide or the flaring of it to sulfur dioxide.

Sieve tray performance, both as to flow and mass-transfer efficiency is crucial to successful plant performance, and extraction is unusually sensitive to small changes in the gas-to-liquid flow ratio.   Thus, even incipient flooding or dumping of trays can cause large losses in production.   Early plant operation was limited in throughput by foaming which caused unstable tray operation.   This problem was resolved by the addition of conventional antifoamers to the feedwater and by tray modification to reduce froth height.   Thus, stable operation at high flow and design extraction were achieved *(39)*.   Surface chemistry studies at WNRE *(41)* showed that multilayer adsorption of hydrogen sulfide onto the water surface at high pressure caused the system to be inherently foamy at process conditions.   Some trace impurities in the water markedly enhanced this foaminess.   Understanding derived from this basic work at WNRE led to the selection of better antifoamers; curiously, what is currently the best one is a foamer at normal atmospheric conditions and only acquires its antifoam characteristics when adsorption of hydrogen sulfide on the water becomes significant (i.e. above 1.5 MPa pressure).

While foaminess had a large effect on tray hydraulics, its deleterious effects were soon successfully controlled.   A more important limitation has been low tray efficiency, lower than the design value derived from U.S.

plant experience. In retrospect, it is clear that the degree of uncertainty in these early plant measurements had been underestimated. An extensive program of basic studies, pilot-plant experiments, and measurements at Port Hawkesbury, Bruce, and Glace Bay, including direct in-tower measurements of froth height and froth density by gamma scanning (42) and sophisticated determination of tray efficiency, has brought us to the stage of a detailed mathematical model of tray performance (43). As a result, tray modifications at each of the plants has produced significant increases in production and further potential for improvement has been identified.

Another important advance in heavy-water plant process engineering has been the development at CRNL of detailed iterative computer simulations of heat and material balances for each plant flowsheet (44). These are unusually complex because of the large number of individual trays and the predominance of many interacting recycle streams: in parts of the plant the recycle flow of deuterium can be more than 1000 times the product flow. Fine-tuning of process parameters, more precise process control, and small improvements in design of the later plants have more than repaid the effort required to develop these programs. Such benefits have been accompanied by an increasing depth of understanding of process subtleties. Exploiting this insight, A. I. Miller and G. Pauluis (45) have developed a new process flowsheet which offers for the next generation of GS plants 5% more extraction for the same feed rate and plant investment as current designs.

Heavy-water production is highly energy intensive—30 GJ/kg D_2O, or in more familiar terms, 5 bbl of oil per kilogram. All of the Canadian plants have their steam supply coupled with large electric power generating plants for good overall thermal efficiency. While Glace Bay and Port Hawkesbury use steam from back-pressure turbines at fossil-fired power stations, Bruce is the world's largest example of the use of nuclear steam for chemical process heat (40).

The major Canadian effort to develop an alternative heavy-water process has concentrated on amine–hydrogen exchange. This process extracts deuterium from a large hydrogen stream such as ammonia synthesis gas—a 1000 Mg/day ammonia plant could produce 70 Mg D_2O/a. Development has reached the stage where a demonstration plant could be built, attached to an ammonia plant, and would be economically attractive (46). Although production from each such plant is limited, the total world potential for deuterium recovery from hydrogen is very large. Another approach to exploiting this process as a major source of heavy water is to link it to a water feed. This can be done best by high-temperature water–hydrogen exchange (47). However, such a complex arrangement may be too expensive.

The amine–hydrogen work at CRNL showed the superiority of methylamine as the exchange medium over ammonia or other amines, and developed an effective homogeneous catalyst for the process—potassium–lithium methylamide dissolved in the methylamine *(48)*. The rate of exchange between the hydrogen gas and the liquid amine is limited both by the kinetics of the exchange reaction and the low solubility of hydrogen. In comparison with the GS process, the mass-transfer coefficient for the methylamine–hydrogen system is an order of magnitude lower *(33)*. Thus, to achieve a reasonable tray efficiency, a special contactor is required having a long gas-phase residence time per tray and an increased interfacial area. At CRNL a deep sieve tray design was developed having an overflow weir of the order of a meter high and with the froth volume filled by knitted mesh packing to reduce bubble coalescence *(49)*. Tray efficiency in pilot-scale tests was an order of magnitude higher than for a simple sieve tray.

Sulzer Bros. had developed an ejector contactor for the similar ammonia–hydrogen process. This achieves an even higher tray efficiency at the expense of a large energy input and complex tower internals. However, since the process operates at high pressure (7 MPa), the resultant smaller tower is an important advantage. A combination of this Sulzer technology and the AECL catalyst technology has provided the attractive plant design referred to above *(46)*.

During the past decade chemists and chemical engineers at CRNL have developed a new catalyst for water–hydrogen exchange *(50)*. Relative to the Trail process arrangement, this new catalyst reduces the exchange tower volume by an order of magnitude. This important development has applications in heavy-water reconcentration (upgrading) and in tritium recovery from light or heavy water, as well as for by-product heavy-water production. These other applications will be discussed later. The heavy-water production process based on this new catalyst is known as combined electrolysis and catalytic exchange (CECE).

Coolant Technology

The water chemistry of CANDU reactors embraces control of corrosion and corrosion-product transport in the coolant system, control of radiolytic decomposition of the moderator *(51)* and control of the concentration of soluble neutron absorbers used to adjust reactivity; and control of boiler-water chemistry to minimize tube corrosion *(52)*. The major chemical engineering effort has dealt with coolant technology and I will confine this review to that aspect of water chemistry.

The important chemical processes which can occur in the coolant are radiolytic decomposition to produce oxygen, corrosion of the system materials, dissolution of the metal oxides so formed, deposition of corrosion products on the system surfaces, and transport of radioactive nuclides generated within deposits on the fuel sheaths. The major system materials are carbon steel, zirconium alloys, and various nickel-containing alloys for the boiler tubes.

Early work (53) in in-reactor fuel test loops showed that radiolytic oxygen can be suppressed by maintaining 5 to 10 cm^3 D$_2$/kg D$_2$O dissolved in the coolant and that operation at pH 10 with lithium hydroxide minimizes deposition of magnetite (Fe$_3$O$_4$) particles on the fuel sheath surfaces. These conditions minimize corrosion and correspond to a minimum solubility of magnetite. With these coolant conditions the fuel surface remains clean and heat transfer is unimpeded—they are the key to the successful use of carbon steel piping, components, and fittings for the CANDU coolant circuit. A simple and effective chemistry control and coolant purification circuit was developed (54).

Despite stringent control of coolant chemistry and very low concentrations of corrosion products in the coolant ($\sim$ 10^{-8} g Fe/g D$_2$O), there is a large potential for transport because of the huge flow of coolant ($\sim$ 10^4 kg/sec) and substantial (50 K) temperature difference from 530 to 580 K. Even a few magnetite particles depositing on the fuel surface (0.1 g Fe/m^2) can yield significant radioactivity—mainly cobalt-60 from the cobalt-59 impurities in the system. Dissolution in-core and precipitation or exchange out-core provide a means to transfer this radioactivity to out-core components. These fields in normally inaccessible areas impede maintenance and cause external radiation exposures to station staff—typically 0.3 rem/MW$_e$. a (17), or about 600 rem/a for the 2000-MW Pickering station.

The importance of corrosion product mass transfer was realized first in the early operation of NRU. Here the solubility of the oxide formed on the aluminum fuel sheathing led to the production of a colloidal alumina floc in the heavy water. The mechanism for its formation, means to control it, and the role it played in transporting uranium and fission products released from failed fuel were studied (55, 56).

Extensive studies (57, 58, 59) defined the controlling processes for activity transport in the power reactors. These are oxide solubility, particle deposition, diffusion through oxide films, and rates of crystallization. Detailed models for activity production in-core and surface activation out-core have been developed (60) that successfully predict the growth of corrosion product fields in each of the CANDU reactors.

Activity transport effects can be minimized by selecting materials with a low cobalt content and by rigid adherence to chemical specifications for the coolant. Because of the important role of corrosion product particles in this transport, filtration has been studied extensively as a means of reducing the rate of growth of radiation fields. High flows are needed to be effective and therefore the filters must operate at full coolant temperature. Two types of filter which have proved successful in pilot tests at the NPD reactor are a deep bed of graphite particles and a bed of steel balls in an electromagnetic field (61).

An important chemical technique to reduce fields due to corrosion products is the CAN–DECON decontamination process developed jointly by CRNL and WNRE *(62)*. This uses a dilute reagent mixture which can be added to heavy water without introducing any ordinary water. By circulation through cation exchange resin the metallic ions are removed and the reagent is regenerated. The radioactivity is retained on the resin for which handling and disposal techniques are already available. The reagent can be removed from the circuit by mixed-bed ion exchange. A full-scale decontamination of the Douglas Point reactor was done in 1975 *(63)* reducing fields associated with carbon steel piping by a factor of six.

This large program on activity transport, involving both AECL and Ontario Hydro, began in response to high radiation fields at Douglas Point due to inadequate chemical control during its early operation. Through close collaboration among developers, designers, and operators, fields have been reduced substantially at Douglas Point, kept lower at Pickering, and even lower at Bruce.

Heavy-Water Management

The success of the CANDU reactor depends on maintaining heavy-water losses at a low level. Experience *(64)* at Pickering and Bruce confirms that losses can be kept to less than 1% of the total inventory per year. Elaborate recovery systems are provided to deal with heavy-water leakage. Most important has been the development of large reliable and efficient molecular sieve drying systems to recover heavy-water vapor from the air in various parts of the reactor building *(65)*.

The recovered water can range from near reactor grade down to a few percent heavy water since it generally becomes mixed to some degree with ordinary water. Reconcentration is done at each large power station by water distillation *(36)*. A large central reconcentration plant which uses water electrolysis is operated at CRNL *(66)*. The CECE process mentioned earlier may be a more attractive reconcentration method; a pilot plant to investigate this recently began operation.

The neutron irradiation of heavy water produces tritium in the form of TDO. After many years of reactor operation the TDO concentration in the moderator can approach 50 ppm (70 Ci/kg). Although tritium produces only a very low-energy beta particle during radioactive decay, ingestion by man will give an internal radiation dose. Therefore reactor operators and maintainers must avoid prolonged contact with tritiated heavy-water vapor or liquid. Total internal dose due to tritium at Pickering has been about 0.2 rem/MW_e.a. Further improvements in heavy-water containment can reduce this dose. An alternative approach is to separate the tritium from the heavy water to limit its accumulation to a few parts per million. The development of this technology by CRNL and OH has begun. The most practical separation method is cryogenic distillation of liquid deuterium. This must be preceded by a process to transfer the deuterium from the heavy water to deuterium gas. The hydrogen–water catalyst mentioned above (50) offers one convenient method to do this; the CECE process is an alternative which also would preconcentrate the tritium in the deuterium feed to the distillation unit.

Alternative Coolants

In the boiling light-water-cooled CANDU it was decided to use carbon steel as the main out-core material in the coolant circuit. Thus, high pH and low-oxygen conditions are necessary. Neither lithium hydroxide nor hydrogen are suitable additives because of the possibility of caustic attack on the fuel sheaths during evaporation and because of volatility of hydrogen during boiling. Ammonia can serve both purposes since it is alkaline and its radiolysis products, nitrogen and hydrogen, suppress oxygen formation. Extensive studies (67) in loops in NRX and NRU, confirmed by operation of the Gentilly-1 prototype reactor (68), defined conditions for good chemical control and the minimum ammonia concentration necessary to avoid forming oxides of nitrogen.

Heavy fuel deposits were expected in boiling systems, and therefore the initial studies of deposition and activity transport for power reactors concentrated on the CANDU–BLW concept until the fields at Douglas Point became a concern. The deposit thickness was proportional to iron concentration in the coolant and to the square of the heat flux (69); deposition was reversible and quickly reached a steady value set by the local conditions. The corrosion products initially deposit by hydrodynamic and electrostatic effects; then boiling accelerates deposition by drawing water and its contained iron into the deposit to replace the steam that leaves. Local alkalinity gradients within the deposit determine whether iron crystallizes to cement the deposit or dissolves to weaken it, and erosion processes then define the equilibrium thickness (70). This model works well in explaining deposition under boiling conditions.

The organic-cooled CANDU concept was proposed by McNelly of CGE in 1958 (71). This began an extensive investigation of coolant properties, decomposition, control of deposition, and many other aspects of coolant chemistry. An organic-cooled, heavy-water-moderated research reactor, WR-1, began operation at WNRE in 1965. It has demonstrated reliable operation with coolant outlet temperatures of up to 675 K. Low corrosion and a low potential for activity transport result in very low radiation fields around the piping.

The coolant finally selected is a partially hydrogenated mixture of terphenyls which is liquid down to 273 K. This advantage outweighs the somewhat higher radiolytic decomposition rate than that of pure terphenyl (72). Radiolytic and pyrolytic decomposition lead to a coolant containing a whole spectrum of compounds from hydrogen and methane through to high-boiling polymers. The composition can be controlled and optimized by degassing to remove gases and volatiles, and by vacuum distillation to separate coolant and high boilers *(54)*.

A major problem with the CANDU–OCR concept was fouling of heat-transfer surfaces by deposits of organic material with a low thermal conductivity. Conditions to minimize fouling were identified: control of oxygen and chlorine content of the coolant, purification by filtration and adsorption on Attapulgus clay to remove particles, and continuous monitoring of fouling potential. Both oxygen and chlorine promote fouling; the latter is particularly undesirable since it complexes iron and causes its transport *(73)*.

Zirconium alloys are used for pressure tubes and fuel sheathing in WR-1. Coolant chemistry control is essential for their long-term life. Chlorine enhances hydriding of zirconium in hot organic coolant and its concentration must be controlled for this reason, as well as to reduce fouling. Most important to minimize hydriding, the oxide film on the zirconium must be kept in good repair by maintaining a water concentration in the coolant of about 200 ppm.

All of the aspects of organic coolant technology—decomposition, purification, physical properties, fouling, heat transfer, materials performance, and flammability—were summarized in 1975 *(74)*.

Irradiated Fuel Processing

Fuel processing to recover plutonium was an important activity from the earliest days of the atomic energy program. A small pilot plant was built at CRNL in parallel with the construction of NRX. It operated from 1949 to 1953 to extract plutonium from dissolved fuel with triethylene glycol dichloride in a batch process. Ammonium nitrate was the salting-out agent *(75)*. Subsequently, the waste solution from this operation was treated with tributylphosphate (TBP) to remove uranium and residual plutonium, and the ammonium nitrate decomposed before the waste was stored as a concentrated fission product solution.

During this period of the early 1950's several other aqueous processing methods were developed to the pilot-plant stage at CRNL and pyrometallurgical processes were investigated on a laboratory scale.

Anion exchange was investigated for application as a small-scale, primary extraction process to recover plutonium directly from dissolved irradiated uranium in $8N$ HNO_3. A 50-kg U/day pilot plant was operated *(76)*. Even with two cycles of ion exchange the fission product activity with the plutonium was undesirably high. Resin stability is another potential problem which was not resolved fully.

The standard TBP (or Purex) process was investigated also and small pilot units were operated. Both packed columns and mixer–settlers were investigated for the solvent extraction steps. An innovative design of mixer–settler in which mixing and pumping was done with air streams was developed (77). It offers the advantages of simplicity and low-energy input. Application of the TBP process to thorium processing was investigated also—the Thorex process.

Pyrometallurgical processes investigated include slagging of molten irradiated uranium, plutonium extraction by silver, plutonium volatilization, and fused-salt extraction (78). Interest in these approaches ended with the selection of uranium dioxide as the CANDU fuel.

All of this fuel processing work was terminated in 1956 with the realization that plutonium recycle would not be needed in Canada for at least several decades. As explained earlier, the once-through natural uranium dioxide fueling scheme gave attractive fueling costs. Studies of recycle costs (8) showed that a large scale was essential for economic operation, and even at large scale the benefits would likely be small if the fuel burn-up forecast for the once-through case was achieved. The situation 20 years later is still that the economics of plutonium recycle are marginal (79).

Starting in 1970 one further processing variant has been investigated—the extraction of plutonium by tricaprylamine dissolved in diethylbenzene (19). Since the irradiated uranium from CANDU reactors has a very low residual uranium-235 content, there is little incentive to recover it. The amine process offers the advantages of small size and a simple, one-cycle arrangement to give the desired decontamination. A bench-scale pilot unit has demonstrated satisfactory performance of the flowsheet, and it is the first time amine has been used to extract plutonium from dissolved irradiated fuel.

Interest now is centered on the thorium cycle (23) and laboratory studies have continued to investigate both an adaptation of the Thorex process to CANDU fuel and the application of the amine process to recovering uranium-233 from irradiated thorium. The program to develop and fully demonstrate the thorium fuel cycle has been outlined, and would require about 25 years to complete. However, the current research level will not be expanded until a decision can be taken by the Canadian Government when the information from the current International Nuclear Fuel Cycle Evaluation has been assessed.

Waste Immobilization

In parallel with the studies of processes for recovering fissile material from irradiated fuel in the early 1950's at CRNL, work began on the treatment of the fission product wastes. It was recognized that a safe and permanent method of disposal would be needed once the nuclear power industry became very large. Immobilization of the fission products in a stable and very insoluble glass was chosen as the best approach.

In 1955 White and Lahaie *(80)* showed that the concentrated acidic waste solution could be incorporated into a glass formed by calcining and melting a mixture of the solution, lime and nepheline syenite. The latter is a silicate rock mined in Ontario and used by the glass and ceramics industry. Development proceeded to the stage of a small batchwise pilot-scale demonstration of process feasibility *(20)*. This unit produced glass hemispheres weighing about 2 kg and containing up to 100 Ci of 6-year-old fission products. Fission product volatility, especially of ruthenium and caesium, during decomposition of the nitrate salts and melting of the glass, required the development of adsorbers and an efficient off-gas treatment system.

The leaching rate of glass samples immersed in water in laboratory tests dropped rapidly in the first month and then tended to level out or decrease more slowly. Rates less than 10^{-8} g glass/cm^2 · day were achieved for many glass compositions.

A disposal experiment was initiated in which 25 hemispheres of glass, each containing 12 Ci of 6-year-old mixed fission products, were placed in the ground below the water table in 1958. After 1 year no activity could be detected in ground water samples 3 m downstream of the mid-plane of the burial. In 1960 a second experiment was initiated with 25 hemispheres of glass each containing about 40 Ci of the same mixed fission products *(81)*. This glass was made less resistant to leaching by the addition of metal oxides so that the interaction of the glass with ground water could be monitored more readily.

Over the first 8 years the leaching rate continually decreased from 4×10^{-8} to 5×10^{-11} g/cm^2. day *(82)*. The integrated release over a period of 17 years has been about one part in 10^6 of the radioactivity initially in the glass. None of this released activity has moved more than 50 m from the hemispheres because it has been adsorbed on the soil. Thus, the released activity has migrated by repeated adsorption–desorption along the path of the ground water which has been moving at the rate of about 70 m per year. In 1978 one hemisphere from each of the experiments was retrieved and found to be in excellent condition.

This experiment provides considerable confidence in the concept of waste immobilization in glass as one step toward isolating the long-lived radioactive by-products of nuclear power from man's environment. Of course, the immobilized material would not be placed deliberately in shallow ground water for permanent disposal. The consensus today is for deep underground disposal in a stable geological formation *(21)*.

In the past few years work has resumed on the development of the process for immobilization of wastes in glass to adapt it to the types of wastes now anticipated *(83)*. Since it is not certain that Canadian irradiated fuel will be processed to recover plutonium, this program also is

assessing methods for immobilization of the spent fuel for final disposal
(21).

Although the irradiated fuel contains over 99% of the radioactivity
produced during reactor operation, the other wastes are important because of their large volume.   These include ion-exchange resins, filters,
combustible materials, and liquids.   Pilot-plant studies (22) began
several years ago of reverse osmosis, evaporation, and incineration as
methods of volume reduction, and of bituminization to immobilize the
concentrated wastes.   Earlier work had developed a process for incorporating low-level wastes in concrete (84); however, this now is considered to be less satisfactory than bitumen.   The demonstration phase of
this program will begin next year with the operation of a Waste Treatment Center at CRNL which will concentrate and immobilize the laboratories' wastes.

Active Fuel Fabrication

In the thorium fuel cycle the recycled uranium-233 inevitably is
contaminated with uranium-232 and its decay products.   The first of
these, thorium-228, will be contained in any recycled thorium.   Thal-
lium-208 in this decay chain emits a very-high-energy gamma ray and for
this reason fabrication of recycle fuels in the thorium fuel cycle will have
to be done remotely in heavily shielded cells.   Conventional fuel fabrication processes may not be the most economical under these conditions.

Therefore some chemical engineering studies of alternatives to pressing and sintering of thoria powder have begun.   One alternative is the
sol gel process in which a fuel consisting of several sizes of high-density
microspheres of thorium dioxide is produced.   Another is the extrusion
of thoria gel in the form of long pellets ready for sintering to high density.

Overview

In the past three decades nuclear chemical engineering in Canada
has spanned a wide variety of activities throughout the nuclear power
program.   Most important have been the contributions to uranium milling and refining operations and to the production of heavy water.   The
fields of Canadian preeminence are heavy-water process technology,
reactor radiation field control, and organic coolant technology.   An early
key contribution was immobilization of fission products in glass.

Total employment of chemical engineers in the nuclear industry in
Canada is about 600 out of a total work force of about 30,000.   The
largest fraction is in operations, followed by design, research and development, and manufacturing, in that order.   Chemical engineers often
are surprised at the range of opportunities available to them in the

nuclear industry. And this range will expand in Canada as the waste immobilization and disposal programs grow and more so if fuel recycle is endorsed as part of our domestic energy resource strategy.

Literature Cited

1. Eggleston, W. "Canada's Nuclear Story"; Clarke, Irwin and Co.: Toronto, 1965.
2. Lewis, W. B. *Peaceful Uses At. Energy, Proc. U.N. Int. Conf. 2nd* **1958**, *1* p. 53.
3. Lewis, W. B. "Some Highlights of Experience and Engineering of High-Power Heavy-Water-Moderated Nuclear Reactors," Atomic Energy of Canada Limited Report, AECL-797, 1959.
4. Haywood, L. R. "The CANDU Power Plant," Atomic Energy of Canada Limited Report, AECL-5321, 1976.
5. McCredie, J.; Elston, K. E. "Program Review of Ontario Hydro's Nuclear Generation and Heavy Water Production Program," in *Annual International Conference of the Canadian Nuclear Association, 18th, 1978*, Vol. 3, p. 1.
6. Fortier, P. C. "Progress Report on Four CANDU Nuclear Generating Stations and Three Heavy Water Plants Outside Ontario," *Annual International Conference of the Canadian Nuclear Association, 18th, 1978*, Vol. 3, p. 57.
7. Lewis, W. B. "Low Cost Fueling Without Recycle," Atomic Energy of Canada Limited Report AECL-382, 1956.
8. Rae, H. K. "Fuel Reprocessing and Recycling for Natural Uranium Power Reactors," Atomic Energy of Canada Limited Report, AECL-494, 1957.
9. Williams, R. M.; Little, H. W.; Gow, W. A.; Berry, R. M. *Peaceful Uses At. Energy, Proc. U.N. Int. Conf. 4th*, **1972**, *8*, p. 37.
10. "Canadian Minerals Yearbook 1977," Department of Energy, Mines and Resources, Government of Canada, Ottawa, 1979.
11. Burger, J. C.; Jardine, J. M. *Peaceful Uses At. Energy, Proc. U.N. Int. Conf., 2nd,* **1958**, *4*, p. 3.
12. Berry, R. M. "Eldorado's Port Hope Refinery–1969," *Can. Inst. Min. Metall. Bull.* **1969**, *62*(690), 1093.
13. Traumer, W. E. "Uranium Conversion at Eldorado Nuclear Limited," *Atomic Industrial Forum/American Nuclear Society Conference, Miami, 1971.*
14. Benedict, M.; Pigford, T. H. "Nuclear Chemical Engineering"; McGraw-Hill: New York, 1957; p. 440.
15. Rae, H. K. *Chem. in Can.* **1955**, 7 (10) 27.
16. Lumb, P. B. *J. Br. Nucl. Energy Soc.* **1976**, *15*, 35.
17. LeSurf, J. E. *J. Br. Nucl. Energy Soc.* **1977**, *16*, 53.
18. Hart, R. G.; Haywood, L. R.; Pon. G. A. *Peaceful Uses At. Energy, U.N. Int. Conf. 4th*, **1972**, *5*, p. 239.
19. Rae, H. K. "Chemical Engineering Research and Development for Fuel Reprocessing and Heavy Water Production," Atomic Energy of Canada Limited Report," AECL-3911, 1971, p. 47.
20. Watson, L. C.; Aikin, A. M.; Bancroft, A. R. "The Permanent Disposal of Highly Radioactive Wastes by Incorporation into Glass," *I.A.E.A. Panel Proc. Ser. STI/PUB/18* **1960**, 375.
21. Boulton, J., Ed. "Management of Radioactive Fuel Wastes: The Canadian Disposal Program," Atomic Energy of Canada Limited Report, AECL-6314, 1978.
22. Charlesworth, D. H.; Bourns, W. T.; Buckley, L. P. "The Canadian Development Program for Conditioning CANDU Reactor Wastes for Disposal," Atomic Energy of Canada Limited Report, AECL-6344, 1978.

23. Critoph, E. *Nucl. Power and Its Fuel Cycle, Proc. Int. Conf.* **1977,** *2,* p. 55.
24. Gow, W. A.; Ritcey, G. M. *Trans. Can. Inst. Min. Metall.* **1969,** *72,* 361.
25. Tremblay, R.; Bramwell, P. *Trans. Can. Inst. Min. Metall.* **1959,** *62,* 44.
26. Vezina, J. A.; Gow, W. A. "Some Design Aspects of the Pressure-Oxidation Acid Leaching of a Canadian Uranium Ore," *Can. Mines Branch, Tech. Bull.-110.* Ottawa, **1969.**
27. Gow, W. A.; McCreedy, H. H.; Ritcey, G. M.; McNamara, V. M.; Harrison, V. F.; Lucas, B. H. *Recovery of Uranium, Proc. I.A.E.A.* Symp. **1971,** 195.
28. Williams, R. M. *Can. Min. J.* **1979,** *100,* 143.
29. Ritcey, G. M.; Joe, E. G.; Ashbrook, A. W. *Trans. Am. Inst. Min. Metall. Petrol. Engrs.* **1967,** *238,* 330.
30. Chalder, G. H.; Bright, N. F. H.; Patterson, D. L.; Watson, L. C. *Peaceful Uses At. Energy, Proc. U.N. Int. Conf. 2nd,* **1958,** *6,* 590.
31. Rae, H. K. "A Review of Heavy Water Processes," Atomic Energy of Canada Limited Report, AECL-2503, 1965.
32. Rae, H. K. "Chemical Exchange Processes for Heavy Water," Atomic Energy of Canada Limited Report, AECL-2555, 1966.
33. Rae, H. K. In "Separation of Hydrogen Isotopes," *ACS Symp. Ser.* **1978,** *68,* 1.
34. Bebbington, W. P.; Thayer, V. R. *Chem. Eng. Prog.* **1959,** *55* (9), 70.
35. Bancroft, A. R.; Rae, H. K. *Can. J. Chem. Eng.* **1957,** *35,* 77.
36. Wartenweiler, M. *Sulzer Tech. Rev.* **1970,** *52,* 84.
37. Thayer, V. R. Canadian Patent 924080, 1973.
38. Blake, L. *Nucl. Eng. Int.* **1976,** *21,* (248), 65.
39. Davidson, G. D. In "Separation of Hydrogen Isotopes," *ACS Symp. Serv.* **1978,** *68,* 27.
40. Petrie, R. I. "Design Developments Bruce Heavy Water Plants," *Annual International Conference of the Canadian Nuclear Association, 16th, 1976,* Vol. 2, p. 27.
41. Sagert, N. H.; Quinn, M. J. "The Coalescence of H_2S and CO_2 Bubbles in Water," Atomic Energy of Canada Limited Report, AECL-5494, 1976.
42. Fulham, M. J.; Hulbert, V. G. *Chem. Eng. Prog.* **1975,** *71* (6), 73.
43. Bancroft, A. R. "Heavy Water GS Process R&D Achievements,"Atomic Energy of Canada Limited Report, AECL-6215, 1978.
44. Miller, A. I. "Process Simulation of Heavy Water Plants—A Powerful Analytical Tool," Atomic Energy of Canada Limited Report, AECL-6178, 1978.
45. Pauluis, G. J. C. A.; Miller, A. I. Canadian Patent 1006337, 1977.
46. Wynn, N. P. In "Separation of Hydrogen Isotopes," *ACS Symp. Ser.* **1978,** *68,* 53.
47. Wynn, N. P.; Lockerby, W. E. "Heavy Water Processes Using Amine--Hydrogen Exchange," *Annual International Conference of the Canadian Nuclear Association, 18th, 1978.*
48. Holtslander, W. J.; Lockerby, W. E. In "Separation of Hydrogen Isotopes," *ACS Symp. Ser.* **1978,** *68,* 40.
49. Bancroft, A. R.; Rae, H. K. "Tecnica ed Economia della Produzione di Acqua Pesante," Comitato Nazionale Energia Nucleare, Rome, 1971, 47.
50. Butler, J. P.; Rolston, J. H.; Stevens, W. H. In "Separation of Hydrogen Isotopes," *ACS Symp. Ser.* **1978,** *68,* 93.
51. Rae, H. K.; Allison, G. M.; Bancroft, A. R.; Mackintosh, W. D.; Palmer, J. F.; Winter, E. E.; LeSurf, J. E.; Hatcher, S. R. *Peaceful Uses At. Energy, Proc. U.N. Int. Conf. 3rd,* **1964,** *9,* 318.
52. Balakrishnan, P. V. "Effect of Condenser Water Inleakage on Steam Generator Water Chemistry," Atomic Energy of Canada Limited Report, AECL-5849, 1978.
53. Robertson, R. F. S. "Chalk River Experience in 'Crud' Deposition Problems," Atomic Energy of Canada Limited Report, AECL-1328, 1961.
54. Hatcher, S. R. "The Chemical Engineer's Role in Nuclear Power Reactor Design, Development and Operation," Atomic Energy of Canada Limited Report, AECL-3911, 1971, p. 41.

55. Hatcher, S. R.; Rae, H. K. *Nucl. Sci. and Eng.* **1961**, *10*, 316.
56. Rae, H. K. "The Behaviour of Uranium and Aluminum in the NRU Heavy Water System," Atomic Energy of Canada Limited Report, AECL-1840, 1963.
57. Lister, D. H. *Nucl. Sci. and Eng.* **1975**, *58*, 239.
58. Lister, D. H. *Water Chemistry of Nuclear Reactor Systems, Proc. Brit. Nucl. Energy Soc. Int. Conf.* **1977**, 207.
59. Burrill, K. A. *Can. J. Chem. Eng.* 1977, *55*, 54.
60. Lister, D. H. "Predicting Radiation Fields Around Reactor Components," Atomic Energy of Canada Limited Report, AECL-5522, 1976.
61. Moskal, E. J.; Bourns, W. T. "High-flow, High-temperature Magnetic Filtration of the Primary Heat Transport Coolant of the CANDU Power Reactors," Atomic Energy of Canada Limited Report, AECL-5760, 1977.
62. Montford, B. "Techniques to Reduce Radiation Fields," Atomic Energy of Canada Limited Report, AECL-5523, 1976.
63. Pettit, P. J.; LeSurf, J. E.; Stewart, W. B.; Strickert, R. J.; Vaughan, S. B. *Materials Performance* **1980**, *19*(1), 34.
64. Kee, K. J.; Woodhead, L. "Progress Review of Ontario Hydro's Nuclear Generation and Heavy Water Production Programs," *Annual International Conference of the Canadian Nuclear Association, 17th, 1977*, Vol. 2, p. 1.
65. Rae, H. K. "Heavy Water," Atomic Energy of Canada Limited Report, AECL-3866, 1971.
66. Morrison, J. A.; Thomas, M. H.; Watson, L. C.; Woodhead, L. W. *Peaceful Uses At. Energy, Proc. U.N. Int. Conf., 3rd* **1964**, *12*, 373.
67. LeSurf, J. E.; Bryant, P. E. C.; Tanner, M. C. *Corrosion* **1967**, *23* (3), 57.
68. Allison, G. M.; LeSurf, J. E. *Nuc. Tech.* **1976**, *29*, 160.
69. Charlesworth, D. H. *Chem. Eng. Progr. Symp. Ser.* **1970**, *66* (104), 21.
70. Burrill, K. A. *Corrosion* **1979**, *35*, (2), 84.
71. McNelly, M. J. *Peaceful Uses At. Energy, Proc. U.N. Int. Conf. 2nd* **1958**, 9, 79.
72. Tomlinson, M.; Smee, J. L.; Winters, E. B.; Arneson, M. C. *Nucl. Sci. Eng.* **1966**, *26*, 547.
73. Bancroft, A. R.; Charlesworth, D. H.; Derksen, J. H. "Impurity Effects in the Fouling of Heat Transfer Surfaces by Organic Coolants," Atomic Energy of Canada Limited Report, AECL-1913, 1965.
74. Smee, J. L.; Puttagunta, V. R.; Robertson, R. F. S.; Hatcher, S. R. "Organic Coolant Summary Report," Atomic Energy of Canada Limited Report, AECL-4922, 1976.
75. Hatfield, G. W. "Reprocessing Nuclear Fuels," Atomic Energy of Canada Limited Report, AECL-259, 1955.
76. Aikin, A. M. *Chem. Eng. Prog.* **1957**, *53*(2), 82.
77. Mathers, W. G.; Winter, E. E. *Can. J. Chem. Eng.* **1959**, *37*, 99.
78. Aikin, A. M.; McKenzie, D. E. "The High Temperature Processing of Neutron-irradiated Uranium," in "Progress in Nuclear Energy Series III"; Pergamon Press: London, 1956; Vol. I, p. 316.
79. Banerjee, S.; Critoph, E.; Hart, R. G. *Can. J. Chem. Eng.* **1975**, *53*, 291.
80. White, J. M.; Lahaie, G. "Ultimate Fission Product Disposal—The Disposal of Curie Quantities of Fission Products in Siliceous Materials," Atomic Energy of Canada Limited Report, AECL-391, 1955.
81. Merritt, W. F.; Parson, P. J. *Health Physics*, **1964**, *10*, 655.
82. Merritt, W. F. *Nucl. Tech.* **1977**, *32*, 88.
83. Tomlinson, M. In "Chemistry for Energy," *ACS Symp. Ser.* **1979**, *90*, 336.
84. White, J. M.; Lahaie, G. "Ultimate Fission Product Disposal II—The Disposal of Moderately Radioactive Solutions in a Cement Mortar," Atomic Energy of Canada Limited Report, AECL-1085, 1960.

RECEIVED May 7, 1979.

The Contribution of Chemical Engineering to the U.K. Nuclear Industry

N. L. FRANKLIN—Nuclear Power Company, Limited, Warrington Road, Risley, Warrington, Cheshire, WA3 6BZ

J. C. CLARKE and D. W. CLELLAND—British Nuclear Fuels Limited, Warrington Road, Risley, Warrington, Cheshire, WA3 6AS

K. D. B. JOHNSON—Atomic Energy Research Establishment, Harwell, Didcot, Oxon, OX11 ORA

J. E. LITTLECHILD—British Nuclear Fuels Limited, Springfields Works, Salwick, Preston, Lancashire, PR4 OXJ

B. F. WARNER—British Nuclear Fuels, Ltd, Windscale and Calder Works, Sellafield, Seascale, Cumbria, CA20 1PG

The earliest engineering decisions on the nuclear fuel cycle in the United Kingdom were made in the late 1940's when texts on chemical engineering were few and nuclear engineering was entirely classified. Since that time in the main fields of feed materials and fuel element production, isotope separation, and active reprocessing, three generations of plants have been constructed in Britain. This chapter examines the different objectives, constraints, and achievements in these three fields with examples from the experience of U.K. chemical engineers over the last 30 years. In conclusion it contrasts what was accomplished in the past with limited financial and human resources but a friendly social environment with the different, present position.

It is 33 years since a small group of people led by the now Lord Hinton and his deputy, the late Sir Leonard Owen, established themselves at Risley in February, 1946 to form what became the Industrial Group of the UKAEA, the Production Group of the UKAEA, and subsequently British Nuclear Fuels Ltd. (BNFL). They designed and built the factories that now make up the Nuclear Fuel Industry of the United Kingdom. They worked in an atmosphere of excitement and improvisation

0-8412-0512-4/80/33-190-335$07.75/1

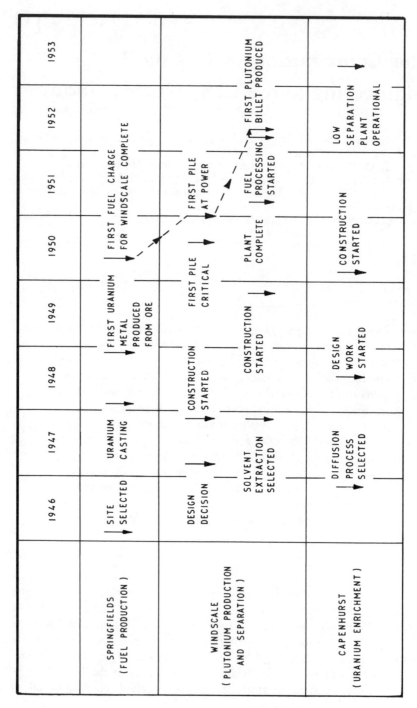

Figure 1. Construction and operation of early fuel plants

and in six years produced the first billet of plutonium.   The chronology, though not the full flavor of this early achievement, is shown in Figure 1.

Since that time the objects of the industry have changed from military to civil applications and three generations of plant have been constructed and operated.   The United Kingdom reasonably can claim to have technical and industrial competence in all of the aspects of the nuclear fuel cycle, although the experience of the final disposition of highly radioactive waste is limited and affected necessarily more by social and political factors than by technology.

The history of the industry runs parallel to the development of the chemical engineering profession in the United Kingdom.   In the late 1940's texts on chemical engineering were few and nuclear engineering was security classified.   There were of course existing chemical factories embodying chemical engineering but design was the province of applied chemists and engineers tending to work within their own provinces.

The nuclear fuel industry brought new constraints to the design of the chemical plant, the normal hazards inherent in the use of reactive chemicals being supplemented by the need to allow for the effects of irradiation and neutron criticality.   This demanded a multi-disciplinary contribution to the technology, with the leadership depending more upon the accident of personality than scientific discipline, because none of the existing disciplines was developed in an appropriate way.   The role of the chemical engineer was to integrate the individual contributions and to build on the general principles of his own subject in order to develop scientific information into industrial-size plants.

This chapter examines this form of development with examples from the experiences of U.K. chemical engineers over the last 30 years. Behind the listing of technical accomplishments there is a history of human achievements in an environment of limited financial and manpower resources but considerable public support which contrasts strangely with the climate of today.

Uranium Processing

Objectives.   Uranium processing in the nuclear industry in Great Britain is carried out mainly at the Springfields factory of BNFL where extraction from uranium ores and ore concentrates, purification and conversion to intermediates and fuel materials has been carried out since 1946.

Uranium enters the factory in the form of:
- ores and ore concentrates containing varying levels of impurities from many parts of the world;
- recycled uranium from the reprocessing plant at Windscale which may be depleted in uranium-235 if from the Magnox

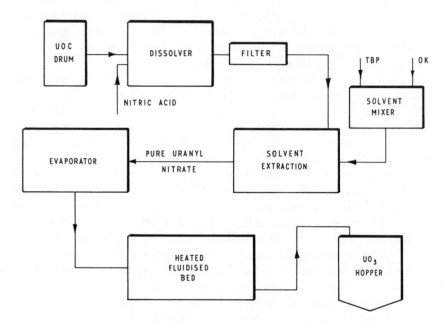

Figure 2. UO_3 *production flow sheet*

reactors or still slightly enriched if derived from the reprocessing of a gas-cooled-reactor (AGR) or water-reactor
fuel;
- enriched or depleted uranium as UF_6 from the enrichment
plants.

The pattern of processes carried out at Springfields is described best
by reference to simplified flow sheets:
- uranium ore concentrate through solvent extraction purification to uranyl nitrate solution and UO_3 (*see* Figure 2);
- fluorination of UO_3 by a dry process to UF_4 (*see* Figure 3);
- reduction to metal with magnesium turnings (for fabrication
into fuel elements for the Magnox reactors);
- fluorination of UF_4 to give UF_6 (*see* Figure 4);
- conversion of enriched UF_6 to UO_2 powder by a dry
process, pelleting, and fuel element assembly (*see* Figure
5).

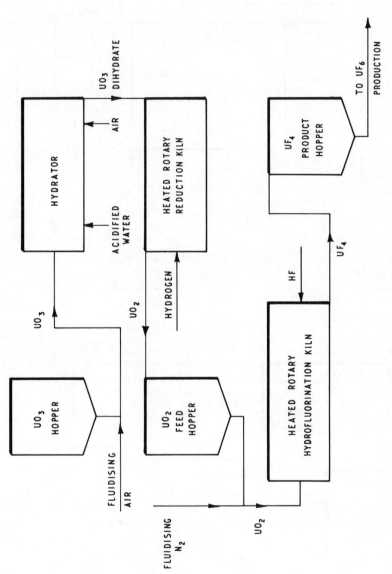

Figure 3. UF₄ production flow sheet

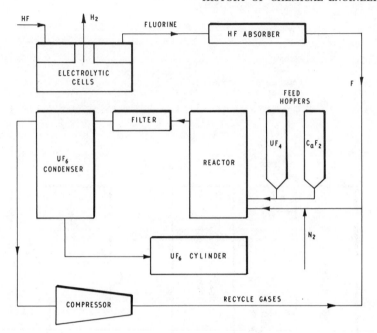

Figure 4. *UF$_4$ to UF$_6$ flow sheet*

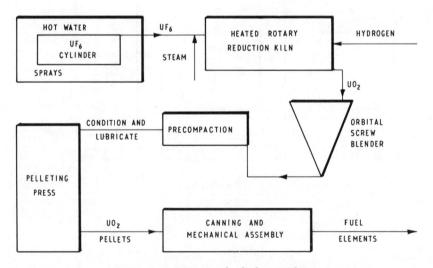

Figure 5. *Conpor oxide fuel manufacture*

In all of these operations various residues such as metal reduction slags are generated and a considerable variety of ancillary processes are operated to recycle these materials. Effluents are produced also, principally aqueous with some solid wastes and a little gaseous effluent. These also require treatment before discharge in a form which satisfies stringent health and safety requirements.

Constraints of Plant Design. The potential hazards encountered in the uranium feed materials processing industry include many that are common to the heavy chemicals industry. However special problems present themselves owing to direct radiation, the possibility of inhalation and ingestion of radioactive dusts and gases, nuclear safety, and more unusual chemical hazards.

RADIATION. The external radiation hazard from uranium is caused chiefly by beta radiation and is significant only if the uranium surface to which people are exposed is large. The problem presents only minor restraints to the designer, localized shielding normally proving adequate without the extreme measures necessary in fission-product-handling plants.

In addition to its chemical action as a heavy metal poison, natural uranium is also an alpha emitter with a long half-life and it can present a serious hazard with respect to irradiation of tissues if inhaled or ingested. Inhalation is a greater problem than ingestion as soluble uranium compounds generally are removed by the normal body metabolism. Insoluble materials, notably uranium dioxide, within certain particle size ranges can accumulate in the lungs and reach maximum permissible burden levels unless rigid measures are taken to exclude it from the atmosphere.

Therefore, an essential requirement in plant design is the stringent application of dust control techniques. Specifically designed enclosures enabling materials to be handled in closed systems and extraction and plenum air systems of sufficient capacity to maintain normal working areas free from contamination are essential. Special plant finishes to facilitate good housekeeping are an added requirement.

NUCLEAR SAFETY. The increased use of enriched uranium in nuclear fuel has meant that the designer of the materials processing plant has had to accept a further constraint that no quantity of uranic material capable of forming a critical assembly will be present in any section of the plant and that an accidental accumulation of such a mass inside or outside of the plant is not possible. This may be achieved by:

1. Ensuring that the mass is always below the critical level by limiting the quantity present; this may be as small as a few tens of kilograms for low enriched uranium. Such a system has the disadvantage of requiring an elaborate system of managerial control combined with a system of mechanical interlocks to ensure that specified quantities never are exceeded.

2. Limiting the dimensions of the equipment so that it is impossible at a given level of uranium enrichment to achieve the critical configuration. This is the so called "safe geometry" technique and is more acceptable as it relies far less on the competence of the operator. However, it places greater constraints on the designer.

It now is considered mandatory that a nuclear processing plant should be designed to prevent accidental achievement of a critical nuclear reaction. Some residue recovery operations had not had this characteristic in the past and one low-level criticality accident has been reported in the United Kingdom (1) in such a plant.

HAZARDOUS MATERIALS. Hazardous chemicals are, of course, a feature of many chemical plants. Materials used in quantity in the uranium industry include hydrogen, hydrofluoric acid, elementary fluorine, pressurized hydrocarbon oils at elevated temperatures, and uranium which is itself fairly highly toxic. Satisfactory safe techniques for the storage, handling, and containment of these materials, some of which are used at elevated temperatures, have been essential precursors to the development of acceptable processes.

Process development. The need for uranium to be made available at nuclear purity and in forms suitable for fueling the early reactor designs required the development of production-scale uranium conversion processes. The ores available in the late 1940's were crude and largely unpurified. Pitchblende, which then was mined principally as a source of radium, was the only readily available uranium source. The early processes for purification were based on precipitation of a crude UF_4 from the solution obtained by leaching the crushed ore in mixed nitric–sulfuric acid followed by solvent extraction with diethyl ether. The pure uranyl nitrate was back extracted into water and a compound, ammonium diuranate, was precipitated from the solution by the addition of ammonia. By drying, calcining, and reducing this compound UO_2 was obtained which could be treated further with HF gas to produce UF_4. This product then was reduced to uranium metal in a thermic reduction with pure calcium, the billets of metal being remelted and cast into uranium rods that could be machined to final fuel element dimensions.

Improvements to this route were made largely to eliminate using some of the hazardous or costly reagents such as high-strength peroxide, ether, and calcium and to permit the use of a flow sheet more suitable for large-scale production.

The present process still depends on the production of UF_4 as a pure intermediate which may be reduced to metal for fueling the Magnox reactors or further fluorinated with fluorine gas to produce UF_6, the essential feed material for all of the uranium isotopic-enriching processes.

UF$_4$ Production. The techniques and methods for producing pure uranyl nitrate solution by dissolving an ore concentrate and solvent extraction of the product are not dissimilar from those used in the reprocessing side of the industry. The greater part of chemical engineering development has been associated with the development of equipment and techniques for carrying out the subsequent gas–solids reactions. Initially these reactions were carried out by passing the appropriate gas over small static beds but these were superseded in the late 1950's by a succession of fluidized-bed reactors. However the batch reactions proved to be lengthy and with long reaction "tails" in order to obtain conversions to the extent required (i.e. 99% of the fully converted product).

A parallel development on a smaller scale of a continuous rotary kiln gas–solids contacting process has led to the introduction of large rotary kilns to replace the batch fluidized beds. These have proved successful in enabling large continuous throughputs to be achieved with very efficient utilization of process gases by virtue of the countercurrent nature of the gas and solid flows, and in reducing residence times from 25–60 hr to 3–5 hr. The thermochemistry of the principal reactions is as follows:

Reaction 1 $UO_3 + H_2(g) \rightarrow \quad UO_2 + H_2O(g)$ –26 kcal

Reaction 2 $UO_2 + 4\ HF(g) \rightarrow UF_4 + 2\ H_2O(g)$ –44 kcal

The required temperature for Reaction 1 ranges from 450°C to 550°C and that for Reaction 2 is 500°C to 600°C. Much depends however on securing a suitably reactive feed material for Reaction 1 which is achieved by hydration of the UO_3 to the β-dihydrate form.

UF$_6$(Hex) Production. The requirements for the conversion of UF$_4$ to UF$_6$ (hex) are quite different in that UF$_6$ has a triple point of 64°C at 0.5 bar gauge pressure and hence is removed from the reaction system in gaseous form. It then is liquified and fed to storage cyclinders where it solidifies.

The earliest hex plant used at Springfields used the batch reaction between UF$_4$ and chlorine trifluoride, the latter being transportable but requiring extreme care in handling. Elementary fluorine is preferred now but this has involved the development and linking of a system of electrolytic fluorine cells using a matrix of molten alkali–fluorides salt, fed continuously with HF with the necessary design and operational precautions against fluorine leaks and explosive recombination with hydrogen.

The UF$_4$–fluorine reaction is carried out in large-diameter fluidized-bed reactors; reaction is very fast, and the product is removed continuously as a gas. It is necessary to control temperature and pressure conditions in order to confine the reaction to $UF_4 + F_2 \rightarrow UF_6$ and to prevent the formation of less volatile intermediate fluorides, UF_5, U_2F_9, and U_4F_{17}.

The feed UF_4 for this process must be exceptionally pure since volatile fluorides of lighter elements pass rapidly up to the enriched product collecting point of an enrichment cascade and cause undesirable contamination of the enriched uranium product. Uranium reprocessed after reactor use must be purified sufficiently well to prevent transuranic elements from passing into the plant.

Uranium Dioxide Production. The majority of the world's nuclear reactors are fueled with slightly enriched UO_2 prepared in the form of dense sintered pellets that are encapsulated in small bore tubes of zirconium alloy or stainless steel. Hex at the required enrichment(s) is produced specifically for a given reactor charge and the first process step with the UF_6 is to convert it to UO_2 having the desired ceramic-grade quality. This means an oxide which after granulation and pelleting can be sintered quickly and uniformly to pellets of near stoichiometric density.

Early conversion processes invariably involved dissolving the UF_6 in water to form an acid solution of uranyl fluoride. Subsequent steps included neutralizing with ammonia and precipitation of the ammonium diuranate compound (ADU) which then could be filtered, dried, calcined, and finally reduced to UO_2. All of the fluoride value of the UF_6 is lost, however, and an undesirable aqueous effluent containing ammonia and fluoride is produced. In the United Kingdom a dry process has been developed involving the continuous reaction of steam with UF_6 vapor as a "plume" to give finely divided UO_2F_2. This then is allowed to grow to give particles of the required size and structure which are passed down a rotating kiln linked to the primary reaction chamber. Reduction conditions are controlled carefully to give a continuous gas–solids reaction which produces UO_2 having the desired properties. Many hundreds of tons of this oxide have been produced to fuel many of the European nuclear reactors.

Isotope Separation

Early Work. In the United Kingdom the earliest work aimed at large-scale separation of isotopes concentrated on uranium and deuterium. There were in the United Kingdom, before the Second World War, relatively few professionally trained chemical engineers, and early isotope separation thinking was done mainly by physicists in the universities. However, industrial experience was essential and ICI became involved and provided staff with industrial engineering, applied chemistry, and chemical engineering skills. The flow sheet theory of isotope separation using multistage methods was described first in physics terms following the prewar publication by Furry, Jones, and Onsager (2).

The relationship of this theory to multistage chemical engineering theory was realized soon and the first plant design flow sheets were developed by Kearton (now Lord Kearton) and others within the design teams at Imperial Chemical Industries Limited (ICI) (Billingham) in 1942–1944 and similar U.S. work was published by Cohen (3) in 1951. The theory of diffusion plant design led naturally to experimental work on vacuum technology, seal materials, construction materials, the theory of plant control instrumentation and stability, and the production of experimental diffusion membranes. ICI (Billingham) built a pilot diffusion plant assembly near Mold in North Wales in 1944 and ICI (Metals) produced membranes on a pilot scale during 1944, which became the basis for the U.K. production in later years.

In parallel with this, dual temperature processes were described by Rideal and Hartley in 1942. The earliest chemical engineering contributions therefore were made by men who would at that time have described themselves as applied chemists but would, in their later careers, describe themselves as chemical engineers. During the period 1941–43 they had the advantage of exchange of information with the United States, until the latter withdrew their earlier cooperation.

The history of the Atomic Energy Program in Britain by Gowing (4) records that the first reaction of Hinton, the man responsible for the industrial program, when faced with the task of building a diffusion plant, was to invite ICI, who had done the pilot-plant work, to operate an industrial factory. The company declined the request in the face of

Table I. Isotope Compositions in Natural Hydrogen, Lithium, Boron, and Uranium

Element	Isotope	Atom (%)
Hydrogen	H_1^1	99.985
	D_1^2	0.015
Lithium	Li_3^6	7.5
	Li_3^7	92.5
Boron	B_5^{10}	20.0
	B_5^{11}	80.0
Uranium	U_{92}^{234}	0.0055
	U_{92}^{235}	0.720
	U_{92}^{238}	99.28

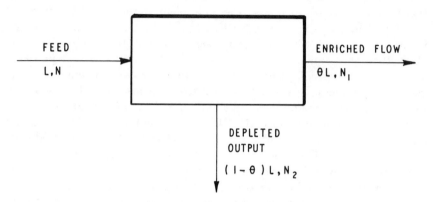

Figure 6. A simple separating element

other pressing post-war ventures. This led to a continuing dominance in the U.K. diffusion plant program by physicists and engineers.

Theoretical Development. Separating isotopes on an industrial scale, where the components are chemically identical and differ physically only in mass, represented a new chemical engineering challenge. As shown in Table I, the challenge was made more difficult in that the desired isotopes deuterium[2], lithium[6], boron[10], and uranium[235] were of low concentration in their respective naturally occurring parent elements.

Historically, the possibilities for isotope separations were discussed first by Lindemann and Aston *(5)* in 1919 who identified four distinct principles on which separations could be based: (a) distillation; (b) diffusion; (c) density difference; and (d) positive rays. Only electromagnetic separation (d) afforded the opportunity to effect an isotope separation of reasonably high purity in a single step but this was considered to be impractical for an industrial plant. All of the other principles related to techniques in which a small change of isotope concentration occurred in a single step. The problem of obtaining high-purity products therefore could be solved only by repeating the fundamental step many times.

The basis of an isotope separation flow sheet starts with a separating element as shown in Figure 6. This divides an incoming two-component isotope feed flow of L mol/sec having a mole fraction N of the desired isotope into two streams, one stream of Θ L mol/sec enriched to a mole fraction N_1 of the desired isotope and the other of $(1 - \Theta)$ L mol/sec depleted to a mole fraction N_2.

A simple process separation factor for the separating element defined as α is given by

$$\alpha = \frac{N_1}{1-N_1} \bigg/ \frac{N}{1-N} \qquad (1)$$

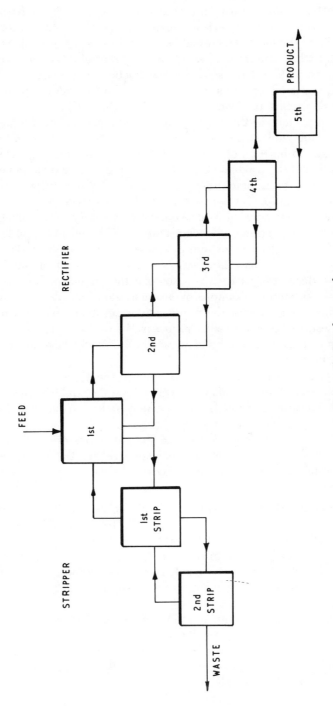

Figure 7. Simple cascade

An isotope separation plant, ideally designed, consists of a large number of separating elements formed in a cascade. Figure 7 illustrates a simple cascade layout showing series connections of elements by combining enriched and depleted streams as feed flow to stages. Also shown is the feed point which enters the cascade at the point of maximum paralleling of elements. The relationship to reflux fractional distillation can be seen in that there is effectively a countercurrent flow of material between the waste and product positions.

The mass flow to the stages progressively tapers from the feed point to the product and waste take-off positions. This results from the need of any stage to have a net isotope transport towards the product take-off equal to that of the product rate. In designing a cascade for an industrial plant it would be impractical to taper at every stage. However, if the stage process equipment is expensive and/or the stage energy consumed is excessive, then tapering is economically necessary. The practical design solution to tapering is a compromise. This consists of splitting the cascade from the feed point into a number of squared-off sections where in any section the mass flow A to the stages is constant but between sections A decreases progressively towards the product and waste take-off positions. A simple example of a squared-off cascade is a fractional distillation column where the reflux ratio is constant as distinct from the ideal cascade where the reflux ratio varies from stage to stage. These concepts of ideally tapered and squared-off cascades have been detailed fully by Pratt (6).

Table II lists the elements separated in the United Kingdom on an industrial scale, the methods investigated, and the equilibrium separation factors. These separation factors are only indicative in that they vary with actual process parameters.

All of the processes used fall in the categories originally listed by Lindemann and Aston; chemical exchange is related to distillation in that the separation effect is determined by the differences in energy states of the isotopes.

Uranium Isotope Separation. The successful industrial separation of uranium isotopes results from (a) UF_6 having a substantial vapor pressure at temperatures below 100°C, and (b) fluorine being monoisotopic. However UF_6 is a toxic, highly corrosive gas which decomposes on contact with water. As a result its use, on an industrial scale, gives rise to plant problems in selecting materials for construction and containment.

The first U.K. uranium isotope separation plant, built in the early 1950's, was based on gaseous diffusion. This phenomenon depends on the observation that the rate of passage of molecules through a membrane is inversely proportional to the square root of the molecular weight. The application of this principle in a cascade stage requires a compressor, membrane, control valve, and cooler.

Table II. Separation Methods and Process Separation Factors

Isotope	Separation Method	Separation Factor
U_{92}^{235}	1. Gaseous diffusion of UF_6	1.004
	2. UF_6 gaseous centrifuges	1.055
B_5^{10}	1. Chemical exchange between BF_3 and	1.03
	$Et_2O - BF_3$	1.006
	2. fractional distillation of BF_3	1.006
Li_3^6	1. molecular distillation of lithium metal	1.08
	2. chemical exchange between aq LiOH and Li/Hg	1.07

A high-enrichment UF_6 cascade based on gaseous diffusion requires approximately 3000 stages. Thus, since the basic separation process is irreversible and therefore requires high-energy inputs at each stage, it is imperative to have squared-off tapering. This was the main reason for developing different sized compressors. However one major benefit from tapering is that having smaller stages towards the product-end alleviates the safety design problems of criticality which became progressively more severe as the U^{235} concentration progressively increased towards the product stage.

An important plant constructional specification was that it should be built to hold a vacuum. This was necessary to prevent inleakage of moist air which, on reacting with UF_6, would give solid plugging deposits on the membranes.

The main development requirement was for the production of a satisfactory membrane. A typical specification calls for large numbers of uniform pores per unit area, a pore size the region of 100 A, a high permeability, and materials of construction inert to UF_6 at the operating pressure and temperature. Membrane development work was initiated at Oxford in 1940 and soon after in Imperial Chemical Industries who, concentrating on porous nickel, produced a satisfactory product for the plant.

The second U.K. uranium isotope plant, built recently to provide low enrichment for nuclear civil needs, uses UF_6 gaseous centrifuges. This process depends on the phenomenon that when a gas is subjected to high gravitational forces the pressure distribution is a function of the molecular weight. Spinning UF_6 isotopes in a cylinder establishes a radial isotopic concentration gradient. The radial separation factor achieved depends on the peripheral speed of the cylinder. As shown in Table II, values of α - 1 are at least an order of magnitude greater than those for gaseous diffusion. This means that the number of stages required are one tenth of those needed in a gaseous diffusion plant of equivalent duty. Similarly the mass flows through stages are reduced much.

While the capital costs for a centrifuge plant are similar to those of a diffusion plant, the energy consumption for centrifuges is at least a factor ten below that of an equivalent output gaseous diffusion cascade. A centrifuge plant has similar constraints to the diffusion plant with respect to construction materials and containment of UF_6. The problems of vacuum integrity are however more stringent since the centrifuge rotor must run in a high vacuum to reduce drag losses and skin heating.

Boron Isotope Separation. The B-10 isotope has a very high cross section for the absorption of neutrons and was separated for use in the fast neutron reactor control rods and shut-off rods. Two methods were used, one using a countercurrent distillation column in which the vapor phase was BF_3 and the liquid-phase BF_3 ether complex. The second, larger, and more effective plant was based upon the low-temperature distillation of BF_3 itself, $^{10}BF_3$ being the less volatile fraction, (Nettley et al. (7). This combined good chemical engineering with good physics. Separation by multiplate distillation (with a single-stage separation factor of 1.006 and therefore a very high reflux ratio of 5,500 through over 1000 theoretical plates) used high-efficiency column packing. Even so, the total column length had to be 30 m and two distillation columns in series were required. Three novelties may interest chemical engineers:

1. The height of an equivalent theoretical plate and the single stage separation factor were both uncertain for this system. They were both determined simultaneously by using two simultaneous equations, one for the condition of total reflux and the second for reflux with product recycle to the waste end of the column.
2. The reflux condensers were operated to condense BF_3 at $-115°C$ using as a coolant liquid nitrogen boiling at $-196°C$, at which temperature BF_3 is a solid. The solution to this problem was a heat break using an interspace containing an isolated pool of boiling BF_3, condensing onto a cold solid layer of BF_3 (solidified at liquid nitrogen temperature).
3. Transfer of material between the two columns was automatic by virtue of two intercolumn vapor-transfer lines. The mass flow in these transfer lines was provided by using appropriate lengths of column packing generating appropriate pressure drop to drive the flow (see Figure 8).

Lithium Isotope Separation. In the 1950's two routes were investigated to separate lithium isotopes. The first was based on molecular distillation of molten lithium at 500°C. The process, which operates at a few microns pressure has a theoretical separation factor of 1.08, the square root of the ratio of the atomic weights. While high separation

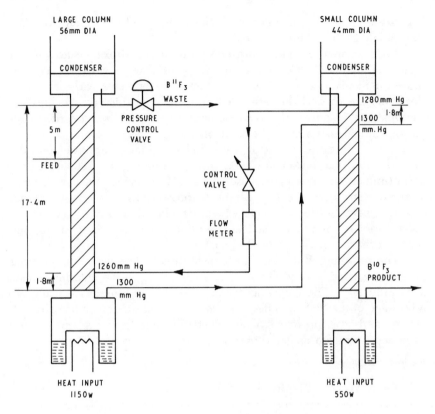

Figure 8. The two-column system

factors were practically demonstrated in cascades, severe technical problems arise when establishing interstage flows with molten lithium at 500°C in a vacuum-tight plant. This route was abandoned ultimately in favor of chemical exchange.

Lewis and McDonald (8) in 1936 showed that isotope exchange occurred between lithium amalgam and lithium chloride dissolved in ethanol with the light isotope concentrating on the amalgam phase. Although the system was not suitable to establish a refluxing countercurrent cascade, development work quickly established that the isotope exchange occurs equally well between lithium amalgam and lithium hydroxide dissolved in water and a practical flow sheet was evolved from this exchange.

$$Li^7/Hg + Li^6OH \rightleftharpoons Li^6/Hg + Li^7OH$$

The three major development problems that arose were: the development of a suitable contactor; the turnaround of lithium from lithium hydroxide to lithium amalgam; and the turnaround of lithium from lithium amalgam to lithium hydroxide.

Lithium amalgam in the presence of lithium hydroxide constitutes an electrical half-cell. Consequently the contactor of whatever concept must be nonconducting. As with many developments early laboratory decisions are carried through into the plant. In this case Perspex was the material used in the development phase and became the major constructional material of the final plant.

In selecting the contactor a number of options were available: dispersed water phase in a continuous amalgam phase; dispersed amalgam in a continuous aqueous phase; use of packed columns; and use of mixer settlers. The early work to establish the separation factor was carried out in a mixer settler which had been developed around the centrifugal glass impeller commonly used in laboratories for mercury cleaning, and mixer settlers were used in the main plant.

The recycling of the lithium from lithium hydroxide to lithium amalgam was carried out by electrolysis, as practiced in the chlorine–caustic soda industry. The recycling of the lithium from lithium amalgam took advantage of the half-cell effect by simply flowing water countercurrent to amalgam in the presence of an electrical conductor. A cascade was designed based on an equilibrium separation factor of 1.07.

Reprocessing

Early Work. The irradiated fuel, upon discharge from the reactor, comprises the residual unburnt fuel, its protective cladding of magnesium alloy, zirconium or stainless steels, and fission products. The fission process yields over 70 fission product elements, while some of the excess neutrons produced from the fission reaction are captured by the uranium isotopes to yield a range of "new" elements—neptunium, plutonium, americium, and curium. Neutrons are captured also by the cladding materials and yield a further variety of radioactive isotopes. To utilize the residual uranium and plutonium in further reactor cycles, it is necessary to remove the fission products and transuranic elements and it is usual to separate the uranium and plutonium; this is the reprocessing operation.

The first successful large-scale separation of plutonium from fission products and uranium was achieved by the Manhattan Project in the United States and was based on the multiple coprecipitation of plutonium with $BiPO_4$ and LaF—a scale-up of an analytical procedure. The project was successful in meeting its targets in a remarkably short time scale and gave an insight into the techniques necessary for the safe handling of radioactive solutions on a large scale.

This process and the experience derived therefrom was not available to the United Kingdom and when it was decided that an atomic program should be initiated, a joint team of Canadian and U.K. scientists was set up at Chalk River, Canada in 1947. This team, which was led by R. Spence and included the French chemist, B. Goldschmidt, surveyed the possible techniques and concluded that plutonium and uranium could be extracted by a selective solvent and then separated by reducing the plutonium to its inextractable trivalent state. The U.K. team chose Butex which combines the desired extraction capability with a high flash point, low solubility in the aqueous phase, and adequate chemical stability in nitric acid. The chemical engineers (A. S. White and C. M. Nicholls) pointed out the long-term advantages of Butex in that it was not necessary to use a salting-out agent and hence treatment of the highly active fission product wastes would be simplified greatly. The consequence of this early choice has been that all of the U.K.'s highly active waste is compatible with a borosilicate glass vitrification process.

The first plant was based on the Butex solvent extraction process, operated in packed countercurrent columns, and used gravity flow for the highly active cycles which in turn led to a 250-ft-high building.

The second separation plant used tributyl phosphate in kerosene (TBP/OK) as the solvent and mixer settlers with simple interstage lifts between cycles. The choice of this solvent enabled the further elimination of salting agents used in the later Butex cycles of the first plant with consequent benefit to effluent disposal. This plant had five times the output of the first yet was accommodated in a low, smaller building. Despite these marked changes between the two plants, many of the principles found to be satisfactory in the first plant were retained in the second, a tribute to the firm foundations laid, in a short time, by the original designers.

The Second Windscale Separation Plant.   A commercial reprocessing plant, as exemplified by the second Windscale Separation Plant, comprises the fuel-decladding units, fuel dissolution, the solvent extraction process to separate uranium and plutonium from fission products and to separate uranium and plutonium from each other. These primary processes are integrated with facilities for the storage of solid wastes (e.g. cladding), the concentration by evaporation of aqueous wastes, and the recovery and recycle of nitric acid. The ventilation system is designed to provide extracts from the process vessels, the cells housing these vessels, and the operating area and to maintain a flow from inactive to highly active areas. All ventilation air must be filtered by primary and secondary high-efficiency particulate air filters (HEPA) and additionally special treatments given to remove nitrous gases, volatile radioiodine, and, in the future plants, krypton.

The products uranium and plutonium are required to be recovered in a high degree of purity and with high efficiencies while the fission products left in the uranium and plutonium must be less than 1 part in

10^7 or 10^8 of the feed concentration. Furthermore, the permitted atmospheric discharges of the fission products must be less than 1 part in 10^{10} to 10^{12} of the amounts fed.

Formidable as these problems may appear at first sight, the development of the industry has shown that by careful evaluation of the principles of design they may be solved by applying conventional chemical engineering technology, albeit in equipment specially designed to take account of the constraints of radioactivity.

From the experience of the first plant, it was decided to base the design of the second plant on the so-called "direct" maintenance system in which the plant within the highly active areas is designed for a long, continously operating life and to duplicate the highly active sections with standby units. The medium- and low-activity sections of the plant are not duplicated and can be decontaminated to a level permitting maintenance or replacement in an acceptable period of time; however, even here there should be no reduction in the standards of design and construction.

The alternative to the direct-maintenance philosophy is the "canyon" using remotely removable equipment as used in the early U.S. plants at Hanford and SRP. This has proved successful but requires more space and is of a more complex design.

The adoption of the direct-maintenance philosophy sets criteria for the design engineer that affect not only the plant design but that of the process as well.

The process should be continuous, capable of control by metering the inactive feeds, free from the necessity of relying on process control instrumentation operating feedback loops, and have a slow response time to variations in conditions to enable the necessary monitoring instruments (temperature, concentration, flow rates and radioactivity) to establish trends so that corrective action may be taken carefully before the product and waste qualities are affected significantly.

The installed plant should have a minimum of moving parts and those that are essential must be capable of replacement without loss of containment. The materials and methods of fabrication must be chosen and controlled carefully to ensure longevity. Because of the inaccessibility of the plant, careful supervision must be exercised over the siting of sample points and instruments, and adequate provision must be made for supplemental equipment of this nature to be introduced as needed.

The very nature of these requirements leads to a conservative design philosophy and yet, as the second Windscale Plant has proved, the reward is high utilization and longevity, with consequent economic benefits.

A necessarily brief outline of some of the process features of the second Windscale Plant will illustrate the application of the above prin-

ciples and demonstrate the broad spectrum of chemical engineering technology used in such a plant.

Process Features. The Magnox elements vary in length and cladding structure but are based on uranium metal rods approximately 1 in.
in diameter. To achieve the plant throughput, one element has to be
decanned every few minutes. The choice, between batch chemical decladding, in which the Magnox is dissolved and the uranium metal left
substantially unattacked, or mechanical decladding in which the element
is forced through a die and the cladding stripped off, was made in favor of
the latter, partly because of successful previous experience, but largely
because of the difficulty of removing activity from the large volumes of
aqueous waste generated by the former process.

However, it was decided to house the equipment required in shielded
cells, equipped with manipulators and hoists rather than in ponds as was
done in the first plant. The machinery was to be capable of being
removed and replaced without direct contact. This departure from the
principles of simplicity and low maintenance requirements, was a necessary concomitant of the choice of mechanical decladding for high throughput. Experience has shown that the process is acceptable but the equipment has required evolutionary development to deal consistently with
the fuel at the required rate.

The dissolution of irradiated metallic uranium in boiling nitric acid to
give a product substantially $3N$ in nitric acid, $1M$ in $UO_2 (NO_3)_2$, and with
plutonium and fission products present, has proved to be a most satisfactory unit operation. Yet for the designer of the first plant it appeared
a formidable problem—corrosion, the risk of fire, foaming, and surging of
the   overflowing   product   all   required   detailed   study   and
modeling. The second plant, with a unit capacity of 5 tons of uranium
per day also was based on continuous dissolution but the larger size
necessitated a welded fabrication. Surging was eliminated by a rede-
signed baffled off-take. Both the first and second plant used fumeless
dissolution in which the evolved nitrous gases were reconverted to nitric
acid by adding oxygen and countercurrent scrubbing by the incoming
feed of nitric acid—in this manner all of the nitric acid fed to the process
was converted to the nitrates in the product solution.

The fabricated dissolver, built in high-grade stainless steel, quality
controlled from cast to installation, has processed 13,000 tons of uranium
in 14 years of near-continuous operation and is being retired now owing
to end-grain corrosion of some forgings used to connect feed and instrument pipes to the dome of the vessel. To indicate the accessibility of
such a vessel to inspection, it has been possible to monitor this corrosion
by means of television cameras inserted through a 4-in. access port
during plant maintenance at yearly intervals without significant radiation
problems. The philosophy of a direct maintenance plant is illustrated in

this instance by the ability to maintain plant operation by bringing into use the second standby dissolver unit, built in its own cell. At the time of writing, the provision of a third cell as a standby and the decontamination and reconstitution of the first cell are being studied as alternatives.

The Separation Process. The heart of a reprocessing plant is the solvent-extraction process and the success of all subsequent operations, including waste disposal, depend upon the soundness of the design of these unit operations. For this reason, a large part of the early R&D work was devoted to solvent extraction and the design principles are well understood. This is not to say that the fundamental chemistry and chemical engineering have advanced to the stage where a design can be computed from basic physical and chemical data and it still is considered advisable to conduct process studies in miniature, fully active pilot plants and to test the performance of the major items (mixer settlers, pulse columns, and instrumentation) on full-scale units operating inactively but with uranium as a key element.

The feed to the solvent-extraction plant may be characterized as plutonium and uranium nitrates and fission product nitrates. Although the latter comprise some 70 elements, there are only a few that partially extract into the TBP/OK solvent, and of these, Zr, Nb, and Ru are the most important. Of the transuranic elements, Np is the only one to extract with the plutonium and uranium to any significant degree.

The design of the separation process starts from the measurement of the partition of the product uranium and plutonium nitrates, the latter in the prevailing tetravalent and hexavalent states and also in the trivalent state to which it is reduced in order to effect the separation of uranium and plutonium. From this data, derived under a variety of conditions of acidity and temperature, the process for extracting scrubbing and partition was computed by iterative calculation using the program SIMTEX developed by Burton.

Because of uncertainties in the data, the extremely low concentration of the fission product impurities, and the possible effects of unidentified solvent degradation products, it was essential to substantiate the process. To this end, the miniature pilot plant, scale 1:5000, was built and the process and its variants tested at full activity.

By combining the data from selected full-scale prototypes and the 1/5000th scale fully active pilot plant, the design features were fixed and the soundness of this work has been demonstrated by the performance of the plant, which achieved full output within one week of going on stream and subsequently was shown to have a capacity 50% in excess of the formal design. During the 14 years of operation, it has had a high availability, exemplified by two campaigns of continuous, uninterrupted operation, each of two-years duration.

Control. The fundamental control of the plant is achieved by setting the aqueous and solvent flows to each of the cycles, and adjusting the flow of uranium rods to the dissolver to match the preset condition. Between the dissolver and the solvent plant, a small tank fitted with a simple rotor equipped with buckets is used to control the aqueous flow to ±1%—comparable with the accuracy of the inactive feeds. To allow this simple technique to be used, we eschewed the use of flow sheets in which the uranium content of the solvent approached that corresponding to a pinch point on the extraction diagram, thus reducing the risk that aqueous raffinate would contain unacceptable amounts of product. The instrumentation for the inactive feeds comprise commercial flow controllers, duplicated, and in some instances triplicated, to measure the flow of acid and water and the combined feed of dilute acid, acidity and density meters, pneumatic level gauges, and resistance thermometers. However, additional on-line instrumentation was essential for the safe and efficient operation of the plant. A novel instrument, which simultaneously measured and displayed the uranium and nitric acid concentration in the feed to the solvent plant was developed. It comprises a density meter and a conductivity cell, the outputs of which are fed to an analogue circuit which calculates the uranium and HNO_3 concentrations.

The Management of Wastes from Reprocessing Operations

Objectives. Reprocessing, like all other industrial operations, results in the formation of a large variety of waste materials. These have to be treated, each according to its form and composition, to ensure that human health and safety is protected and that the impact on the environment is controlled to an acceptably low level. The standards of management required for radioactive wastes are particularly stringent, not only because some radioactive materials are particularly hazardous but because radioactive operations often are connected in the public mind with the extreme results of military applications and, being associated with a new technology, lack of widespread knowledge has retarded the development of a realistic public attitude towards radiation hazards.

The U.K. nuclear industry has used, from the start of its operations, extremely high standards of containment of radioactive materials and radiation protection, and consequently, now has a safety record which is second to that of no other comparable industrial activity. That waste management operations in the nuclear field share this exemplary record is the result of the cautious development of sound waste management principles and practices that have been introduced carefully at a rate which has ensured that the techniques employed have been tested and proved adequately before being operated (9). Chemical engineering principles and equipment have been used extensively and have made a major contribution to the achievement of safe and acceptable practices.

The following objectives of radioactive waste management have been developed in a recent report (10) by an international expert group under the auspices of the Nuclear Energy Agency:

- to comply with radiological protection principles for present and future generations;
- to preserve the quality of the natural environment;
- to avoid preempting present or future exploitation of natural resources
- to minimize any impact on future generations to the extent practicable.

With particular regard to the conditioning, storage, and disposal of radioactive wastes, a number of basic criteria may be identified that now are accepted widely as being of major importance.

1. The amounts of waste produced should be minimized.
2. The best means of utilizing resources, including recycle of valuable components, should be used to conserve the environment.
3. The best practicable means should be used in waste conditioning and disposal operations.
4. Solid waste for disposal and isolation from the biosphere should, as far as is practicable, be:

- chemically stable;
- noncombustible;
- monolithic and therefore having a low specific surface area (not in a powdered form;
- insoluble in natural waters;
- chemically compatible with the disposal environment;
- in such a form that it will not produce combustible or explosive gases in the disposal environment;
- resistant to radiation damage; and
- should not constitute a critical hazard.

The main chemical engineering principles and processes used in the United Kingdom over the past 25 years to achieve these objectives are discussed below under the headings of gaseous, liquid, and solid radioactive wastes.

Treatment of Gaseous Waste. Gaseous wastes arise from the ventilation of process vessels and the concrete cells that house the plant and equipment used for reprocessing spent nuclear fuel. This gaseous waste is largely air contaminated with small entrained liquid or solid particles containing radioactive components. Some ventilation streams also are contaminated with oxides of nitrogen.

The techniques used to clean these gases before discharge, in accordance with government authorizations, through stacks to the atmosphere are wet scrubbing; impingement on irrigated ring packings; electrostatic precipitation; and absolute filtration. In each case the gas-cleaning equipment used has been developed or adapted specially for use in plants that are handling substantial amounts of radioactive materials. The avoidance or ease of maintenance and the ease of cleaning are particularly important in radioactive operations that have to be controlled remotely behind concrete walls that are often several feet thick. The simplest equipment—without moving parts and operating at normal temperature and pressures—is therefore generally favored.

One gaseous waste stream from reprocessing operations meriting special mention, is that which arises from ventilation of the vessel in which, at the start of the wet processing operations, intensely radioactive spent fuel is dissolved in nitric acid. The off-gas from this dissolver is largely air, with some oxides or nitrogen, contaminated with entrained radioactive aerosols and some of the volatile fission products, I^{121}, Kr^{85}, and H^3. The oxides of nitrogen are removed by nitric-acid scrubbing after the addition of oxygen, and aerosols are removed by wet scrubbing and absolute filtration. Wet scrubbing with sodium hydroxide or acidic mercuric nitrate solution and adsorption on packed beds of zeolite or silica gel impregnated with silver nitrate are effective for iodine removal. No processes for Kr^{85} or H^3 removal on an industrial scale are necessary because of the very low hazard, but cryogenic techniques or molecular sieves could be used in the future.

Treatment of Liquid Waste. A number of liquid-waste streams arise from operations. These vary greatly in volume and composition but may be grouped into three classes according to activity content— highly active, medium active, and low active.

Since, with only a few exceptions, the fission-product nitrates are nonvolatile, distillation of radioactive liquid wastes is a very effective method of separating water and nitric acid as an almost pure condensate, while at the same time retaining the fission products in a concentrated solution of such small bulk that long-term storage is practicable. By this means a comparatively large volume of highly active liquid waste can be converted into a small volume of concentrate for storage and a low-level liquid waste can be discharged safely to sea in accordance with government authorizations. Very high separation factors for activity (10^4–10^6) are achieved readily in a single-stage batch distillation. The reprocessing systems in the United Kingdom have been operated in such a way that the highly active liquid-waste streams are of low-salt content which allows high concentration factors to be achieved.

Distillation units which have been developed for treating high-level liquid waste have a number of special features required for operation with reprocessing wastes:

- exceptionally thick construction (to high fabrication standards) in a special stainless steel;
- operation at low pressure and therefore low temperature and corrosion rate for long life;
- remote operation inside concrete cells with walls about 5 ft thick; and
- liquid transfers accomplished by vacuum lift, siphons, or air pressure (no moving parts).

These distillation units, with capacities of typically up to 50 m^3/d, have played a major role in the treatment of radioactive waste since their steady and reliable operation over many years has ensured that the high-level liquid waste from the nuclear fuel cycle routinely has been reduced in volume for safe storage prior to further treatment to condition it for disposal and long-term isolation from the biosphere.

At present a total of about 750 m^3 of high-level waste concentrate, containing about 4×10^8 Ci (beta) and having a radioactive decay heat of about 2 MW, is contained in special storage tanks at Windscale. This is the cumulative total from about 25 years of reprocessing operations.

These tanks, each with a capacity of about 150 m^3, are constructed of stainless steel to very high standards and are contained in concrete cells. The tanks are fitted with water-cooling coils and a water jacket to remove radioactive decay heat, and a cooling tower system is used to reject this heat to the atmosphere. In order to ensure the high reliability required, diversity, segregation, and redundancy are used in the cooling system, both external and internal to the tank. The insides of the concrete cells that house the tanks are lined with stainless steel to act as a third line of containment.

Although it is considered that this storage system will safely contain the high-level waste concentrate for at least several decades, a process is now under development for the incorporation of this waste into a glassy solid which will be suitable for disposal into a deep geological formation on the bed of the deep ocean.

This process, which is named HARVEST and has been described in a number of papers (11, 12, 13) will draw on basic chemical engineering technology at present used in other industries, e.g. in the construction and operation of high-temperature furnaces, glass making, and gas cleaning.

In the HARVEST process, highly active liquid waste concentrate, together with a slurry of glass-making chemicals, is fed into a special steel container (18 in. in diameter and 9 ft high) which is preheated in a furnace capable of operating up to about 1000°C. By the action of heat, the waste and glass-making chemicals are dried, calcined, and converted into a glassy solid which is very insoluble in natural waters, including sea water.

When sufficient glass has been produced the container is removed from the furnace, cooled, and placed inside a storage vessel which is sealed by welding.   The sealed storage vessel, with its contents, then is stored under water in a pond for a cooling period prior to final disposal.

The cooling period in the pond will depend on the disposal route chosen.   If disposal onto the bed of the deep ocean is chosen, where removal of radioactive decay heat is efficient, the cooling period might be 10–20 years; whereas if disposal into a geological formation or into the bed of the ocean is planned, where heat removal will be more difficult, then a cooling period of 50 or more years might be required.

Medium-active wastes of low-salt content can be treated by distillation in the same way as described above for highly active liquid waste. The small volume of concentrate produced from medium-active waste distillation can be combined with high-level waste arisings and the very low low-active condensate can be disposed of to sea in accordance with government authorizations.

Medium-active waste also can be treated by floc precipitation processes similar in character to those widely used in water treatment. Precipitants are required which will remove the unwanted radioactive species from the waste solution and quickly settle, carrying the radioactivity into a small bulk of sludge.   The supernate can be treated as a low-active waste and discharged locally to the environment.   Typical precipitants for common cations are as follows.

Precipitant	*Main cations removed*
ferric–aluminium hydroxide	Pu, Am
barium sulfate	Sr
nickel cobalt cyanoferrate	Cs
nickel sulfide	Ru

The efficiencies of such coprecipitation–adsorption processes cannot, in practice, be predicted accurately because they are sensitive to pH and the presence of competing ions and are dependent upon the chemical behavior of extremely small amounts of radioactive materials in large volumes of waste.

Floc treatment processes are generally three to four orders of magnitude less effective than distillation in the separation of radioactive components but are particularly valuable for waste streams of high-salt content that cannot be reduced to a small bulk by distillation, and for large volumes of wastes the distillation of which would entail excessive energy costs.

Treatment of Solid Wastes. The solid wastes generated in the nuclear fuel cycle vary greatly in character and composition but the major wastes can be categorized as follows:

- contaminated spent fuel cladding removed from the fuel at the start of reprocessing operations (low α, high β, γ);
- sludges and ion-exchange resins from pond operations and the treatment of medium-active liquid wastes (low α, medium β, γ);
- plutonium-contaminated materials mainly from the maintenance of plants handling plutonium (high α, low β, γ).

Magnox cladding, which has been removed mechanically from spent fuel rods at the start of reprocessing operations, is contaminated with small pieces of fuel and will require treatment before disposal to the environment. At present, this waste is stored under water (to eliminate any fire risk) in large concrete silos and processes are now under development for the conditioning of this waste to make it suitable for disposal. The favored processing route comprises the following operations:

- selective dissolution of the Magnox component of the waste in dilute nitric acid with recycle of the undissolved spent-fuel component to the reprocessing system;
- treatment of the Magnox solution by one or more of the following routes depending upon its activity level:
  1. direct solidification in concrete or bitumen;
  2. solvent-extraction to remove uranium and plutonium;
  3. ion exchange and floc treatment to remove radioactive components, followed by disposal of the purified solution to sea.

It is unlikely that the sludges and waste resins from the exchangers will require having any of their chemical components removed before disposal and the treatment envisaged is therefore a direct incorporation of the waste into a cement or bitumen matrix followed by packaging for disposal. These techniques, now often referred to as waste solidification, have been developed specially in the nuclear industry to aid in the disposal of radioactive wastes.

Plutonium-contaminated wastes, which mainly comprise paper, gloves, and contaminated, unserviceable equipment from the maintenance of plants handling plutonium, can be treated in the following ways: combustible waste—incineration and acid digestion; and noncombustible waste—decontamination by electropolishing or ultrasonic washing.

Although incineration is used widely for volume reduction of industrial and household combustible waste, the incineration process for plutonium-contaminated waste has required very special development to ensure protection of the operators, efficient off-gas cleaning to protect the environment, and the avoidance of criticality owing to the concentration of plutonium in the ash. An incinerator for this duty now has been operated successfully in the United Kingdom on a pilot-plant scale and it is planned to extend this to industrial-scale operation when required.   A process for the recovery of plutonium from incinerated ash by acid leaching and solvent extraction is also under development together with a system for the incorporation of treated ash into a ceramic matrix for final disposal.

Acid digestion is being considered as a means of treating combustible waste.   It has the advantage that plutonium is recovered relatively easily from the acid solution, whereas it often can prove somewhat intractible as a component of incinerator ash.

Discussion

The organizational practices that have been evolved during the post-war years for undertaking major process projects are essentially multi-disciplinary, the composition and balance of the teams being determined by the content and novelty of the process.   In the United Kingdom during the same period a complete nuclear fuel cycle industry has been established.   It has involved the wide variety of processes outlined in this chapter and others relating to the manufacture of fuel elements that have not been described.   The balance of contribution by the main engineering and scientific disciplines differs from process to process but chemical engineers have been fortunate in the opportunities which the new industry has afforded them to apply and extend their professional skills.

These opportunities have arisen for a number of reasons of which two main ones can be identified.   The first is associated with the need of the nuclear fuel cycle industry for separation processes.   Nuclear reactors and weapons depend on the properties of the atomic nucleus in relation to neutron reactions.   Such properties, unlike chemical characteristics, differ markedly between isotopes of the same element and provide a strong incentive to separate isotopes on a large scale.   The separation factors involved are usually very small, the capital investment and operating costs involved are correspondingly large, and the premium for skillful process selection and design is great.   Furthermore, isotopes of a few of the nonfissile elements have very large neutron cross sections, e.g. boron, cadmium, and gadolinium.   The resulting capture of neutrons by small concentrations of these elements is entirely parasitic and even can be damaging to the materials of the nuclear reactors.   To compensate for such losses the proportion of the isotope, uranium 238, in the uranium fuel must be reduced.   Therefore the cost of high-cross-section parasitic absorbers can be measured by the relatively high cost of uranium isotope separation, and very high standards of purification from such absorbers are sought.

The second main field for the application of separation processes arises in the treatment of irradiated fuel.   Such fuels contain of the order of 1% of mixed fission products involving 60 or more different elements, and a similar but smaller quantity of plutonium. To produce product streams of uranium and plutonium that are suitable for further processing and effluent streams that can be disposed of, decontamination factors of more than 10,000,000 are required. This is a consequence of the high specific activity of the fission products and the same property leads to the need for remote and highly reliable operation of the first of the separation cascades in the reprocessing plant. Thorough development, demonstration and design of such processes is particularly cost-effective because it is often extremely difficult to make modifications once the plant has been committed to active operation. These requirements have called for a strong bridge between the complexities and uncertainties of fission product chemistry on the one hand and the performance and reliability of mechanical plant on the other, and the chemical engineer has proved particularly effective in the construction of this bridge.

It seems likely that his contribution will be no less important in the future. The technology for the manufacture and installation of gas centrifuge plants for the separation of uranium isotopes has been brought to an advanced stage by the cooperative efforts of teams in West Germany, Holland, and the United Kingdom acting within the framework of URENCO. Capital investment in plants of this type during the next decade will depend on the rate of increase of demand for enriched uranium but it is likely to be many hundreds of millions of pounds, and the opportunities for advancing the basic technology and for improving the details and efficiency of plants will be correspondingly great.

Planning permission has been granted also for a major new reprocessing plant by the Secretary of State for the Environment after a public enquiry conducted by J. Parker (14) which lasted one hundred days, and two parliamentary debates each of which led to support of the project by a substantial majority (15). The development of the Windscale site to accommodate this plant and the main ancillary facilities, together with replacement capacity which may be required for existing plant, will represent one of the largest land-based projects in the United Kingdom during the next decade and will provide demands and opportunities for skilled engineers and applied scientists of many disciplines, but particularly for the chemical engineer.

Literature Cited

1. Hughes, T. G.   "Criticality Incident at Windscale," *Nucl. Eng. Int.* **1972**, *17* (189).
2. Furry, W. J.; Jones, R. C.; Onsager, L. *Phys. Rev.* **1939**, *55*, 1083.
3. Cohen, K.   "The Theory of Isotope Separation as Applied to the Large Scale Production of U_{235};" McGraw Hill: New York, 1951.
4. Gowing, M.   "Britain and Atomic Energy 1945–1952;" The McMillan Press Ltd: London, 1974.

5. Lindemann, F. A.; Aston, F. W. *Philos. Mag.* **1919**, 37, 221.
6. Pratt, H. R. C. "Countercurrent Separation Processes," Elsevier: New York, 1967.
7. Nettley, P. T.; Cartwright, D. K.; Kronberger, H. Proceedings of the International Symposium on Isotope Separation (1957) pp. 178–197.
8. Lewis, G. N.; MacDonald, R. J. *J. Am. Chem. Soç.* **1936**, 58, P2525.
9. Roberts, L. E. J. *Brit. Nucl. Energy Soc.*, in press.
10. "Objectives, Concepts and Strategies for the Management of Radioactive Waste Arising from Nuclear Power Programmes;" Nuclear Energy Agency Organization for Economic Cooperation and Development: Paris, Sept. 1977.
11. Clelland, D. W., Presented to the *Int. Symp. Management of Wastes from the LWR Fuel Cycle, Denver, Co, July 1976.*
12. Corbet, S. D. W. Problems in the Design and Specification of Containers for Tritrified High Level Liquid Wastes, IAEA Symposium paper IAEA/SM/207/3, Vienna, 1976.
13. Hall A. R. "Development and Radiation Stability of Glasses for Highly Radioactive Wastes, IAEA Symposium Paper IAEA/SM/207/24, Vienna, 1976.
14. Parker, Justice "The Windscale Inquiry;" Her Majesty's Stationery Office: London, 1978.
15. Debates in the UK House of Commons, March 22nd 1978 and May 15th 1978.

RECEIVED May 7, 1979.

The Role of Chemical Engineering in Providing Propellants and Explosives for the U.K. Armed Forces

R. P. AYERST and M. McLAREN—PERME Waltham Abbey, Essex, England

D. LIDDELL—ROF Bishopton, Renfrewshire, Scotland

The decline in the importance of gunpowder and its replacement by nitrocellulose and nitroglycerine led the chemist and chemical engineer to play an increasingly dominant role in the production of materials upon which military capability was, and still is, critically dependent. The vital contributions of chemical engineering fall into five epochs from the early provision of plant and processes for cordite and picric acid before World War I to complex plant design for the development of TNT, RDX, plastic propellant, and nitroguanidine. A major role in the founding of the Institution of Chemical Engineers was played by those engaged in explosives manufacture. The importance of adopting modern trends in plant design for explosives is emphasized and the manner in which this is achieved in the United Kingdom is examined.

For 500 years gunpowder served the Armed Forces as a propellant for guns and as an explosive. The art of manufacture was highly developed especially at the Royal Gunpowder Factory, Waltham Abbey in England but its preparation was largely a grinding and mixing operation attended by great hazard (1).

Associated with the start of the Industrial Revolution early in the 19th Century was an intense interest in chemical matters and by 1847 both nitrocellulose (NC) and nitroglycerine (NG) had been prepared. The industrial scene was fertile ground for the exploitation of these materials on a manufacturing scale but the conversion to large-scale production led to tragedies. In England a disastrous NC plant explosion at Faversham in 1847 killed 21 people and similar events occurred at Le Bouchet and Vincennes in France (2). Nobel set out to develop Sobrero's invention of NG but his work was set back temporarily by the violent explosion at the Heleneborg works in Sweden in which his brother was killed.

0-8412-0512-4/80/33-190-367$06.25/1
Published 1980 American Chemical Society

These events delayed but did not prevent the adoption of NC and NG. Their great potential was realized and more science and ingenuity was brought to bear on their manufacture. Nobel expanded his activities to other European locations including Ardeer in Scotland. The British and French government supported work on the new propellants and explosives. In 1884 Vieille, a French engineer, invented Poudre B, a NC propellant which was adopted for the Lebel rifle and about this time work commenced on NG and NC manufacture in the Royal Gunpowder Factory, Waltham Abbey. During this period the vital importance of processing, especially stabilization, and its influence on the quality of NC became known. Meanwhile, pursuing applications of NG in 1887 Nobel invented and produced ballistite, a mixture of 12.6% N NC with 50–60% NG. The British government set up a committee which included such eminent scientists as Dewar, and, under Sir Frederick Abel's leadership, in 1889, the committee formulated the idea of cordite. This also combined NC with NG but used a more highly nitrated cellulose than Nobel's ballistite, with 13.1% N (guncotton) to give a more powerful propellant which was extruded in the form of cords. The name cordite, derived from the extrusion process, distinguished it from the ballistite made by Nobel at Ardeer.

Ample supplies of acids and other chemicals became necessary and, in company with the expansion of work on NC and NG, chemical manufacture became a rapid-growth industry; as the size of the acid plants increased it became important to design equipment of all kinds far more carefully than in earlier times. It was evident that a new era of large-scale fine chemical manufacture had begun. Even in 1895 the Royal Gunpowder Factory at Waltham Abbey was producing 500 tons/year of cordite and at a private cordite plant run by Kynoch Ltd at Arklow in Ireland 300 people were employed making a similar quantity. The new materials for the armed forces depended on the availability of chemicals and on the design by competent chemical engineers of safe and efficient plants.

This presentation examines how the chemical engineer translated proposals into manufacturing plants and so enabled the armed forces to obtain their essential supplies of explosives and propellants on a massive industrial scale. It is convenient to divide the history of the industry into five epochs broadly defined as: the period leading up to World War I and including the years of the Boer War; the 1914–1918 War; the post-war years up to 1935 when rearmament began; from 1936 to the end of World War II and the post World War II period up to the present day. The five epochs and the products prepared during them are outlined in Figure 1.

The first period was largely a NC era and towards the end saw the introduction of trinitrotoluene (TNT). The second period saw the intense activity produced by war conditions and shows how vitally impor-

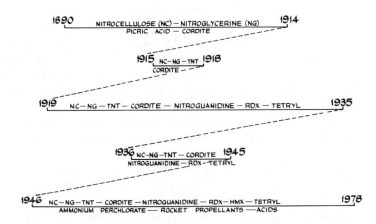

Figure 1. The five epochs in explosives and propellants manufacture

tant the provision of high explosives and propellants (especially TNT and cordite) was.   After the war the tempo relaxed and until 1935 there was only a small effort devoted to the manufacturing side.   But the change occurred in 1935 when the possibility of war again loomed over Europe. For the ten years from 1936 there was again an acceleration of effort, and plant design for the production of new materials as well as the established explosives became important.   Cyclotrimethylene–trinitramine (RDX) and nitroguanidine became vital needs of the services.   After the 1945 armistice until the present day the technology of defense has undergone extensive change.   The requirements have become more exacting and more chemical engineering expertise has been needed; nevertheless, for various reasons there has been an acute shortage of chemical engineering expertise within government establishments so that insufficient advantage has been taken of the considerable advances in the application of science to process design.   The few chemical engineers available have worked only in restricted fields and have relied on the importation of expertise from private industry to compensate for the small numbers.

1890–1914: The Nitrocellulose and Nitroglycerine Era

In 1890 Sir Frederick Abel's committee recommended using cordite for the British Services in place of gunpowder as a propellant and in 1891 its production began at the Royal Gunpowder Factory at Waltham Abbey —this was a somewhat more rapid development than could be hoped for today.

Cordite was 58% NG, 5% mineral jelly, and the remainder was 13.1% N NC; it was fabricated by mixing in acetone as solvent.   The two nitro compounds used the natural raw materials, glycerine and cotton, and required extensive quantities of concentrated nitric and sulfuric acids, which in turn needed the recovery or disposal of the spent acids after nitration.   The acetone was obtained by the destructive distillation of wood via the thermal decomposition of calcium acetate.   The production of cordite required the support of a considerable chemical factory.   Even in 1872 Waltham Abbey was producing 250 tons/year of guncotton (3) which would have required the handling of several tons per week of acids.   By the time the expansion of the new works there, at the Quinton Hill site, had been completed for cordite manufacture in 1905 the capacity had been increased fivefold.

For one of the last of his 15 worldwide sites to manufacture NG Alfred Nobel had chosen a suitably remote location in Scotland at Ardeer where production started in 1873.   P. A. Liedbeck became Nobel's chief chemist and he both designed and operated the plants, the products of which were principally dynamite and blasting gelatin for mining purposes.   There is an apocryphal story that the miners at nearby Kilwinning highly prized Nobel's NG because when burned in their lamps it gave a far better light!   Nobel's exploitation of NG was towards the safer use of the material and it was perhaps a natural extension of blasting gelatin to make a product that did not explode but gave controlled burning.   Nobel mixed larger quantities of 12.6% N NC with NG and invented ballistite (4).   Lundholm and Sayers working in the Ardeer chemical research department devised a safe procedure for mixing large amounts of the soluble 12.6% N NC with 50% dry weight of NG by using an aqueous slurry process.   The slurry was dewatered to form a paste which then was treated in heated rolls to consolidate it into a sheet.   The sheet was chopped into small squares to form the propellant ballistite which is suitable for small arms.   This processing has been retained up to the present day in a modified form as a basis for mixing many propellants. It is worth pondering on the effect of the prompt patenting action taken by Nobel to forestall any imitators.   His patent may have led Abel to avoid using the safer wet-mixing process and use the highly dangerous dry guncotton (13.1% N) to avoid contravening the patent in government factories.   Abel found that the guncotton mixed with NG formed a stiff, horny material.   The dry guncotton was placed in lead trays and NG was poured onto the fibrous powder.   To convert the mixture to a workable dough suitable for extrusion acetone and mineral jelly were added in a Werner Pfliederer and Perkins mixer.   The dough was extruded into cord from a hydraulic press to form Mark I cordite.   The mineral jelly was added originally with the intention of lubricating the gun barrel but in fact it acted as a diluent and stabilizer.   Aromatic components absorbed nitrous gases thus contributing to the safe storage of British ammunition.

Abel's method of nitrating cotton was not elegant.   The plant was a series of water-cooled cast-iron pots containing mixed acid into which cotton hanks were dipped.   J. M. Thomson from Ardeer and his brother came to work at Waltham Abbey and in 1905 devised the displacement process.   William MacNab describes the operation:

"In this beautiful process the acid is contained in a series of shallow circular earthenware pans provided with perforated false bottoms.   The charge of cotton is forced under the acid and sectional perforated earthenware plates laid on top to keep the cotton down and a thin layer of water is run on top of the acid to prevent the escape of fumes.   When nitration is finished the spent acid is allowed to run off slowly from the bottom while water is distributed at the same rate over the surface and displaces almost completely the acid from the nitrocotton" (5).

MacNab points out the need for extensive chemical engineering knowledge to enable this process to be effected since there is much associated equipment in the way of pipes, pumps, and valves to control correctly the corrosive liquids.   At this time Sir Frederic L. Nathan had been Superintendent at the Royal Gunpowder Factory, Waltham Abbey for five years.   He was the first of the scientific superintendents and was destined later to participate in the formation of the British Institution of Chemical Engineers.

Four years earlier in 1901 Dr. Robertson (later Sir Robert) and Rintoul devised scrubbing towers for acetone recovery and the payoff for this work was to be seen early in World War I when acetone was in short supply.   Acetone was prepared by the destructive distillation of wood via calcium acetate and in 1915 the United Kingdom was burning 1400 tons of wood each week for the manufacture of this essential commodity. Rintoul's work therefore had an important effect many years after his foresight had provided the process.   Rintoul, as chemist-in-charge of the NG plant at Waltham Abbey was not satisfied with the hazard presented by the stoneware cocks used on the Nobel-designed nitration plant and so he had, with Nathan, devised the displacement nitrator for NG which eliminated the moving part which had contributed so much danger to manufacture in the earlier years.   His design of a cockless batch nitrator enabled 1633-kg batches of NG to be made, in much greater safety than with the old design using stoneware cocks.

It is not altogether surprising to learn that in 1909 both Nathan and Rintoul, the leading explosives technologists of their time, left government service to become General Manager and Chief Chemist, respectively, of the Nobel's Explosives Company at Ardeer, Scotland.   At Ardeer the manufacture of NC and NG were well established and there is every indication that the newcomers were most welcome and set a new course of collaboration between government and industry. This enabled the Ardeer factory to turn rapidly to the manufacture of cordite in 1914 and for other areas of cooperation to be established which persist to the present day.

When considering the developments over this period the words of F. D. Miles in his history of the Nobel Division of Imperial Chemical Industries Ltd are worth quoting.

> *"Nearly always the fundamental development was the work of one enterprising man. It has been the fashion to ascribe the industrial development of the last 150 years to the 'advance of science,' meaning by this, the growth of pure science in the hands of scientists not interested first of all in the applications of their discoveries. To do this is to ignore the fact that industrial developments as often as not owed comparatively little to pure scientific knowledge or academic work. The industrial revolution was the combined work of ingenuity and business ability!"* (6).

To include men like Nathan and Rintoul, the acknowledgment should include the qualities of drive and duty that undoubtedly have motivated many excellent men in government service.

Several other factories were set up for making propellants and Kynoch Ltd of Birmingham operated the largest. Their plant at Arklow in Ireland was designed by an engineer, John Morrison. At this time Kynoch's was a competitor of Nobel's for government contracts to supply cordite and their production helped to supply the British army in 1914. Kynoch's, who eventually was to be absorbed into the ICI Ltd complex, used well designed plants for many operations including solvent manufacture.

Up to the start of World War I in 1914 picric acid, trinitrophenol, had been used as a high-explosive shell filling. It largely had replaced black powder and was termed melinite by the French and lyddite by the British. Trinitrophenol was a relatively easy batch-reaction nitration which had been carried out as a nonexplosive operation in the dyestuff manufacturing industry until a disastrous explosion took place in Manchester when a chemical factory caught fire.

Although picric acid was in use during the years up to 1914, the first stirrings of a revolutionary change were taking place. Sanford (7) noted in 1896 that trinitrotoluol may be prepared by nitrating toluol from coal tar sources and refers to its properties as a high explosive, what seems to be undiscovered at that time was the greater complexity of the nitration and the consequences in terms of plant and process. The need for chemicals involving much more complicated processes greatly changed the scene and underlined the contribution of plant and process design tailored to fit the chemical reactions.

From 1900 onwards some work commenced on TNT in private industry and a batch process was available producing up to 10 tons/week but it was not adopted officially as a high explosive by the U.K. government until 1914.

Picric acid and TNT both require using "boosters" for conveying detonation to the main charge from the detonator and tetryl (2, 4, 6-trinotrophenylmethylnitramine) came into use for this purpose mainly because of the availability of diphenylamine in the dyestuffs industry. The Woolwich Arsenal Research Department had developed a batch process for its preparation by 1910 and it continued to be made by a similar method until World War II.

We see the end of this era as one where both complexity and scale were increasing in the explosives and propellants industry, calling for more sophisticated designs with even greater attention to the economics and safety of the operations.

1914–1918: World War I Era

In 1914 Britain was in an unprepared state for war. H. Levinstein (8) in his Presidential Address to the Institution of Chemical Engineers said that a nation makes war with the whole of its resources including industry. Because industry was not expecting war in 1914 it took an immense effort to mobilize the chemical and engineering capacity of the country to supply the needs of the greatest force the Army and Navy had ever fielded. Shells, guns, and cartridges were needed urgently. Private industry was the only producer of TNT at the rate of only 20 tons/week. Private cordite manufacture amounted to about 200 tons/week and the Royal Gunpowder Factory at Waltham Abbey was making 57 tons/week.

The position was critical but eventually a tremendous expansion of production took place and in a short time many new explosives and propellants plants were designed and built. Waltham Abbey raised its cordite output first to 140 tons and then to 200 tons/week while private industry expanded production to about 1300 tons and the great new government factory at Gretna was making a further 1000 tons/week by 1917. TNT manufacture had to be increased to match the total of 42,000,000 shells produced during the four war years and 230,000 tons of TNT in all were produced by the national explosives factories.

How was the expansion achieved so rapidly? Undoubtedly by the most intensive plant design effort, completely new designs were produced for TNT, and some new designs were produced for propellants. In the case of TNT the problems of scaling up were serious obstacles and chemical engineering contributed largely to the achieved success. Levinstein was at the heart of the action as General Manager of the Dyestuffs

Company in Blackley where the expertise on nitration of aromatics resided. His Presidential Address in 1937 includes a valuable summary of events and he records the assistance rendered by Du Pont in the United States who supplied 31,000 tons of TNT in 1915 to bridge the gap in supplies. Somewhat tardily the government recognized the principal task after 3½ months of war and in November of 1914 set up a powerful committee under Lord Moulton. It included William MacNab who later helped to found the Institution of Chemical Engineers. The government realized that factories had to be built and men of suitable training to design and run them had to be found. Unfortunately, unlike Germany, who had 5000 trained chemists, mostly in industry, Britain had only 500 chemists, mainly teachers. Teaching was reckoned in some circles to be a more honorable occupation than industrial activity for intellectuals. Lord Moulton brought out the importance of the national interest and, to the chagrin of the dyestuff industry, pushed the chemical industry in the direction of TNT. The paramount importance of supplies was recognized eventually by the setting up of the Department of Explosives Supply under Lloyd George who was made Minister of Munitions in June, 1915.

Events in 1914 had included the despatch of Colonel Sir Frederic Nathan to London to assist in the setting up of a large scale TNT plant at Pembrey in Wales. Reader's (9) history of the Imperial Chemical Industries Ltd describes how the expansion of TNT production was assisted by every means. Nobel's provided expertise, Messrs Chance and Hunt, acid manufacturers, built a factory at Oldbury and new sites were sought wherever nitric acid was available. An important feature was the mononitration of a mixture of 55% toluene and 45% petrol—Borneol petroleum. After the mononitration the MNT was separated from the petrol by distillation and used for the DNT and TNT stages. The Oldbury plant is believed to have been the first effective continuous process and it operated on a countercurrent principle. MNT was fed in at one end of a chain of nitrators each of which was fitted with a separator. After passing through 14 such vessels at temperatures ranging from 40° to 100°C, TNT was run off at the far end. This application of a new chemical engineering technique marked a major change in the industry to a higher level of complexity and hazard requiring new scientific ideas. By 1916, 16 private firms and 12 government factories were producing 1000 tons/week. One notable achievement of chemical engineers and others which enabled this expansion to take place was the design and erection of oleum plants which were essential for the nitration. German supplies had been relied upon before 1914 and sulfur burning units had to be put up in record time.

It was the same story of rapid expansion in the field of propellants. In the case of cordite no new processes were devised but better handling of the subsidiary processes became necessary. Acetone was a bottleneck and it was Rintoul's recovery process which helped to extend the utilization of available supplies. The situation became so acute by 1917 that a new form of cordite called RDB was adopted which, with lower nitrogen NC, could be gelatinized or colloided with ether and alcohol. Another method of overcoming the shortage was devised by Weizmann at the government factory at Holton Heath using a fermentation method to produce acetone from starch. Earlier in 1915 fermentation methods of making glycerine had been started with the accompanying need for designing countercurrent extraction plants using isopropyl alcohol as solvent.

The rate of growth of complexity was increasing rapidly and a good example was the need to expand the guncotton drying capacity without increasing the hazard. A chemical engineer, K. B. Quinan, at Waltham Abbey in 1916 designed drying stoves that are still in existence and use today. The principle of preparing warm, dry, filtered air and forcing it through fluidized beds of guncotton restrained from escaping by covers of special fine cloth was a brilliant solution well ahead of its time. It probably increased the output of the cordite factory by 50% by reducing the throughput time.

Ammonium nitrate production became a major factor towards the end of World War I once the merits of amatol, an 80:20 mixture of ammonium nitrate and TNT, had been accepted by the Services as the most expedient method of extending the limited TNT supplies. Nitrogen fixation for fertilizers was the objective of the Haber process for ammonia and the successful chemical engineering of this process had made Germany independent of the Chile nitrate supplies by 1918. The plant at Oppau made 25 tons/day of ammonia. The British requirement for ammonium nitrate was 4000 tons/week by the end of the war, which was met largely with gas works ammonia, Chile nitrate, and some manufacture via a Birkland–Eyre oxidation furnace and calcium nitrate.

The closure of the major plants at the beginning of 1919 gave time to think about the requirements of the chemical industry and chemical engineering became a major topic of discussion including a lecture at the Faraday Society (10) and of course it was given close attention in the explosives industry itself.

1919–1935: The Inter-War Years

Not unnaturally the scene in explosives after the war was one of contraction of manufacturing capacity. The Waltham Abbey plant practically closed down and, of other government plants, only the huge Gretna plant was operational. Attention turned to experimental and investigational work in the government area and to civilian uses of explosives in mining and associated activities for the private industry plants.

Government sought to avoid the lack of preparedness demonstrated in 1914 by concentrating teams of scientists onto the problems of the day. Three important fields were recognized. TNT plants needed improvement to avoid the explosion hazard, guns needed flashless propellants with reduced barrel erosion, and if possible a new high explosive was required. The Research Department at Woolwich Arsenal looked at the last two problems and made excellent progress; in fact their flashless solution is still so important today that actions arising will be described in some detail in both this and subsequent sections. TNT manufacturing methods were reviewed and new plants were devised. A new high explosive was selected for development but in the manufacture of the propellant ingredients, NG and NC, little progress was made—effort being concentrated on interesting chemical investigations.

The development of a flashless, low-erosion propellant neatly demonstrates the substantial contribution made by a small team of dedicated men in Britain to the support of the Armed Forces in World War II and to events even up to the present day. When Marshall's *Explosives (11)* was published in 1917 the reasons for gun erosion and for gun flash were known—it records that erosion was related to combustion temperature. Sir Hiriam Maxim in 1901 advocated the addition of carbon black as a flash suppressant and to reduce erosion. Sir George Beilby knew in 1904 that nitroguanidine was an effective flash suppressant. Vieille found in 1912 that nitroguanidine incorporated into cordite would reduce gun erosion by a factor of two and by a factor of three in ballistite; but he also found that it made the propellant brittle and ruined the ballistics. In 1914 William MacNab suggested using nitroguanidine combined with a tetranitro compound bound with rubber to yield a flashless propellant but the idea was not pursued even though it anticipated by some 30 years the composite propellants used today in rocket motors. It is interesting to note that MacNab became President of the Institution of Chemical Engineers in 1934 and he is commemorated by the MacNab Medal awarded annually for the best Home Paper.

In 1921 Sir Robert Robertson, when he was chemist-in-charge of the Woolwich Arsenal Laboratory, decided to investigate nitroguanidine thoroughly and appointed J. N. Pring to carry out the task. He thus set in motion a tremendous train of events which was to extend over 60 years up to the present day. Nitroguanidine when suggested by Beilby was rejected on cost grounds but in 1921 it was acceptable because advances in chemical manufacture had lowered the price of calcium cyanamide and therefore nitroguanidine. Pring rapidly demonstrated the effectiveness of nitroguanidine as a flash suppressor for small-caliber guns and showed that particle subdivision was the clue to its utilization in cordite. The preparation on a manufacturing scale of the fine material was a difficult chemical engineering task which took some years to achieve.

By 1925 trials had been carried out in 6-in. breech-loading guns with compositions containing 10–70% of the nitroguanidine prepared in a coarse form in an edge-runner grinding mill.   The trials were successful so more attention was focused on the method of manufacture.   In 1925 the factory at Waltham Abbey began to produce nitroguanidine from calcium cyanamide.   The aqueous suspension of calcium cyanamide was treated with carbon dioxide to give cyanamide which was then reacted in an autoclave to yield guanidine nitrate.   The latter was converted to nitroguanidine by treatment with sulfuric acid and the crude product, after filtration, was recrystallized from hot water and dried.   The product after milling in an edge-runner mill was termed petrolite.

At this stage Pring encountered difficulties with ballistic regularity and identified the problem still as one of particle size; therefore efforts were directed towards alternative methods of obtaining finer material. Holden at Woolwich designed a somewhat cumbersome but effective shock-cooling technique using a rotating gunmetal cylinder cooled on the outside with brine acting internally as a crystallizer.   Improvements in both the process and the crystallizer were effected by the chemical engineers at the Royal Gunpowder Factory in Waltham Abbey who polymerized the unstable cyanamide to dicyandiamide thus increasing the yield.   They also introduced a new vortex crystallizer—more suitable for production—which gave the fine crystals required; the product was re-named picrite.   By 1931 several tons per week were being produced and further plants were designed for the Admiralty factory at Holton Heath and Nobel's at Ardeer.   Nobel's, by this time had passed through a transition of being first part of Explosives Trades Ltd and then being absorbed with Brunner Mond and British Dyestuffs into the giant Imperial Chemical Industries complex.

With the manufacture of picrite well established in the early 1930's, Pring continued to promote the case for its use; the events leading up to World War II and the eventual adoption of this important ingredient of service propellant by 1938 will be described below.

The work of the Research Department at Woolwich on high explosives established that cyclotrimethylenetrinitramine, also known as hexogen or cyclonite and termed RDX, was likely to be the most powerful explosive available and only a marginally superior substitute could be foreseen.   This significant prediction has stood the test of time but, more importantly, being made at an early stage it allowed attention to be given to the critical problem of the large-scale production of RDX. Work started in 1922 and three names—Simmons, Forster, and Bowden (two of them chemical engineers)—were associated with the development (12). The process involves the manufacture of hexamine, its nitration with 98% nitric acid, and an ingenious continuous decomposition of the by-products after diluting with water.   The adoption of the novel principle of continuous operation was crucial to the success of the enterprise. De-composition of the by-products produced large quantities of nitrogen peroxide that needed collection, conversion to nitric acid, and recycling. By 1933 a small unit producing 75 lb/hr had been designed and was operating at Woolwich. The further development of this vital material belongs to the subsequent era of 1936 to 1945.

Some effort was directed towards the further development of the TNT process. According to Knapman (3) the factory at Waltham Abbey was allowed to become almost derelict but it was decided to locate some work there on the improvement of the Oldbury continuous TNT plant in the development of which William MacNab (13) had played an important part in 1916. The Chemical Engineering group of the Ordnance Factories examined the whole problem of ensuring adequate supplies of explosives and propellants in the event of war and it was clear that although the Oldbury plant was an outstanding development, considerable modification would be required to produce high-set-point TNT. A new pilot plant was constructed at Waltham Abbey which introduced screw lifts for the nitrobody between nitrators, a reduction in the size of the acid–nitrobody separators, and an increase in the number of nitrators. The plant was the work of Bowden and Smith (14). Their patent is remarkable in that it retains the secret that it is actually applicable to TNT. The pilot plant produced Grade 1 TNT continuously at the rate of 2 tons/week and was probably a world leader in this respect.

The NG manufacturing process also was being made continuous but the work was taking place outside the United Kingdom. The Schmid continuous process appeared in 1927 (15) and eventually was first installed in a British government factory at Holton Heath in 1937.

NC manufacturing in the United Kingdom was considered adequate and little attention was given to it. The mechanical nitration processes developed by Du Pont in 1926 were not favored. Developments did occur in the manufacture of solventless cordite and a feature of this at the RNCF Holton Heath was the overdue adoption of the principle of wet mixing of NC and NG as originally employed by Nobel. The drying of guncotton with its attendant risks therefore was eventually abandoned by the government factories at the end of World War II.

1936–1945: The Hitler War Period

With the gathering of the war clouds in the mid-1930's, plans were put in hand for a comprehensive scheme of rearmament by sea, land, and air. Ordnance Factories were planned for erection in various parts of the country, the necessary labor recruited and a comprehensive production plan formulated. By 1939 23 Royal Ordnance Factories (ROFs) had been planned and officially approved. By the time war came, seven of these were in actual operation, four were engineering factories, two, Holton Heath and Irvine in Ayrshire, were explosives factories, and the only filling factory ready for action at the outbreak of war was Hereford. The factories at Waltham Abbey, Enfield and Woolwich Arsenal had been maintained at readiness during the inter-war period.

Thus ten ROFs out of the eventual 44, together with a limited number of specialist firms, were ready to equip the Navy, Army, and Air Force when war broke out in September, 1939. The location of these factories within the limited confines of the British Isles in the midst of the competing claims of other vital manufactures was a classic chemical engineering exercise in itself, and was carried out by the newly appointed directors of the various manufacturing groups—explosives, engineering, and filling. Propellant and high-explosives manufacturing were allocated to separate factories; likewise were the operations of filling and engineering. The new ROFs started with a considerable inheritance of expertise in production but with so many new establishments to be staffed and managed, experienced staff from the parent factories augmented by specialist staff from private firms with explosives-manufacturing experience could only be spread around very thinly. It says much about the ability and dedication of these men that they were able to pass on their technical and managerial skills so rapidly and effectively to the new recruits coming from the universities, colleges, and other industries in the critical stages of the build-up to full output. In the case of picrite the development assumed a new level of intensity. The need for fine crystals had been established but the method of producing them in quantity had not. The installation of improved pilot plants at Holton Heath and Ardeer had shown that picrite propellants were a viable and essential contribution to the Armed Forces. By 1938 the British Army and Navy were convinced of the advantages of a flashless propellant with lowered erosivity which could be produced in quantity. The production of fine crystals remained the limiting factor until after war had commenced. The original crystallizer produced only 15 lb/hr from three sprays. The Holton Heath design with six sprays had twice the output giving approximately 8 to 15 tons/week.

In 1940 the British Purchasing Commission of the Ministry of Supply arranged with American Cyanamid for the erection of a 1625-ton/month plant at Welland, Canada. The application of American chemical engineering design teams made this a successful enterprise which was in full production by 1941 and its output by 1943 had been raised to 2300 tons/month. The production of cordite N using this picrite was investigated in the United States and Canada and led to the adoption of picrite propellants by the United States in 1944.

Although the crystallization process was studied extensively and at Welland trials were carried out to spray hot solutions upwards into a hot-air stream, the successful process was evolved at Ardeer.   Jackson and Miles demonstrated that picrite could be sprayed from superheated water solutions into a brine-cooled crystallizer to give the correct product.   Shortly afterwards in the United Kingdom the development of this idea, combined with vacuum cooling, led to the modern technique still used today.   This chemical engineering design is the key to modern gun propellants and the difficulty of the task can be judged from the long period taken to solve the problem.   A shorter time than 1925 to 1942 might have been needed if, during those years, more chemical engineers had been employed in government service.

The story of picrite does not end here.   As will be seen later, prompted by the realization of its importance, further work was undertaken in the United Kingdom to ensure the economics and adequate supply of this vital chemical in 1950.

RDX was expanded in production from 75 lb/hr to 225 lb/hr in 1939 by the installation of a double-size Woolwich unit at Waltham Abbey.   The scale-up was by the unadventurous method of putting in two lines instead of one.   The considerable problem of acid recovery and recycling had not been worked out so that there were tremendous pressures on the government factories when the need for production arose.   A chemical engineering approach in the pre-war years would have reduced this difficulty.   There was a collaborative effort between the Research Department and the government factories to design a 90 ton/week plant (1200 lb/hr) for installation at Bridgwater in Somerset and Simmons, Forster, and Bowden (12) described the details.

Scaling-up of the nitrator resulted in a fairly inefficient multiple-stage cascade reactor with severe cooling requirements.   The solid separation used a Dorr–Oliver raking classifier in a novel way.   Considering the extremely restricted time and effort available and the unfamiliarity of those engaged with either chemical engineering, or the hazards of explosives in the case of plant contractors, an excellent result was achieved in the time scale required, but with hindsight effort should have been expended years earlier.   This was less true of the acid, hexamine, and fume recovery plant associated with RDX manufacture.   These were obtained from specialist plant manufacturers.   Bamag Ltd dealt with ammonia oxidation for nitric acid manufacture and nitrous fume recovery while ICI Ltd (Fertilizer and Synthetic Products Division) provided plants on the hexamine side.

During the years 1941–1945 improvements were effected in the nitration plant by careful analysis of the operation by chemical engineering analysis.   Denbigh, Bransom, and others (16, 17) worked out the behavior of the nitrator and optimized the continuous-process operating conditions. The control of the hexamine–nitric acid ratio is fundamental to the economics of the process.   An output of 90 tons/week of RDX requires 405 tons of sulfuric acid to be concentrated daily in 20 Bamag pot stills since this is required to prepare the pure nitric acid.   The alternative Bachman process developed in the United States reduces this problem but has its own difficulties associated with the acetic anhydride cycle.

The situation on TNT production was more satisfactory than in World War I.   The 25 ton/week unit using the Bowden–Smith process incorporating screw lifts for the nitrobody and special separators for easy cleaning was erected in 1936 at Irvine.   After some delays caused by a serious fire it was used in 1939 both as the production unit and as a prototype for 24 units set up in four factories at Drigg, Irvine, Pembrey, and Sellafield. These produced 324,000 tons of TNT during the war and were augmented by batch production of 93,700 tons at the agency factories—Allen, Ardeer, Girvan, and Powfoot.   No major improvements were possible during the war years.

Tetryl production was essential to the war effort and a continous plant was set up in 1941 at ROF Pembrey which satisfied the needs for all of the stores.   Improvements were effected in the control of particle size by "flash" graining which in turn enabled a more reproducible "exploder" pellet to be produced.

In the propellant field new chemical engineering designs were being evolved to improve economics and safety; safety was particularly important in the case of NG which has had the dubious record of at least one major explosion per year since manufacture started (2).   The Schmid continuous process (15), the first successful continuous process which appeared in 1927, was installed at Holton Heath and was adopted also for the Ranskill plant.

The Schmid process embodied some radical improvements on the former plants in that it used mechanical stirring in a calandria-cooled nitrator, an inclined baffled separator of novel design, and a cocurrent washing system composed of glass-sectioned columns through which the NG was passed by jet pumps or eductors.   Air was introduced simultaneously to aid dispersion and improve contacting (18).   Although the holdup of detonable NG was relatively much less than in batch units of similar throughput, it was still considerable and in a 900 kg/hr Schmid nitration plant it normally would amount to some 800 kg.

In 1935 Dr. Mario Biazzi introduced a further method of continuous manufacture based on the use of stirred vessels for the washing stages and characteristic onion-shaped separators with tangential entries to impart a gentle rotary movement to encourage coalescence and separation (19, 20).   Most important perhaps was that this particular arrangement was amenable to remote control from a protected position.   It was with the introduction of the Biazzi plants that lead was replaced by stainless steel for reactors and subsequent stages.

Little alteration took place in guncotton NC manufacture although the process patented by J. R. Du Pont in 1922 was exploited elsewhere.

The United Kingdom adhered to the displacement method until well after the end of the Second World War. However, the Royal Naval Cordite Factory in association with the Research Department at Wool-which developed a method of making 12.2% N "soluble" NC from wood pulp paper in the same displacement equipment as that used for the nitration of cotton linters to guncotton (13.1%N). The Abel boiling and pulping procedure after nitration was retained with effort directed towards improving the efficiency of the stabilizing operation rather than varying the basic process.

A system of continuous boiling of NC during passage through a long serpentine pipe was developed by Milliken (21) in 1930 and this has been re-examined in recent years; however it was not adopted in the U.K. factories because of doubts about the reliability of continuous systems with such unstable materials as NC. The 12.2% N NC was used for solventless cordite manufacture because of its ease of working when mixed with NG. It was during this work that a major change took place in the mixing operation; it was derived from Nobel's original method but mechanized to suit the needs of bulk production. NC was handled exclusively as an aqueous slurry adjusted to a known concentration and NG was added through a spray rose in an open-spiral-tracked trough. The NC–NG mix was dewatered on a paper-makers' sheeting table. It was not until 1943 that this method was adopted for the solvent cordite process in the government factories, probably following the explosions which occurred while handling dry guncotton at Bishopton in 1941 and 1943.

The mixing and finishing processes for cordite remained virtually the same with the same type of machines as those used in 1900.

It is appropriate here to give some attention to the supply of nitric and sulfuric acids. Most of the problems of acid supply involved chemical engineering and their solution lay in improved plant and process design. The First World War gave a considerable impetus to the development of all types of plant for the production and concentration but by the early 1920's the momentum in Britain seemed simply to die away while in the United States and on the continent endeavors continued without interruption (22). The concentration of dilute nitric acid by dehydrating agents was pursued vigorously in the United States and on the continent while the use of nitrogen tetroxide and oxygen was researched mainly in France and Germany (23).

The greatest item in the cost of concentrating nitric acid using sulfuric acid as the dehydrating agent is the cost of reconcentrating the sulfuric acid. The alternative nitrogen tetroxide process therefore was of great interest and led to the choice of the "Hoko" process for ROF Irvine in which dilute nitric acid, nitrogen tetroxide, and oxygen were autoclaved together at a pressure of 50 atm.

Oleum was produced in nearly all of the factories in standard commercial sulfur-burning units while concentration of sulfuric acid was carried out in a variety of plants including pot stills, drum concentrators, or Gaillard towers. The vast experience accumulated over the years of the Second World War has been collected together in a summary volume under the auspices of the Acid Plant Design Committee (24).

The manufacture of picrite posed an extremely difficult problem in sulfuric-acid recovery for the substantial amount of spent acid produced contained only about 18% sulfuric acid. Submerged combustion was applied to this duty at ROF Bishopton and while this procedure functioned well as far as actual concentration was concerned, corrosion of the equipment by the nitrate ion present was so severe that the process could only be run intermittently and eventually had to be abandoned leaving the acid to the dealt with by neutralization with limestone followed by prolonged drainage in settling lagoons.

It now is considered that three- or four-stage vacuum concentration in a glass-lined tantalum plant is probably the best long-term solution but even here there are problems in the disposal of the ammonium sulfate which is unavoidably produced in the course of concentration.

Pot stills are inefficient owing to the high working temperature and the poor heat transfer through the pots which demand a correspondingly higher furnace temperature (900°–1000°C) with a high loss of heat by radiation and in the furnace gases.

Drum concentrators utilize direct heat from the furnace gases which pass through the acid and are consequently more efficient but suffer from the disadvantage that owing to the high acid temperature in the concentrated acid stage much acid spray is evolved which requires mist precipitators to remove and which moreover are costly to maintain.

As a postscript to this important period it is worth noting some remarks made during his address in 1965 by the eminent chemical engineer, the late Dr. A. J. V. Underwood (25). He refers to the policy pursued in 1943 by Lord McGowan, then Chairman of ICI Ltd. This policy boiled down to proposing that large organizations ideally should employ only chemist and engineer teams and, accordingly, before 1958 the ICI Billingham Division employed just ten chemical engineers. Underwood goes on to say that it is only fair to add that some years ago ICI Ltd ceased to hold this view and by 1964 the Division, after splitting into two, employed 100. To some degree this attitude and the subsequent change has been reflected in other areas including government and as will be seen later, the post-1945 years started off with a considerable thrust towards injecting more chemical engineering expertise into the government explosives industry. Although only partially achieved, it has had a small measure of success.

1946–1978: Technological Change

The final epoch, following the end of World War II, has seen rapid changes in the methods adopted in chemical industry for processing, major changes in the requirements of the Services, and above all the alteration of the Defense picture by the advent of atomic weapons.

Continuous operation with more sophisticated control has been a feature of all manufacturing processes and one can no longer deal with the complex problems in the ad hoc manner of the past. Rockets and their attendant demand for special propellants have become a significant factor in the Armed Forces while social changes have modified considerably attitudes to safety and have altered the economics of production.

As before, in 1919, the run down of production from wartime levels had to take place in 1946. There were differences however. Research departments recognized the need to emphasize manufacture and there was an appreciation of the role of chemical engineers throughout industry. At Waltham Abbey the production factory was closed but a chemical engineer, Dr. F. J. Wilkins, was made Chief Superintendent of the Chemical Research and Development Department and he made a break with the old industrial chemist–engineer system adopted by McGowan by putting together a Chemical Engineering and Engineering Service Group under a chemical engineer and by recruiting suitably qualified staff, both new graduates in chemical engineering and experienced plant chemists. There was a very real attempt to avoid the mistakes of the past and this was to bear fruit in promoting the immediate further development of the picrite work. Unfortunately the excellent start was short lived and after a few years following the appointment of more academically oriented staff some of the benefits began to fade as less attention was given to the needs of production. There were, however, areas where the new approach was sustained. A Scientific Advisory Council Committee was set up on which some of the country's leading chemical engineers were invited to serve. Professor F. H. Garner and Dr. A. J. V. Underwood were among the well-known names on the Committee. Eventually, however, after much valuable work, the Committee was disbanded because of lack of support for the chemical engineering projects with which it was concerned.

For picrite development the new approach sustained an activity which was to continue for nearly ten years. It was conceded that it was required in large quantities and several chemical routes were explored in the laboratory followed by pilot-plant design and trials. Nitrolim or crude cyanamide was retained as the starting material but new conditions were worked out to provide for higher yields based on fixed nitrogen than had been obtained previously. The pilot planting of this process to provide a 100 lb/hr continuous unit, operating virtually automatically, marked a substantial advance in plant and process design. The process was called the British Aqueous Fusion (BAF) process and the Waltham Abbey design was scaled up and copied in several locations. A team of a few chemical engineers enabled this to be done and although other processes have been tested on pilot scale the BAF process remains a firm leader.

The crude calcium cyanamide is reacted directly at just above 120°C with concentrated ammonium nitrate to yield guanidine nitrate contaminated with soluble and insoluble impurities. The beauty of the process is the manner in which guanidine nitrate is recovered from this mixture without loss of fixed nitrogen. The product then is treated with either sulfuric acid or oleum in a special plant to form nitroguanidine. The importance of nitroguanidine has fluctuated considerably but it is certain that it now holds an important position in the propellant field as long as the potential threat of conflict remains and conventional weapons are retained to guard the peace. Nitroguanidine has taken its place alongside NG and NC to be joined but not displaced by new materials.

New technology has had its impact since 1946 on both NC and NG production and both are made now in continuous plant with a very high content of chemical engineering design although here the expertise has been derived from private industry with notable contributions from ICI Ltd and other specialist companies.

The problem with all continuously operating explosives plants, particularly those concerned with NG is that of securing and maintaining an effective detonation break between the various processing stages. Schmid solved the problem by using eductors to send the NG through connecting pipelines in the form of a water dispersion. Experience has since shown that the effectiveness of this method is dependent on the material of construction of the pipelines and on their diameter in relation to the strength of the emulsion or dispersion with adequate turbulence over sufficiently long lengths to ensure interruption of detonation, should an explosion take place in some stage of the process. Before these requirements were appreciated fully some multiple incidents occurred which otherwise might have been confined to a single plant or building.

Pursuing means of reducing the holdup of NG at the nitration stage still further, Nilssen and Brunnberg of the Nitroglycerin Aktiebolaget, Gyttorp (26) in 1952 developed a NG process in which conventional processing vessels were dispensed with altogether, a jet pump or eductor serving as the nitrator, the mixed acid being refrigerated before passing to the eductor jet (27). This process also took the bold step of applying DeLaval centrifugal separators, suitably modified, to the separation of NG from the spent acid. The holdup of such separators is of the order of 4.5 kg compared with 50 kg or so for one of the most efficient static separators of equivalent throughput.

Line-washing processes have been developed both at the government factory at Bishopton and at the Nobel's Explosive Company designed to reduce the hazards attendant on the washing processes when these take place in bulk in a single building.

NC manufacture by mechanical nitration gradually became accepted, eventually overcoming the innate conservatism of the system controlling the acceptance of new ideas for manufacture, so that by 1956 a unit was installed, and in due course it was shown that the whole of the nitration of NC could be effected in continuous nitrators of the Du Pont type.

In the 1950 period there was a new area of activity which the explosives industry was called upon to take up. Rocket propulsion had come to stay and the competing merits of liquid and solid systems were to become the center of controversy. While at the time the liquid fuels like hydrazine seemed to have some advantage, the smaller nonspace rocket motor seems now to be settled in the solids field and an extensive development of solid propellants has taken place. The British (28) invented plastic propellant, a composite of ammonium perchlorate and a polyisobutene binder, while the cast propellant system was developed in the United States together with the curable rubber system.

In the United Kingdom the plastic propellant required the manufacture and recrystallization of ammonium perchlorate and while the scale was small compared with the U.S. effort in this field a substantial expertise was accumulated at Waltham Abbey. Crystallizing and milling equipment was devised to give particle-size ranges of 2 to 400 μ since this new brand of propellants showed remarkable sensitivity to particle-size variation. The United Kingdom also imported the expertise from Hercules Corporation in the United States for making the double-base charges cast in situ in a rocket motor case. This process was tried out first at Waltham Abbey under government control and when the technique had been worked out for U.K. conditions staff were sent from Imperial Metal Industries Ltd (at the time a division of ICI Ltd) to be trained in the casting process. Production was transferred then to the Summerfield Research Station and operated by IMI as an agency factory. Although there was little chemical engineering in the casting operation, the manufacture of the special casting powders utilized all of the knowhow and equipment used for cordite and had to be made either at ICI Ltd Nobel Division or, if for research, at Waltham Abbey. At the latter establishment studies are still in progress to improve upon the older mixing techniques and to introduce new methods of controlling the quality of the ballistic modifiers used in these propellants. The exceptional advantages of this cast double-base propellant lie in its smokelessness and increased safety because of the unique ballistic properties. The U.K. plastic propellant does not show these benefits but it has the advantage of cheapness and ease of manufacture.

Both cast and plastic propellants find extensive application for the Services. In both cases the production of the propellants has called for a completely new range of chemical manufacture and in most of these the unit processes of comminution and crystallization rank high in the control of the product.

The manufacturing methods for gun propellants have continued to be largely modifications of existing methods devised in the early cordite days with the attendant disadvantages of high cost and poor reproducibility. Much effort has been devoted to rocket propellant development, no doubt with the expectation of rockets displacing guns in the future although present trends seem to disprove this.   One recent development in gun ammunition, probably the most interesting in the 30 years since the war, has been the combustible cartridge case, developed at Waltham Abbey and now in quantity production.   The original proposals were made by G. I. Nadel *(29)* but R. A. Wallace, a chemical engineer at Waltham Abbey, led the way to a production process.   The operation consists of felting a suitable case from a slurry of NC, kraft fiber, and synthetic rubber latex.   Their manufacture has been taken up by a number of countries.

In the high-explosives field there have been some significant developments.   Tetryl is produced by a continuous process.   RDX manufacture has been improved marginally while a number of other nitrobodies have been made on a pilot-plant scale.

The major changes in the United Kingdom have been in connection with the manufacture of TNT although naturally the requirement for this material in peacetime is negligibly small.

Despite the simplifications and improvements that had taken place, the Bowden and Smith plant was still fairly complicated and maintenance was heavy.   The dwell time required was relatively high so that the output was lower in relation to the plant size.   The main advantage was that the process did not require the use of oleum and therefore could operate on a closed cycle.   The advent of the sulfuric acid cracking process in Germany during World War II however opened up the possibility of using oleum with its economic advantages while still allowing the acid cycle to be closed.   With the end of the war, work on this aspect of the process was discontinued until the early 1950's when the Chemical Engineering Committee of the Scientific Advisory Council reopened the subject and Waltham Abbey demonstrated the feasibility of a three-stage countercurrent process using either 20% oleum or the concentrated oil of vitriol (COV) and offering three times the output for the same volumetric capacity.

About the same time a novel liquid–liquid or solid–liquid countercurrent contacting system was being developed by W. H. Morris *(30, 31)* at ROF Pembrey who proposed its use for the manufacture of TNT as one of a number of applications (including the manufacture of antibodies and other solvent-extraction applications and ion exchange).cThe equipment consists essentially of a rectangular tank divided along its length by cross partitions to form mixer sections agitated by gate paddle stirrers alternating with an agitated transfer section and with a phase-separating section at either end.

The dispersion in the mixer sections is swept under the dividing partitions into the adjacent transfer sections. If the dispersed phase is the less dense of the two its droplets rise up through the transfer sections against the down-coming flow of continuous phase after which they pass to the next mixer section where the process is repeated. Unlike a mixer–settler the ROF contactor does not depend for its operation on the coalescence of the dispersed phase in separating compartments and in its application to the nitration stage of the TNT process nitrobody separates out only at the product end where a smaller unstirred compartment operates as a true separator.

In the remainder of the nitration vessel the nitrobody exists as a dispersion in the spent acid. This may not carry the advantage it was once thought to have, however, and recent American investigations indicate that even a 15% solution of TNT in nitrating acid can be brought to detonation. Be that as it may the ROF contactor is quite clearly a significant advance on the previous plants not only for its simplicity but for its safety, cleanliness, and ease of operation.

It is one of the particular attractions of the Morris contacting system that it is able to function with solid–liquid systems as well as with liquid–liquid systems. Advantage of this is taken in applying it also to the purification stage of the TNT process which no longer handles the TNT in the molten state as in the Holley–Mott system but granulates it and subjects the granules to countercurrent treatment with sodium sulfite solution in the cold with a useful reduction in sulfiting loss.

Acid manufacture continues to be an important activity in making propellants and explosives. Perhaps the most outstanding developmnt in acid manufacture coming from the ROFs was a process for the continuous production of concentrated nitric acid by the action of liquid nitrogen peroxide and air on weak nitric acid under moderately elevated pressures. The process of W. H. Morris was built and run successfully on the pilot-plant scale at Pembrey and has been developed by Humphreys and Glasgow Ltd into a full-scale commercial process known as the "peroxide process" which is both simpler and less costly than the "Hoko" process and operates on a continuous basis under a pressure of only 8 atm compared with the 50 atm required by the latter. It is to be noted that the peroxide process does not impose any penalties with regard to materials of construction beyond those normally met in conventional concentrated nitric acid processes.

The process already has met its commercial test and at least one full-scale unit has been operating since 1967.

Epilogue

The story since 1890 demonstrates a steady progression towards the use of a wider variety of explosives and propellants by the Armed Forces. At the same time the chemical plant to produce them has become more complex and with the introduction of more continuous processes has called for an increasing chemical engineering content in its design.   More pilot-plant work has become essential and scale-up factors have assumed greater importance especially where the size increase from pilot to full scale has been of the order of ten or even 100.   More recent activities in the field of high explosives have shown the need to explore thoroughly all aspects of the plant prior to adoption for large-scale production.   Tetra-nitrocyclotetramethylene tetramine (HMX) was one material examined in this way in recent years and a similar procedure has been adopted for new rocket propellant systems and their ingredients.

Not unexpectedly the industry is hard pressed in times of war to provide adequate supplies for the Services, while in peacetime it is difficult to maintain sufficient effort to keep abreast of world developments and to justify the installation of expensive manufacturing facilities. The lowest cost activity is laboratory research and any movement towards chemical engineering development entails a sharp escalation of cost and an extension of the time scale to several years.   There is clearly a need to compromise between the higher cost of taking a process to pilot- or full-scale development and the unpreparedness which results from excessive economy.

The early chemical engineers—Davis, Mond, Nathan, Quinan, Cullen, and others—enabled laboratory processes to be translated onto an industrial scale.   They were selecting and designing processes, designing plants, choosing suitable materials of construction to resist highly corrosive chemicals, and overcoming the hazards and difficulties of the industrial environment.   These were the functions defined by the British Institution of Chemical Engineers in 1924 as appropriate to the profession.   In the early days recruits to the technology were trained chemists who developed new techniques as they established manufacturing processes.   In the course of time the gap between laboratory chemistry and plant design was widened tremendously with the extensive study of unit operations and the scientific treatment of heat and mass transfer.   Consequently the concepts of the 1930's when Lord McGowan of ICI Ltd encouraged the use of chemist–mechanical engineer teams now has been abandoned almost completely by industry and the need for chemical engineering specialism has been established.

In government service there is an encouraging understanding in the ROFs of the role of chemical engineering in production.   As far as the research departments are concerned, at one, Waltham Abbey, the problem is to attract the number of high-grade chemical engineers required for research and development in the face of the keen competition encountered from the heavy demands of industry. Experience in guiding decisions in the explosives and propellants field can be obtained only by actually working with the materials and their hazards. It really is no solution to award contracts for design work to specialist private firms except for those parts of the plant that are of more universal use, e.g. acid concentration plants. For explosives and propellant manufacture, development, and research in plant design, chemical engineers always have been required "in house" and the need for them is even greater today than it has been ever in the past.

Acknowledgment

The authors wish to thank F. H. Panton, Director of the Propellants, Explosives, and Rocket Motor Establishment for the opportunity to publish and for his encouragement and assistance during its preparation.

Literature Cited

1. Fitzgerald, W. "How Explosives are Made," *Strand Mag.* **1895**, *9*, 307–318.
2. Biasutti, G. S. "Histoire des Accidents dans l'Industrie des Explosifs"; Corday S. A.: Montreux, 1978.
3. Knapman, P. G. "RGPF History of the Factory"; Public Record Office Supply 5/451: London, 1943.
4. Nobel, A. British Patent 10 376, 1889.
5. MacNab, W. *Trans. Inst. Chem. Eng.* **1935**, *13*, 9.
6. Miles, F. D. "History of Research in the Nobel Division of ICI"; ICI Ltd, Nobel Division: Birmingham, 1955.
7. Sanford, P. G. "Nitro-explosives"; Crosby Lockwood: London, 1896.
8. Levinstein, H. *Trans. Inst. Chem. Eng.* **1937**, *15*, 10.
9. Reader, W. J. "Imperial Chemical Industries: a History"; Oxford Univ. Press: Oxford, 1970; Vol. 1, p. 132.
10. Hinchley, J. W. *Trans. Farady Soc.* **1917**, *13*, 87.
11. Marshall, A. "Explosives," in "History and Manufacture"; Churchill: London, 1917; Vol. 1.
12. Simmons, W. H.; Forster, A.; Bowden, R. C. *Ind. Chem.* **1948**, *24*, 429, 530, 593.
13. MacNab, W. *J. Soc. Chem. Ind.* **1922**, *41*, 353T.
14. Witham, G. S.; Bowden, R. C.; Smith, T. R. British Patent 381, 291, 1932.
15. Schmid, A. *Z. Schiess-u. Sprengstoffwesen* **1927**, *22*, 169, 201.
16. Denbigh, K. G. *Trans. Faraday Soc.* **1944**, *40*, 352.
17. Denbigh, K. G. *Chem. Ind.* **1961**, 920.
18. Brown, D. G. *Chem. Ind.* **1947**, 87.
19. MacDonald, J. O. S. *Br. Chem. Eng.* **1956**, *1*, 254.
20. Klassen, H. J.; Humphrys, J. M. *Chem. Eng. Prog.* **1953**, *49*, 641.
21. Milliken, M. G. *Ind. Eng. Chem.* **1930**, *2*, 326.
22. Chilton, T. H. "The Manufacture of Nitric Acid by the Oxidation of Ammonia: the Du Pont Pressure Process," in *Chem. Eng. Prog. Monogr. Ser. No. 3* Am. Inst. Chem. Eng.: New York, 1960.

23. Manning, A. H. *Chem. Ind.* **1943,** *62,* 98.
24. Directorate of Royal Ordnance Factories, Explosives "Report of the Acid Plan Design Committee," Ministry of Supply, London, 1949.
25. Underwood, A. J. V. *Trans. Inst. Chem. Eng.* **1965,** *43,* T302.
26. Nitroglycerin Aktiebolaget British Patent 832 870, 1958.
27. Bell, N. A. R. "Loss Prevention in the Manufacture of Nitroglycerine," in "Major Loss Prevention in the Process Industries"; *Inst. Chem. Eng. Symp. Ser.* **1971,** *34,* 50.
28. Poole, H. J.; James, F.; Runnicles, D. F.; Nicholson, A. J.; Crook, J. H.; Thomas, A. T.; Slack, G. W. British Patent 874 564, 1961.
29. Nadel, G. I. U.S. Patent 2 991 168, 1961.
30. Morris, W. H. British Patent 885 503, 1961.
31. Morris, W. H. British Patent 974 829, 1964.

RECEIVED May 7, 1979.

Chemical Engineering and the Public Image

WILLIAM F. FURTER

Department of Chemistry and Chemical Engineering, Royal Military College of
Canada, Kingston, Ontario, Canada K7L 2W3

*The most outstanding characteristic of chemical engineering
in comparison with the other main branches of engineering
is its broadness, a characteristic which results from its
uniquely wide base in chemistry, physics, and mathematics.
Yet, by and large, the public view of the profession is
almost exactly the opposite. Rather than being perceived
as a broad, general, and versatile form of engineering,
chemical engineering is seen instead in a much narrower
and more specialized sense—the engineering application of
chemistry in the manufacture of chemical products. Part
of the problem lies in the name itself, and the connotations
which it evokes. Yet a more suitable name has not been
forthcoming, so the paradox seems destined to continue.*

C hemical engineering, at more or less the century mark of its founding
as a profession distinct from chemistry, currently finds itself in search
of a public image which accurately reflects its modern-day scope and
nature. This search continues to be confounded by the very name under
which the profession is known, and the resulting connotations which the
name evokes. If the chemical engineer is not perceived as "a chemist
who works in a plant" (presumably instead of in a laboratory), he is at
least seen by many as "an engineer who applies chemistry," or "a person
who manufactures chemicals." In other words, there is a persistent
image of the chemical engineer as an industrial, applied, or engineering
chemist. Yet the image is misplaced; he is first and foremost an engineer
rather than a chemist—applied, industrial, or otherwise, and like other
engineers he is concerned primarily with applying the engineering approach to problem solving in his particular areas of expertise.

0-8412-0512-4/80/33-190-393$05.00/1
©1980 American Chemical Society

The Chemical Engineer: Narrow Specialist or Broad Generalist?

If one were to bet on whether the general public views the chemical engineer primarily as a narrow specialist or as being very broadly based and versatile, the smart money would go with the former. The public image of the chemical engineer, at least if not as a person in a white laboratory coat holding a test tube, has been of a person wearing a construction hardhat holding a test tube. The image continues to represent a valid and important, but ever diminishing and no longer highly representative, segment of the profession. In fact the image of chemical engineering as seen from within the profession could hardly be more different than that with which it is viewed from outside, in that it emphasizes versatility and breadth rather than intensive specialization. The bulletin of the Massachusetts Institute of Technology states it this way:

> *"Chemical engineering is the most broadly based of all engineering disciplines. This breadth results from a very deep involvement with chemistry in addition to the applications of physics and mathematics which are common to all engineering disciplines"* (1).

From the calendar of the Imperial College of Science and Technology in England comes:

> *"Chemical engineering can lay strong claims to being the most broadly based of the engineering disciplines"* (2).

From yet another direction:

> *"If the military were permitted only one type of engineering then it should be chemical engineering, because of the breadth"* (3).

R. L. Pigford, in his treatise *Chemical Technology: The Past 100 Years*, said it this way:

> *"The chemical engineer . . . needs to understand chemistry, physics, and mathematics* in approximately equal proportion *in order that he, apparently better than those from other backgrounds, can assemble and evaluate whatever knowledge is required to "bring things together"* (4).

A similar view is given as:

> *"It is apparent that chemical engineering is not at all the exploitation of a small area of some basic science but rather a study of the strategy and modes of integration by which many underlying fields of science may be brought into play in the solution of problems related to compositional change"* (Anonymous).

From H. A. McGee, Jr., in a recent issue of *Chemical Engineering Education:*

> *"As a profession we are justly proud of our great breadth, for we are the only applied science profession with in-depth training in chemistry as well as in physics and mathematics. Our background and perspective as scientist-engineers makes for flexibility and adaptability that is the envy of our sister disciplines"* (5).

Finally, D. Mackay, in his paper on *The Case for a Chemical Engineering Education,* stated:

> *". . . the chemical engineer has become the engineering generalist through his unique ability (among engineers) to understand and exploit chemical change. He is familiar with all forms of matter and energy and their manipulation. He leaves detailed specialization in one area of technology, such as construction, machine design, or electronics to others"* (6).

It may be concluded, then, that the chemical engineering profession, at least at the present time, does not view itself as just "based on chemistry," but rather as strongly based on chemistry, physics, and mathematics, from whence comes its extraordinary breadth and versatility.

The Misconception Inherent in the Name

A major cause of the misconception about what chemical engineers are and what they can do continues to lie with the term "chemical" engineer. Many of the other one-word designations for engineering disciplines reflect, with not unreasonable accuracy for single adjectives, the nature and scope of the particular profession so named: for example, mechanical engineering, electrical engineering, metallurgical engineering, geological engineering, mining engineering, etc., even the original military engineering and its historical civilian counterpart, civil engineering. Yet the term "chemical" engineering, with its connotations of the manufacture of chemicals and an almost exclusively chemical base, no longer represents, if it ever did at least in North America, more than part of the profession so named.

The difference is that the designator word "chemical" for chemical engineering indicates what is different about it compared with the other engineering fields, i.e. that unlike the others, it is based substantially on chemistry as well as on mathematics and physics; rather than on what the profession actually does. The misconception resulting from the name works to the detriment of the scope of employment available to chemical engineers particularly outside the chemical process industries, to their utilization at full potential by at least some of their employers, and to the attraction of young people into the profession. The chemical engineering profession in North America has been wrangling over this problem for years now, and the unhappiness with the name is growing sharply as the scope of the profession continues to expand.

There have been continuing attempts and proposals from a wide variety of directions to change the name to something more representative, yet so far no one seems to have come up with anything better (or at least less worse). Many of the proposed replacement names center around energy and materials. For instance, one hears alternatives such as energetics engineer, energy engineer, materials engineer, process engineer, and combinations of these and others. Indeed, chemical engineers are unquestionably the experts within the engineering fraternity in certain major areas of both energy and materials, yet such names are hardly more definitive. The problem is that all of the alternative names for the profession that have been proposed so far are either too narrow, lengthy, inaccurate, or otherwise inadequate in comparison with the original. Hence the profession seems destined to continue with its current name, at least for the foreseeable future, with the resulting misconceptions and disadvantages unrelieved. Some examples of these follow.

Twelve years ago, the American Institute of Chemical Engineers (AIChE), in conjunction with the leaders of the process industries, selected the ten most outstanding achievements of chemical engineering (7). These were: production of fissionable isotopes; production of synthetic ammonia; commercial-scale production of antibiotics; production of petrochemicals; establishment of the plastics industry; establishment of the synthetic fibers industry; production of chemical fertilizers; establishment of the synthetic rubber industry; development of high-octane gasoline, and electrolytic production of aluminum. Although all of these developments involve manufacturing chemicals of one type or another, for how many of even these would the general public perceive chemical engineering to be the principle technical discipline involved? All? Some? Let's not be too sure!

Chemical engineering always has had wide areas of overlap with its neighboring professions, and continues to share many new and recent fields with other technical disciplines. Yet it rarely seems to become identified as the primary or dominant discipline in them, either in the eyes of the public, or even in the eyes of the other scientific and technological professions. For instance, environmental engineering and pollution abatement now seem to be associated primarily with civil engineering, while combustion, rocket propulsion, nuclear power production, and energy conversion are thought to be related to mechanical engineering, and process control and simulation are assumed to be associated with electrical engineering, despite the fact that chemical engineering has played a leading role in all of them. Yet guess which engineering profession is associated most frequently with industrial pollution!

In staking out a "home ground" for chemical engineering, a major focus point has been the unit operations. Yet the paradox continues; most of the so-called unit operations are primarily (if not wholly) physical in nature. Hence even on home ground the term chemical engineer has its problems.

Relation with Chemistry

Chemical engineering always has been perceived as being somewhat different from the other engineering professions for a variety of reasons. Much of the uniqueness has come from its association with the chemistry profession, an association particularly strong during the formative years. It is not intended here to explore the relation between the two professions, except to note that the relation continues, although perhaps not quite as strongly as in earlier years, to the benefit of both. This relationship has been the envy (and rightly so) of the other main branches of engineering, none of which have had anything quite comparable. The benefits to chemical engineering of this close association have been many over the years, and two are given here as examples: the much earlier accommodation to the PhD degree than in the other main engineering disciplines and the early founding and sharing in *Chemical Abstracts*. Both have been factors of enormous consequence in facilitating advances in chemical engineering research during the present century to the ultimate benefit of mankind.

Relation with Other Engineering Professions

The "Big Four" of engineering—civil, mechanical, electrical, and chemical—are so called because there is neither a close nor a consistent fifth to add to these four, in terms of enrollment, among the multitude of other engineering specialties.   The order of the first three varies from time to time, with chemical nearly always being fourth.   Not only is this the current case, but it has been for many decades.   Chemical engineering has a built-in constituency among students entering universities—i.e. with the group of students who liked chemistry better than physics in high school and wish to become engineers.

A very important rule that chemical engineering plays in the engineering school is to maintain, by its presence, the integrity and academic respectability of the other engineering degrees offered, particularly those of the remainder of the Big Four.   It does this by bringing substantial representation of the chemical sciences into the engineering program, in support of the thesis that engineering is based on the three cornerstones of mathematics, physics, and chemistry; a representation that might not otherwise exist to a viable extent.   Mathematics, of course, is usually well represented in all of the engineering programs, but of the Big Four only chemical engineering has a science base which is not weighted predominantly toward the physics side and away from the chemical side. Hence the presence of chemical engineering keeps an engineering school honest by helping to keep the other engineering programs credible. Without it, the academic credibility of the other main engineering degrees offered could be cast in serious doubt.

In short, chemical engineering fills the gap that would exist otherwise between the natural sciences on the one hand, and the primarily physics-based disciplines of civil, electrical, and mechanical engineering on the other.

Conclusions

In many ways chemical engineering has been its own worst enemy by failing to ensure that its image represents it fairly.   Saddled with a name which reflects its differences with other engineering professions more than its own nature, it has failed to keep itself from being perceived in the public mind as a relatively narrow specialty—the engineering application of chemistry—rather than as the broad, versatile, wide-ranging, polytechnical discipline perceived by its current practitioners. The chemical engineer, while an expert in chemical processing, also manages to be the true generalist of the engineering profession because of the uniquely wide (for engineers) spectrum of science represented in his base.   Indeed, of all of the major types, he comes the closest to being an all-purpose engineer.   The correction of the misconceived public image to reflect current reality is surely one of the most, if indeed not "the" most important single challenge that chemical engineering now faces, if it is to continue to survive and flourish in the years to come.

Acknowledgment

Acknowledgment is made to the Donors of the Petroleum Research Fund, administered by the American Chemical Society, for the award of a Special Educational Opportunities Grant to the author.

Literature Cited

1. "M.I.T. Bulletin, 1975–76 General Catalog Issue"; Massachusetts Institute of Technology: Cambridge, 1975; 154.
2. Calendar of Imperial College of Science and Technology, University of London, London, England 1972, 10.
3. Sawyer, W. R., Director of Studies, Royal Military College of Canada, Kingston, Ontario; quotation from an address, 1966.
4. Pigford, R. L. *Chem. Eng. News* **1976**, *54*(15), 190.
5. McGee H. A., Jr. *Chem. Eng. Education* **1975**, *9*(2), 94.
6. Mackay, D. *Chem. Can.* December, **1971**, 22.
7. *Chem. Eng. Prog.* **1967**, *63*(12), 29.

RECEIVED May 7, 1979.

The Big Future Program for Chemical Engineers: Fuel and Energy Conversions

DONALD F. OTHMER, Department of Chemical Engineering, Polytechnic Institute of New York, 333 Jay Street, Brooklyn, NY 11201

History is being made and will continue to be made as the chemical engineer converts valueless solids into fuels for producing energy, by his unit processes, unit operations, and process and equipment design. The solutions of the energy problem will be on a magnificently grand scale because of the tremendous masses of materials to be processed. Vast economic, social, and political values will be involved. Many other sources of energy are being converted — also by chemical engineering methods — into usable heat and electric power. The petroleum industry became a chemical process industry a half century ago. This doubled the program, work load, and opportunities of Chemical Engineers. These will be doubled again through energy conversion plants which will require by the end of the century greater amounts of capital than all of the other chemical process industries combined — and will promise the largest opportunities ever for chemical engineers.

O urs is an expanding profession, both as to the numbers of those who practice it and especially as to the fields, disciplines, and industries of our interest and occupation. There may be two or three hundred thousand of us, world wide; and by now we affect or control at some stage the materials and/or the production of almost every article manufactured on an industrial scale. And in these industries, classification of the duties and responsibilities of chemical engineers is difficult because their tasks are of such great variety and scope.

Indeed it may be difficult to identify all of the tasks and influences of even any one individual chemical engineer. There are, of course, the usual general and approximate job descriptions, the boundaries of which are nebulous and cross each other like the sticks in a pile of jack straws.

0-8412-0512-4/80/33-190-401$05.00/1
© American Chemical Society

Some of the paths may be labeled research and development, pilot plant operation, equipment design, project engineering, equipment fabrication, plant construction, operation, supervision, and management. All of these rather empirical labels may be bundled together in parallel as were the fasces, the symbol of the ancient Roman official. His sticks were bound by two cords—and here, the basic one is process development and design. The other equally important one is Economics, meaning profitable operation.

Certainly the chemical engineer always has his kit of tools at hand— the unit physical operations and the unit chemical processes. These accomplish among so many other tasks—the movement, sizing, and separation of materials, the diffusion of molecules and heat, and the understanding and use of the interrelation of the thermodynamic possibility and the reaction velocity of chemical change. The chemical engineer shows his skill in the use of these individual tools in their assembly, arrangement, and fitting together to develop an optimum process; one which on the bottom line demonstrates a desirable profit.

No definition of the work of the chemical engineer is adequate to identify all of his many capabilities and roles. These, however, may be generalized and integrated by a simple expression—he converts cheaper raw materials into finished products which are more valuable to man's use through processes, usually involving physical and/or chemical transformation. These processes involve a series of steps skillfully chosen and arranged to give products which may be sold at an acceptable profit.

For over a hundred years the name chemical engineer has been used and his skills have been practiced long before its academic recognition as a profession to be encouraged through specialized education (1). Certainly, in the inorganic chemical industries, and in many extractive metallurgy productions, the amounts of materials reacted chemically have long been much larger than those conveniently worked with by a chemist. The organic industries processed the smaller quantities of materials available as by-products from the carbonization of piles of coal and wood and the fermentation of carbohydrates, as well as other conversions of biomaterials from both the animal and vegetable kingdoms.

The vast industrial expansion of the last decade of the past century and the first decade of this century magnified the earlier industrial chemistry as to the number of raw materials available and used, the number of products made, and in particular, to the size and scale of their production. Great indeed were these expansions, and great was the need for skill in developing new, better, and more profitable processes and equipment to be used in them. Thus the need for recognizing a separate profession of the chemical engineer, who worked with these assignments, culminated in the organization in 1908 of the American Institute of Chemical Engineers (AIChE). However it was not until 1925 that the AIChE recognized the curricula of 15 schools, including that of the

Polytechnic Institute, by accrediting them to give a collegiate degree in chemical engineering.

During World War I the profession quantified with new skills what had been the art of the processes and the products of many chemical productions, and of the petroleum refining industry. The ease and low cost of obtaining the widely distributed and vast resources of paraffin hydrocarbons compared with any other source of organic materials led in the 1920's to their use as starting materials for derivatives, petrochemicals. (This writer's first patent application in 1927 used for many years ethylene dichloride (2) one of the earliest petrochemicals, first available commercially in about 1926). This use has grown mightily until in 1978, 10% of America's natural gas and oil is used for chemical feedstocks; and America's largest chemical company, DuPont, derives 80% of its products from them.

It is probable that the petroleum and petrochemical industries, and their downstream productions, and chemical uses and conversions now employ about one half of all of the world's chemical engineers. Thus the total number world wide has been doubled through the advent of petroleum processing, including its chemical conversions.

Fossil Energy Sources and Conversions

Material resources often have controlled a nation's rise—and fall. Ancient Egypt developed bronze from its copper and tin, and weapons therefrom enabled it to dominate its sphere; when copper ran out, Egypt succumbed. Similarly a silver mine gave Athens this noble metal for trading; with its exhaustion Athens deteriorated. Energy resources today are equally important to national attainments, although the world's economics and politics now are very much more complicated than those of the ancients.

The principal supply of both the world's fuels for generating energy and its organic chemicals has for two generations been petroleum liquids and gases. Suddenly it has become evident that these can be obtained no longer in the amounts needed by simply drilling holes—from meters to kilometers deep—in the earth's crust or the sea bottom.

Fuels for energy and chemical feedstocks must come soon from sources other than petroleum by conversions requiring very involved physical and usually chemical processing. This is particularly true since our national energy consumption is expected to grow (3) by 1.7 to 2.9% per annum.

Solid fuels—coal in all of its ranks from peat to anthracite as well as oil shales and oil sands—will be the abundant fuel materials for the next 100 years. Many chemical engineers are and will be required in the researching of processes for these conversions.

New and often elaborate processing, sometimes in quite novel equipment, must be developed; and many plants must be built and worked in sizes that are tremendous compared with almost any other chemical production or mineral extraction. These huge plants will require in their building stupendous resources of capital, labor, and materials, as well as the talent and time of all available chemical engineers, which is important to us.

As agricultural wastes can be utilized in comparatively smaller plants, these and wood stoves for home heating will contribute energy presently supplied by oil. Large-scale biomass utilization may be possible, if (a) land is available for growth of special crops after food production, and (b) if labor costs for collection of waste or virgin vegetation to central plants can be met. Almost every system for production of energy thermally or otherwise will require chemical engineers in the research, development, design, building, and operation of plants.

Factored into these requirements for energy world wide, must be the expected increase in the world's population from that now estimated at somewhat over 4,000,000,000 by some 50% to the 6,000,000,000 who will be entering the 21st century. (Hopefully it will reach a steady state at not over 10,000,000,000.) Fortunately the population of the United States and of most of Europe and Japan—the industrial nations which use and depend most on energy—will not increase by anything like that percentage but the desires and uses for energy of all of us may continue upward nevertheless.

With the abrupt change in the supply and costs of the world's major source of energy most of the world must look for new sources or conversions from available sources. Most of these conversions will be designed, built, and operated by chemical engineers.

A recent report to the U.S. Congress gave a shopping list of plants for energy conversion which are needed to be added in this century (4). If the averages of the ranges of the numbers are taken, this list includes:

- 1350 electric power plants of 1000 MW equivalent, of which 650 might be fired with fossil fuels or their derivatives, 625 with nuclear fuels, and 75 with geothermal energy. Now there is an equivalent of 400 such plants using fossil fuels, 37 nuclear plants, and also 1 geothermal of 500 MW;
- 105 plants making liquid or gaseous fuels from coal and 60 making oil from shales, each equivalent to 50,000 bbl of oil daily—to date none;
- 3 times as many operating coal mines;
- 15,000,000 electric automobiles in the consumer sector and 17,500,000 buildings electrically heated and cooled—practically none to date.

These are only the largest items which are catalogued readily; there are many more. Meanwhile there must be spent also huge amounts of money and technical manpower—much from chemical engineers—in bringing into practicality those great promises of nearly unlimited cheap power: solar electricity, the breeder reactor, and fusion power. Many of the developments necessary for their realization and also for the realization of the possibilities of many other suggested systems must go beyond the horizons of present human knowledge and technology. Many of these developments and their operation if successful, will be in, or verge on, the chemical process industries.

As with every new industrial development the competitive economic factors will be all important to their success. One fundamental axiom is that with every dollar increment in the price of the standard, crude oil (at mid 1979 at about \$20/bbl) additional systems of energy production "break even." They then can be added progressively to our sources more and more as the price escalates and as a brake on its escalation *(5)*.

In introducing the results of the new scientific discoveries and the essential and very sophisticated technology which must be infused into the old arts of energy conversion, one recalls that the expansion of cheaply available energy has come not so much by evolutionary growth as by quantum jumps. The important ones have been: (a) from the power of human—and other animal—muscles; (b) to the steam power fueled by wood and coal which energized the industrial revolution; (c) thence to internal combustion engines which used petroleum liquids and gases to put man on wheels and wings; and (d) thence via all of these fossil and now nuclear fuels, and water falls, to electricity which has illuminated the world and worked many other miracles. Each jump has exanded greatly the work and hence the demand for many engineers.

Scientists and engineers are always optimistic in setting goals, every one hopes, and they strive for an early attainment of those goals which may come by some breakthrough or quantum jump. This would unfold a new source of low-cost and abundant energy via one of the routes which, as yet, is possible only in theory. Meanwhile while striving for the accomplishment of these magnificent but probably distant energy sources, chemical engineers must cultivate those which are less dramatic and more readily achievable. All of man's technology will be researched; and its pertinent points will be developed.

Consider the magnitude of this task from both the throughput capacities of materials to give energy and the size and number of installations, and also the sophistication and complications of the entirely new processes which will be required. It becomes evident that the total army of the world's chemical engineers may have to be doubled in number to accomplish the magnificent assignment of supplying the world with the energy required to roll in the next century *(5, 6)*.

Certainly these stupendous conversions of material resources into energy will more than equal all of the conversions of the world's present chemical plants, for which chemical engineers are responsible. Just as certainly, the research and design of the processes and plants necessary for these conversions also will require a stupendous input of chemical engineering talent. However, there is a tremendous magnitude required of these operations. Thus, due to the factor of scale up, a somewhat lower input of time of chemical engineers per million dollars of capital investment in plant may be required sometimes than for conventional chemical manufacture. The same may be true per million dollars of costs of operation.

Chemical engineers, like all others, are builders. They will be stimulated by the many opportunities to design and build new processes and plants for converting energy. However, equally important will be the redesign of processes and the rebuilding of present plants in the chemical process industries to minimize and conserve the use of energy. This is the task of every chemical engineer. To paraphrase the English poet Milton, "They also serve who only stand and [save]" (7).

Already many processes and their production plants have been reworked to save from 10 to 50% of the energy used because of its higher cost. Sometimes other benefits have accrued, partly because of the new and critical examination of all aspects of the processing. Here is a great need for chemical engineering talent and man hours which almost invariably has paid off handsomely in the resulting savings—hence profits—through reduction in energy use.

Physical Conversions to Obtain Energy

Many of the newer means of obtaining energy, and especially of conserving energy, fall in the interdisciplinary areas between the clearly understood functions of chemical engineers and other engineers or scientists. Particularly these may be in mechanical, electrical, hydraulic, aeronautical, naval, nuclear, and related engineering fields; also the scientists may be physicists, biologists, bacteriologists, geologists, and others in related scientific fields. As this list indicates, almost every branch of engineering and science must be concerned with the major materialistic problem of our time—the provision of energy for the future.

The direct utilization of solar energy to date for thermal supply to buildings for heating and cooling has been almost entirely through the design function of heat transfer and storage, as well as equipment interconnection. All of this is based on simple astronomical and climatological data. However, large arrays—hectares—of lenses or mirrors focus the dilute radiant energy of the sun on a much lesser area for electric power production or other industrial use. Not only is much land required but also much expensive hardware (8).

By comparison, the almost limitless areas of tropic seas receive the same solar constant of heat as do contiguous land areas; and the water with the heat it has collected, at no cost for hardware, may be flowed or pumped into flash evaporators under vacuum.   The vacuum steam generated in cooling the warm water is passed through a turbogenerator set to produce power.   This steam, now at an even lower pressure is condensed in giving up its latent heat to cold water drawn from the sea bottom and circulated through the condenser.   Fresh water is obtained as   condensate,   an   equally   valuable   product   in   many   tropical lands.   The cold sea water brought up from the depths is rich in nutrients and may be used to produce algae as single cell protein.   The algae may be the start of a mariculture food chain to feed bivalves such as oysters or clams, shellfish and other fish.   Sea weed also is valuable as a source of various chemicals and also as biomass for conversion to energy in some other manner *(9)*.

Chemical engineers will contribute much to such power production based on ocean temperature differences, which up to the present use projected plants costing entirely too much to be profitable even at present energy rates.   Additional revenues may change this picture.   These may be available from other products—here fresh water and mariculture products, which may vary from single-cell protein as the simplest to expensive seafood *(10)*.

Other energy sources depending less directly on the sun, such as windmills and devices to use the energy in sea waves, may be almost entirely mechanical.   However, some of these depend on chemical or electrochemical conversion processes for storing the energy produced.

However, the chemical engineer will be a partner with mining, hydraulic, and power engineers, geologists, and in geothermal energy developments and productions.   The hot water which is the carrier of the heat from the earth's depths usually carries minerals, most often as salts of greater or lesser value.   The separation of these solutes may give values to add to, or possibly even dominate the revenues from the power generated from the heat in the water *(11)*.

Only chemical engineers have been active in the use of osmotic pressures—forward or reverse—dependent on the difference of concentration of two adjacent solutions of salts.   At the point where the fresh water of a river flows into the salt water of the sea, the difference in osmotic pressure equals about 24 atm, which is equivalent to the pressure of 240 m (750 ft) of water behind a dam of that height.   It has been calculated that theoretically for each cubic meter flowing from the river into the sea per second at least 2 MW of power are involved *(12)*.   The osmotic pressure difference is greater when fresh water flows into the Dead Sea (500 atm) with its high concentration also of divalent $MgCl_2$ as well as NaCl.   The relative theoretical power here would be more than 30 MW *(13)*.

Several mechanical and electrochemical methods for converting this energy into a usable form have been suggested (14, 15). These may be used also in the dissolution by sea or fresh water of the salt in salt domes, such as those in Texas near the coast (14). The saturated brine, so formed, then is passed to be diluted by sea water to its normal concentration of about 3.5% salt. Texan salt domes have been the site of very large productions of petroleum. Now it has been calculated that, for all but the most productive of the oil wells, there is more latent energy available from the saline gradient through the dissolution of the remaining salt than that in the petroleum which has been removed, or which still may be removed.

Fermentations to Obtain Fuels

Many combustible products can be made by fermentation of vegetation—chiefly of the carbohydrates, sugars, starches, or cellulose in the various structures of the plant. The only fermentation product ever commercially considered as a fuel is alcohol. This was used as a motor fuel in the early days of the automobile, in the 1930's in many countries to encourage agriculture, and again during W.W.II when gasoline was unavailable. Raw materials varied from bamboo and bananas to potatoes and wood chips. Many chemical engineers were concerned; notable achievements included the development of economic processes to make anhydrous alcohol (16).

The competition of fuel vs. food in agriculture with respect to land and labor always will be important. Under conditions as of 1978 the pump price of motor fuel alcohol per gallon has been shown (17) to be for sugar cane—$1.19, corn—$1.27, and wood waste—$2.03. With its lower heat content and mileage per gallon compared with gasoline, this would be expensive motor fuel in this country even without any taxes.

Quite aside from the energy—and expense—the growing, collecting, processing, and transporting of the sugar cane, corn, or wood chips to the distillery, the energy cost for fermenting and distilling the alcohol is almost invariably greater than the energy it delivers in combustion (18).

In older plants, this processing energy cost may be 1.8 to 2.0 times as much. There is hope that chemical engineers will reduce this energy cost to about one half of that which the alcohol can deliver (19). However, considering the other energies required, alcohol probably always will cost more in energy than it produces. In late 1979 its price for chemical grade was about 3 times that of chemical-grade methanol.

Because of the political desire to increase agricultural revenues, the production of alcohol for blends of about 10% in gasoline may be expected to increase. Necessarily there will be considerable development work to try to improve the economy of the gasohol so produced. These programs, and the design, building, and operation of plants will require many chemical engineers.

New Ways to Burn Fuels

Probably all of the recently suggested modifications of the systems of combustions using conventional fuels are based on techniques usually used by chemical engineers. One example is the use of fluidized beds—either atmospheric or pressurized—as developed by chemical engineers. Usually such beds have assisted chemical reactions which are more involved than simple combustion. Such fluidized combustion of coal particles under boilers has significant advantages in some instances.

Fuel cells using methanol, hydrogen, or other fuels usually are a means of combustion of specialized fuels based on electrochemical engineering. The production of the fuel also may be within the responsibilities of the chemical engineer. Many look to these fuel cells for much of future energy. More recently these have operated at temperatures up to 650°C and with efficiencies up to 50%.

Similarly the generation of electric power by passing an ionized gas in a channel between the poles of an electric magnet invariably requires chemical reactions and processes in very specialized reactors of unusual materials of construction. (An early program in magnetohydrodynamics—over 40 years ago directed by this author and privately financed—failed to produce electricity at competitive prices, as have all of the other programs since, even at the present, much higher energy prices.)

Combustion in Large Excess of Water. Numerous other systems of burning fossil fuels to produce power will require chemical engineering principles, processes and equipment, as for example wet-air oxidation (WAO). Because of its quite different system of combustion, its close relation to other chemical reactions, and its usefulness, otherwise, to the chemical industries it may be discussed with an example *(20)*. This commonly is called ZIMPRO—i.e. the Zimmerman Process.

Almost any organic material in water, as a solid or in solution, when heated in a closed vessel to 175° to 250° C with oxygen or air introduced against the pressure developed, will oxidize or burn the solids autogenously. The pressure is controlled by the rate of supply of air and of withdrawal of the steam and gases formed by the combustion. The organics may be combusted either partially, and with some preferentially to others, or completely by the supply of adequate oxygen as such or in air. Over a hundred installations have been built for partial WAO of sludges from sewage treatment plants *(21)*. These and some other suspensions having over 95% water cannot be dewatered mechanically. Many such solid–water mixtures have colloidal characteristics. Simply heating them to 200° C breaks the colloid and thermally conditions them

to allow mechanical dewatering of the resulting solid particles.   Sludges
containing up to 400 tons/day are combusted by WAO in units of heavy
construction to withstand the corrosive conditions and high pressures.

An early use of partial WAO was to break the colloid of bog peat with
up to 19 parts water per pound of solid by this thermal conditioning
(22).   Then mechanical dewatering allows the separation to drop down to
0.8–1 part water per pound of solid.   Some 5–10% of the fuel value is
used in the burning by WAO or otherwise destroying the colloidal components.   The heat generated accomplishes the direct heating without
heat-transfer surfaces, which soon would be fouled.   The resulting peat
solids are denser and have a higher unit heating value than those resulting from simple drying of the peat; and in some cases the heating values
are equivalent to some bituminous coals.   There is an increase in the
drained water of water-soluble, sugar-like solids, due probably to hydrolysis simultaneously with the WAO.   These solids will give alcohol and
other products when the liquid, drained from the peat solids, is fermented.   The thermal conditioning "coalifies" peat in a few minutes to
give the same results due to pressure, heat, and some oxidation as that
accomplished during epochs of geological time (22).

Large expanses of many hundreds of square kilometers of surface
peat, from 2–8 m deep, are in at least several parts of the United States.
After removal of the peat and draining, valuable agricultural land results.
Equally large amounts of peat are available in many other countries.   In
Russia, Finland, Sweden, and Ireland a thin surface layer of the bog peat
is turned over to dry in the sun—using, for this drying, up to twice the
energy of today's sun as was absorbed by the vegetation many years ago
and stored until now.   To this extent the peat bog may be regarded as
merely a specialized solar collector.   In each of these countries, the dry
material is collected and burned in power plants having capacities totaling
hundreds of megawatts.

Peat may be gasified to form synthesis gas, as may almost any solid
fuel, and a demonstration plant of the Institute of Gas Technology and the
State of Minnesota showed that it is converted to hydrocarbon gases
much more readily than lignite or coal (23).

If the WAO is carried to completion through the addition of sufficient
oxygen as such or in air, over 95% of the heating value of the peat may be
obtained in the steam–gas mixture leaving the reactor, compared with
85% or less when air-dried peat or coal is burned under a boiler to
generate steam.   All of the sulfur is oxidized to form the sulfate radical,
nitrogen oxides are not formed, and there are no particulate emissions.
Ash, even from coals with very high ash or of low melting points, presents
no problem as it discharges in a slurry with the blow down water, along
with the sulfur as sulfates.

The WAO is extremely fast at the high temperature (up to 350° C) and pressure (up to 240 atm) which may be maintained in the reactor. Solubility of oxygen in water is many times as great as at ambient temperatures, as is also its diffusion in the reactor to every particle of fuel. The reaction velocity is also extremely high.

Complete WAO produces a mixture of steam and gas at pressures of hundreds of atmospheres which can be expanded through turbines to give power; whereas by heat interchange, pure steam is generated at a somewhat lower pressure. The cost of a WAO plant for power production is competitive with that of a modern coal-fired steam boiler unit, as is also the cost of its operation *(24)*.

High sulfur coal may have all of its sulfur—both organic and inorganic—preferentially oxidized to the sulfate radical in a WAO unit *(20)*. If the coal or lignite is to be ground and put in an aqueous slurry for pipeline transport, it is prepared completely for feeding to a WAO unit for desulfurizing it at a minimum cost. This may cost less in equipment and operation than the scrubbing of sulfur oxides out of the large volumes of stack gases.

All of these newer systems for burning fuels tend toward chemical processing and require chemical engineers to develop, design, and operate.

Gasification and Liquefaction of Solid Fuels. Many thousands of chemical engineers have been working on gasification of coal. This is an intermediate step in its combustion which uses 25–45% of its heating value so that the balance may be transported and used more conveniently. Carrying forward the processes and know-how developed over 150 years in making town gas, much of the work is devoted to the production of fuel gas below 500 Btu/cu ft for use near the gasifier. Many more sophisticated processes will produce substitute natural gas (SNG) with heating values in the 950–1000 Btu/cu ft range. Approximately 60% of the heating value of the coal goes out in the gas.

Our Department of Energy estimates that the costs of conversion of coal to SNG will be in the range of $4–$5/1,000 cu ft. The numerous necessary plants each will cost about $5,000 daily 1,000 cu ft or 1,000,000 Btu *(5, 28)*. These plants will be designed and built, then operated largely by chemical engineers. However their total bill for the capital costs, also that for the costs of their operation over and above that of the fuel used, are almost incommensurably larger than the costs associated with the present industries which chemical engineers have built and operated.

Some of these processes which are in the stage of pilot plants or design of demonstration plants are the HYGAS, BI-GAS, CO_2 Acceptor, Synthane, and Self-Agglomerating Ash gasification processes *(5)*. These processes are designed to fill the pipeline distribution systems and to mix with and ultimately take the place of natural gas as its supplies diminish. With SNG produced by the HYGAS process of the Institute of Gas Technology, as an example, at the coal mine and transported 1500 km by existing pipelines, the cost at the point of usage would be between $4 and $5/1,000,000 Btu or 1000 cu ft based on coal costing 60¢–80¢/1,000,000 Btu. The capital cost of the plant might be some $4,000–$5,000/1000 cu ft/day of SNG capacity. Both capital and operating costs are very much cheaper, and thermal efficiency is much greater than for coal–steam–electric plants with power transported to the user by high-tension wires.

Since coal gasification plants must be large, in the range of 250,000,000 cu ft/day, the battery-limits cost of a plant is over $1,000,000,000. Some recommendations have been made for at least 60 such plants. The number of chemical engineers required for the process and equipment design, construction, and operation of so many plants would be huge (5).

An almost equal amount of development work is being carried on by hundreds of chemical engineers to make liquid fuels from coal. Again demonstration plants are being built following 8 or 10 pilot plants which make syncrude, then by refining the usual liquid fractions. The syncrude may cost $25/bbl and investment cost may be $25,000/daily bbl or about $5,000/1,000,000 Btu/day. Several processes also are being developed which produce both gases and liquid fuels, with somewhat more processing required for their separation.

Synthetic Crude (Syncrude) for Oil Shales and Tar Sands

Two other more dilute and much less easily worked sources of solid fossil fuels are oil shale and tar sands. Each of two relatively small areas of the North American continent have amounts of oil bound in shales and sands which, when retorted out, could produce more oil than the total of all of that of the Middle East (5). Billions of dollars have been spent and are being spent to implement the work of a thousand chemical, mechanical, mining, and civil engineers, geologists, and other scientists working for 15 to 20 companies. Climate problems add to the material problems of having to mine, process by extraction and/or retorting, and then replace ten times as much solids as products. The oil shales are in an almost barren semidesert and the tar sands are in a very wet (in summer) area within 10° of the Arctic Zone.

Methanol—A Synthetic Liquid Fuel

There are so many suggested conversion processes and plants that it is impossible even to scan them to note the fascinating technical problems to be solved. Particularly it is impossible to examine and tabulate the number of chemical engineers required to develop the processes and to design, build, and operate the required equipment and plants. Thus it may be more helpful to look at the chemical production of one fuel of wide interest and use and of such large possibilities of commercial success that it has required no government financing in its development. The billion dollar, projected plants also will be built without government help. Under present energy costs, certainly under those of the future, it will be profitable; and although the capital requirements are huge, they can be handled readily by private enterprise. This project has great flexibility in the raw materials and the uses of the produced fuel, and it may be taken therefore, as exemplary. Methanol is the only true synthetic of the so-called "synfuels" that have been proposed. It has been in large-scale production for many years, thus no research and development is needed; also it requires no refining as do the others. Yet it will require many chemical engineers.

The product fuel material, methanol, is very well known (26). It may be used as produced to supply the two largest requirements for energy; i.e. the turning of the armatures of our generators of electric power and the wheels of our automobiles. Also it is one of the best fuels developed to date for use in fuel cells (27) The steps for making methanol have been operated in the chemical industry in units of the size now determined as optimum for a fuel-grade methanol plant. Furthermore, long used processes will convert any solid, liquid, or gaseous fuel from either fossil or biomass sources to synthesis gas, thence to methanol (28). Thus there is great flexibility and versatility as to both the sources of the raw materials and the methods of using the finished product.

Numerous studies (26, 27, 28) have demonstrated methanol's high thermal efficiency as a fuel for cars and diesels (trucks and locomotives) and for generating electricity (29). Other successful developments in processing, reactors, and catalysts have been operating for many years in large methanol-production plants. Further improvements, based on the long history of methanol in the chemical industry, are possible (28) in plants specifically designed to make fuel-grade material, such as that trademarked Methyl Fuel (30). Unfortunately there is no similar experience in designing and operating other synthetic fuel processes, which have been tested on only a relatively small scale.

Any petroleum liquid or gas or any of the available solid fuels may be used—from peat and lignite through coal of any rank including anthracite; also wood and other vegetation can be used if available in the necessary quantities. High-grade coal is usually more valuable for direct firing; but high-sulfur and low-rank materials such as subbituminous and lignite, often with large amounts of water, are suitable feedstocks. Peat, which has been air dried as it is harvested or preferably thermally conditioned, may be used also. The solid fuel is ground to a suitable particle size and slurried in water to feed to a coal gasification reactor which is supplied also with oxygen produced on site.

The synthesis gas, after the customary shift conversion and purification, is delivered at synthesis gas specifications to the compressors for the methanol synthesis loop. This operates as a series of catalytic converters, which have advantages over the systems used conventionally both as to the flow arrangement and catalyst used in the Wentworth process. Sulfur is removed in the elemental form; but the crude methyl fuel retains the higher alcohols and other impurities produced in small amounts (2–3%). They contribute to its value as a fuel.

The Methyl Fuel produced has a higher heating value (HHV) of 20,000,000 Btu/ton; and it requires an input of from 1½ to 2 times as much heat based on the coal's HHV. Thus the overall energy conversion is in the range of 50 to 65%; and the lower efficiencies in this range are for poorer coals. About 2 tons of a subbituminous coal as mined with 8600 Btu per pound HHV will produce 1 ton of Methyl Fuel at an overall conversion efficiency of 58%.

Several major plants for producing Methyl Fuel which are immediately adjacent to a mine for coal or lignite are now in the design stages after extensive feasibility studies (30). These are necessarily very large and have had firm sales contracts made in advance with major electric-power complexes. The capital cost will be in the range of $80,000/daily ton of Methyl Fuel and there may be an additional $20,000 cost for off-site facilities including those for coal mining. The total operating and capital costs in 1979 dollars may be in the range of $2.50–$5.00/1,000,000 Btu, i.e. $50–$100/ton of Methyl Fuel or about 16¢–33¢/gal.

Methyl Fuel may be delivered readily as a liquid in pipelines or standard tankers and stored in usual tanks. Environmental problems are practically nil in production, use, and transportation. Since it is water soluble, spillage at sea would not be disastrous.

Extended operation in large power plants has shown that Methyl Fuel can be burned under boilers to generate steam with major advantages over usual oils such as: no sulfur oxides, no particulates, and very low nitrogen oxides. The only necessary retrofit may be a different orifice in the fuel injection nozzle.

However Methyl Fuel allows a much more efficient generation of electric power than by the conventional steam boiler. This uses the modern, combined-cycle gas turbine, which requires about 10 tons/day of Methyl Fuel to generate 1 MW. The installed cost of such large, modern turbine units is in the range of $550 to $600 per kilowatt generated in 1978 dollars, while that of a large (700 MW) coal-fired steam electric generating unit is in the range of $1000 to $1200 per kilowatt, including the electrostatic precipitators and stack gas scrubbers that are required today. Moreover the time required for construction is about twice as long for a coal-fueled steam power plant. The gas turbine plant will have a thermal efficiency almost a third greater than the steam plant (46%

against 32%). This reduces substantially the loss of fuel value in the production of Methyl Fuel.

The cost of electric power will be the sum of the cost of Methyl Fuel at the mine-mouth conversion plant, plus the cost of its transportation, and its power generation with gas turbines. With all of the capital costs included, one very large (3000 MW) complex will be about 4.0¢ per kilowatt hour based on 1978 dollars, coal at 57¢/1,000,000 Btu at the Methyl Fuel plant, and about 4500 km of ocean transport.

Methanol has long been used in automobiles as a special, premium fuel; and Methyl Fuel may be used either by itself or in mixtures with gasoline. Mixtures have the disadvantage in storage and use that small amounts of water may cause separation into two layers. Widespread use will require only minor modifications of individual automobiles, but will require the major modification of the distribution system for automotive fuel now based on gasoline. Recently one of the major oil companies (Mobil) demonstrated a simple process of converting methanol to gasoline. This could be an early source of motor fuel from coal without waiting to rebuild the automotive fuel distribution system. However, methanol has been used in fleets of cars and gives a much better and cleaner performance with an octane rating of 106; also it gives much better mileage than would be expected from its LHV compared with gasoline.

Methyl Fuel may be made from excess natural gas now being flared to waste in numerous petroleum-rich countries. It may be transported then at a fraction of the costs for liquified natural gas (LNG). Costs of both systems of overseas delivery of the heating value of the natural gas have been determined under wide variations of the several parameters. LNG may provide the cheaper system from North Africa to Europe and the East Coast of North America, possibly also from the Persian Gulf to Europe. A gigantic 50,000 tons/day Methyl Fuel plant was being engineered for installation in Saudi Arabia just at the time of the escalation of energy prices when the Egyptians invaded Israel. From the Persian Gulf, it had been designed to supply clean energy to either coast of North or South America and also to Japan more cheaply than an LNG system (28).

The efforts of the numerous chemical engineers who have developed Methyl Fuel and similar systems from the established practice for producing industrial methanol total many man years. The several groups involved in such projects have completed various feasibility and related studies. Many more man years of chemical engineers will go into the definitive project and design engineering for each individual synfuel plant. Such engineering must include hundreds of related legal papers for the numerous applications for permits to many federal, state, and local agencies such as the Environmental Impact Statements, etc. Then the equivalent amount or even more time of chemical engineers will be spent in each phase when the actual (and big) money is being spent in the procurement, equipment construction, installation, and finally the initial operation and running in of the various components of a final plant. It will have many units of equipment and trains of producing units in parallel.

Because of the large size of each unit of methanol and other synfuel plants, and the fact that each unit will be duplicated in each of the several trains of processing in the very large plants which will be required, the value of the equipment per man year of chemical engineering will be large indeed. This is true both before and after initial operation and also for the size of the operating staff compared with usual chemical or even pretroleum-production plants. This means, reciprocally, that the dollar value of plant investment and material throughput as well as personal responsibility will be very high indeed per chemical engineer.

The costs of engineering of plants vary from 5 to 10% of the total capital costs. Even at a low 5% of investment, the numerous multibillion dollar plants for synfuels and for the many other energy conversions will pay the salaries of many chemical and other engineers. The operation of all of these facilities will require another army of chemical engineers.

Conclusions

The work of chemical engineers has been most important in the conversion of nature's raw materials into the consumer products and services which are the essential fabric of modern civilization, culture, and comfort. The greatest material threat to these ever has come with the realization that energy is such an important contributor and necessity, and must be obtained and supplied even at much greater costs than thought previously. Chemical engineers must and will supply the means for producing energy from other sources hitherto overlooked in the presence of an abundance of readily available petroleum liquids and gases. This task in the next 20 years will equal all of those of the chemical industry; also the needs and work of chemical engineers will thus double. The related changes will make energy production a part of the chemical process industries and make it another one of the responsibilities of chemical engineers.

Literature Cited

1. "Encyclopedia of Chemical Technology," 1st ed.; Kirk, R. E., Othmer, D. F., Eds.; Interscience Encycl. Co.; New York, 1949; Vol. 3, p. 653.
2. Othmer, D. F. *Trans. Am. Inst. Chem. Eng.* 1933, *30*, 299.
3. Report of Committee on Nuclear and Alternative Energy Systems, *Nat. Res. Council, Supporting Paper #2, 1978.*
4. Seamans, R. C., Jr. (Adm. ERDA) *Chem. Week*, Dec. 14, 1977, 22.
5. Othmer, Donald F. *Mech. Eng.* 1977, *99*, 30.

6. Othmer, Donald F. *Chem. Ind. (London)* March 18, **1977**, 177.
7. Milton, John Sonnet XV, 1652.
8. Balzhiser, R. E. *Chem. Eng. (N.Y.)* **1977**, *84*, 88.
9. Othmer, Donald F. *Mech. Eng.* **1976**, *98*, 27.
10. Othmer, Donald F. *Desalination*, **1975**, *17*, 193.
11. Christopher, H., Armstead, "Geothermal Energy", Halsted Press: 1978. 8.
12. Bromley, L. et al. *J. Am. Inst. Chem. Eng.* **1974**, *20*, 326.
13. Loeb, S. *Science*, **1975**, *189*, 654.
14. Wick, G. L.; Isaacs, J. D. *Science*, **1978**, *199*, 1436.
15. Olsson, M.; Wick, G. L.; Isaacs, J. D. *Science* **1979** *206*, 454.
16. Wentworth, T. O.; Othmer, D. F.; Pohler, G. M. *Trans. Am. Inst. Chem. Eng.* **1943**, *39*, 565.
17. Prebluda, H. J.; Williams, R., Jr., presented at the 177th National Meeting of the American Chemical Society, Honolulu, HI, April, 1979.
18. Marion, L. *Chem. Eng. (N.Y.)* **1979**, *86*, 78.
19. *Chem. Week* Oct. 31, **1979**, p. 45.
20. Othmer, Donald F. *Mech. Eng.* **1979**, *101*.
21. Pradt, L. A. *Chem. Eng. Prog.* **1972**, *68*, 72.
22. Othmer, Donald F. *Combustion* **1978**, *50*, 44.
23. Institute of Gas Tech. "Reports to Dept. of Energy," April 1978, Chicago, Il.
24. Pradt, L. A. "Paper at Kentucky Coal Conference," Univ. of KY, May 2, 1979.
25. Lee, B. S. *Chem. Eng. (N.Y.)* **1978**, *85*, 3.
26. Stinson, S. C. *Chem. Eng. News* **1979**, 28.
27. Reed, T. B.; Lerner, R. M. *Science*, **1973**, *182*, 1299.
28. Othmer, D. F. "Important for the Future," 1979; Vol. 4, p. 7.
29. *Power* **1979**, 14.
30. Anon. *Syn Fuels*, January 25, *1980*.

RECEIVED May 7, 1979.

INDEX

INDEX

Jacket design by Carol Conway.
Editing and production by Candace A. Deren.

The book was composed by Carolina Academic Press, Durham, NC,
printed and bound by The Maple Press Co., York, PA.

660.209
F984h

76600

660.209
F984h

76600

    Furter, William F.,ed
    History of chemical engineer-
    ing